Energy and Motorization in the Automotive and Aeronautics Industries

Energy and Motorization in the Automotive and Aeronautics Industries

Tomasz Krysinski
François Malburet

ISTE

WILEY

First published 2020 in Great Britain and the United States by ISTE Ltd and John Wiley & Sons, Inc.

ISTE Ltd
27-37 St George's Road
London SW19 4EU
UK

www.iste.co.uk

John Wiley & Sons, Inc.
111 River Street
Hoboken, NJ 07030
USA

www.wiley.com

Library of Congress Control Number: 2020935303

British Library Cataloguing-in-Publication Data
A CIP record for this book is available from the British Library
ISBN 978-1-78630-572-5

MIX
Paper from responsible sources
FSC® C013604

Contents

Foreword

The future of the planet cannot leave anyone feeling indifferent. Environmental problems and more particularly those related to global warming concern us all and require general mobilization. Industries must be particularly active in reducing their greenhouse gas emissions and finding innovative solutions that will enable sustainable, environmentally friendly growth.

The automotive and aeronautics industries have become fully aware of these challenges. Modern vehicles have already made a lot of progress in reducing fuel consumption. The new CAFE (Corporate Average Fuel Efficiency) regulation sets extremely ambitious emission reduction targets for car manufacturers, combined with possible financial sanctions. For aviation, States have made commitments to the ICAO (International Civil Aviation Organization) to stabilize emissions from the sector from 2020 onwards and to even go beyond within the framework of the Paris Agreement, with the objective of reducing CO_2 emissions by half by 2050.

Manufacturers today view these topics as an exceptional opportunity to offer innovative technologies, both in terms of vehicle structure and propulsion systems. All potential new energy sources are being explored: biofuels or synthetic fuels, electric or hybrid engines, hydrogen engines, etc.

At least three major challenges must be met in this research for both the automotive and aeronautics worlds: optimizing propulsion efficiency – including engine technology and thrust or traction generation – weight reduction and reduction of forward resistance, both in terms of aerodynamic drag and ground friction.

Throughout the 20th Century, both the automotive and aeronautics industries inspired each other in terms of components and materials as well as means of production. The same type of internal combustion engine, with pistons, has been used in particular in automotive and aeronautical applications for certain small aircraft. Autopilot was first introduced in the aeronautics field and its gradual integration into modern cars with new autonomous driving functions can now be observed. Another example of a similarity between the two industries is the increasing use of composite materials.

This book provides a detailed comparison of energy challenges, with a particular focus on the issue of energy storage – whether it is electric energy or hydrogen – which is one of the main issues for both industries to date.

It shows the major challenge of achieving zero emission cars, helicopters and aircraft. It also allows us to hope that these extremely innovative industries will master the necessary technologies and mutually enrich each other with the experiences and progress made by each other to achieve this goal, thus awakening the minds of the pioneers who have always been able to meet the challenges facing them and enabling these two modes of transport – both of which are definitely part of our modern lives – to build a future for themselves, for future generations, while respecting the planet and its environment.

Guillaume FAURY
CEO of Airbus

Preface

"Never believe in any theory until it's confirmed by experimentation in various conditions and scales."

Prof. Alexander A. Nikolsky,
Department of Aeronautical Engineering,
Princeton University

The transport sector represents a significant part of the world's energy consumption. The technologies used have an impact on the depletion of non-renewable resources and on the environment, whether it is the air quality (CO_2, fine particles, NO_x, etc.) or noise pollution.

Beyond organizational and behavioral solutions, the improvement of existing technologies or the development of new technologies is a major challenge for the coming years. After decades of almost exclusive use of fossil energy, the need for energy transition has been recognized. Several solutions, not necessarily new ones, are being developed or considered for the automotive and aeronautics industries: this is the case for electric or hydrogen solutions. It appears that no single solution is likely to offer sufficient potential in the short or medium term to address both the problem of energy transition and sustainable development as well as the issue of mobility, a major challenge for the development of tomorrow's society (social, economic). It must be considered that each solution will have to be used in the coming decades in a complementary way.

In the above context and on the basis of their professional experience and culture, the authors decided to write this book on energy and the engine power of transport systems in the automotive and aeronautics sector by linking science and technology in an industrial context.

This book is intended for students of engineering schools, students in mechanical faculties, young engineers involved in companies in the automotive or aeronautics sector and people wishing to have a more global vision on the subject.

A large number of books have been written on the field of energy, each in its own specific field. Similarly, there is a number of structures dedicated to the automotive and aeronautics industries. The authors wanted to write a book that would put the automotive and aeronautics industries into perspective, the needs of which are sometimes similar and sometimes opposite, and whose technical solutions may be similar, even common or, on the contrary, very different. The objective was to present the entire propulsion chain from the product's energy requirement to energy storage by integrating the engine(s) and power transmission elements.

This book was written with the aim of transmitting a technical culture and know-how in order to support future generations in the development of future solutions. It is the result of a long collaboration between industry and university.

Introduction. This introductory chapter first describes a historical overview of the main technological leaps in the energy field that have made it possible to develop the automobile, the aircraft and the helicopter in recent decades. This overview makes it possible to suggest that – in the coming years – similar transformations such as electrification or the development of urban mobility will represent new challenges facing manufacturers in the aeronautics or automotive sectors. A second part, through a functional approach, introduces Breguet-Leduc's formulation which at first glance links the characteristics of the structure (aerodynamics and mass) and propulsion energy (electric or fossil fuel) to the distance that can be travelled by the product (a plane, a helicopter or, by extension, a car). This Introduction provides an initial reflection on the product's structure in terms of its energy consumption as well as on the productivity of the product according to its use.

Chapter 1: Motorization and Reflection on Ideal Engines. This chapter summarizes the methods for obtaining the order of magnitude of the engine power associated with the product and its mission. The first part presents Froude's theory, necessary for sizing useful hover power of rotary wings, and then justifies the development of useful forward power in flight. The second part details the case of the aircraft by distinguishing the power required for take-off and then for forward flight. The third part introduces the case of the car. This chapter defines an ideal type of engine to meet each product's power requirements.

Chapter 2: Engine Technologies. As a logical follow-up to the previous chapter, this chapter aims to describe different engine technologies used in aeronautics or automotive applications. The first part describes the operating principle of gas turbines and their application to turbo propellers or turboshaft engines that equip aircraft or helicopters. In this context, their thermodynamic cycles and the expression of their yields in relation to the constructive elements are presented. Then, possible improvements of these engines are described by means of exchangers to recover exhaust gases or by an electric hybridization of a turboshaft engine. The second part is devoted to electric motors. After describing different technologies used, several examples of use on cars, aircraft or rotary wing aircraft are proposed. The third part concerns piston internal combustion engines. The thermodynamic cycles of these different types of engines (diesel or gasoline) are presented, as well as a comparison of the performance of these technologies and an introduction to the different technical means of possible improvement.

Chapter 3: Power Transmission Elements. This chapter justifies the need to introduce a fixed or variable reduction in the ratio between the engine and the propulsion element (rotor, propeller or wheels). The first part deals with the case of conventional helicopters and then examines other cases such as multi-rotor systems. The second part presents the case of propeller and then turbojet aircraft. The third part introduces the case of the automobile. For internal combustion engines, gearbox technologies and their staging rules are described. The case of electric energy cars is also discussed.

Chapter 4: Storage. This chapter describes energy storage technologies for transport systems linked to the automotive and aeronautics industries. The first part presents different energy sources available and introduces the notion of an energy carrier. In the second part, the forms of storage used or that are potentially usable on cars, aircraft and helicopters are described. Electrochemical storage (battery), supercapacitors, storage in the form of liquid or gaseous hydrogen, storage in the form of compressed gas, inertial mechanical storage and energy storage in thermal form are thus detailed.

Chapter 5: Hybridization. The purpose of this chapter is to present hybridization and different existing formulas. The first part deals with the use of hybridization to increase the autonomy of the initial propulsion energy (Range Extender). Examples associated with automobile and/or aeronautics are proposed. The second part deals with hybridization of internal combustion engines for the automotive and aeronautical industries. Several specific cases such as parallel or mixed hybridization are developed and given as examples. The chapter concludes with a reflection on hybridization structures.

Acknowledgements

The authors would like to thank:

– Airbus Helicopters for allowing us to use the knowledge, experience and know-how developed by its employees for the purpose of this book;

– the Airbus Helicopters innovation department team, which provided effective support for some of the carried-out studies;

– the management of the *École nationale supérieure des arts et métiers* and its development subsidiary AMVALOR, for their collaboration;

– the teachers and students of the *École nationale des arts et métiers d'Aix-en-Provence* who were able to participate in some of these studies;

– Pierre Rougier, flight engineer of the Eurocopter Tiger helicopter, Christian Mercier, chief chain engineer of the Airbus Helicopters hybrid propulsion system, Jan Krysinski, university professor at the Lodz University of Technology, and Philippe Malburet, secondary mathematics teacher, for their careful proofreading and relevant advice in writing this book.

Tomasz KRYSINSKI
François MALBURET
April 2020

Introduction

Energy Issues Linked to Transportation Powertrains

The main role of aircraft or land vehicles is transporting a payload (passengers or freight) over a certain distance. This role can be identified according to different mission types: commercial passenger or freight transport missions, military missions with a very specific spectrum or more general use that includes all work activities such as surveillance or rescue as well as individual use for work, leisure or sport.

The choice of the propulsion system's energy – in terms of being thermal, electric or hybrid – and the calculation of the range or autonomy according to the system's characteristics (mass, aerodynamic quality, type of power transmission, etc.) are important analyses when comparing modes of transport. Their different adaptations (cars, helicopters, aircraft or compound helicopters) are also important. To do this, it is important to take into account the following aspects:

– **The environmental impact** of different modes of transport. For aviation, States are committed to the International Civil Aviation Organization (ICAO). The ICAO's objective is to stabilize emissions from the sector beginning in 2020 and to even go beyond this in the context of the Paris Agreement, by halving CO_2 emissions by 2050. In terms of the automobile, the requirements of the Corporate Average Fuel Economy (CAFE) regulations, Euro 6c, or Euro 7, are very strict and can go as far as financial penalties for manufacturers. The new CAFE standard for 2017 to 2025 predicts a reduction in fuel consumption of 5% per year until 2025. These regulations encourage manufacturers to look for new technologies, especially linked to the powertrain electrification.

– **Transport safety and security:** development policy must lead to a permanent reduction in accident rates and integrate cybersecurity.

– **The market and competitiveness:** the proposed solutions must be accessible to a large number of customers.

– By 2030, the development of megacities (e.g. London) and congestion in large cities leading to mega-regions (e.g. Johannesburg and Pretoria) or mega-corridors (e.g. Hong Kong or Shenzhen-Guangzhou) will require **sustainable mobility strategies** to be implemented. The development of new means of transport such as autonomous vehicles or flying taxis based on electrical energy is one of the envisioned solutions (Figure I.1).

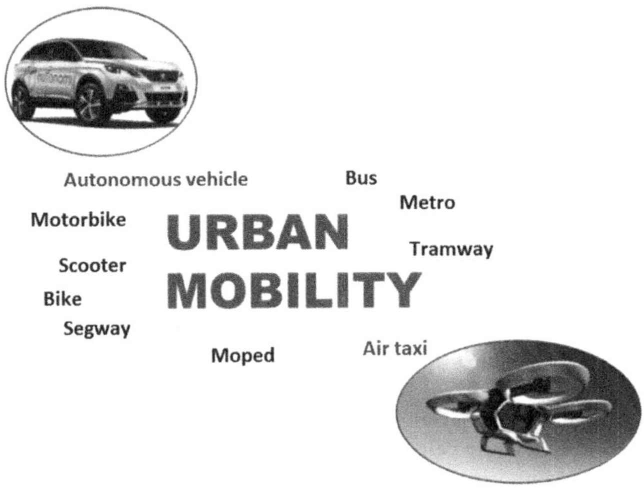

Figure I.1. *Means of transport within the urban mobility framework (source: Airbus, PSA). For a color version of this figure, see www.iste.co.uk/krysinski/automotive.zip*

I.1. History of energies in automotive and aeronautical transport

The availability of energy and the technological maturity of storage systems and engines are the key points that make one technology take precedence over another. Throughout history, different phases have been observed where certain technologies were required and then disappeared in favor of other solutions.

I.1.1. *Steam engine*

The invention of the steam engine, particularly James Watt's invention in 1769, led to the development of large ships and railways, such as George Stephenson's Locomotion in 1825. In terms of use, water vapor required heavy equipment and often numerous personnel. In fact, for aeronautical or automobile transport, these elements were an obstacle for this type of development, even if Cugnot's *fardier* (steam dray) (1781) can be considered as the stepping stone to the automobile.

I.1.2. *Trial and error*

The technologies of electric and internal combustion engines were developed almost alongside each other. Among the different historical dates, we highlight the following for internal combustion engines:

– the first single-cylinder combustion engine designed by Eugenio Barsanti and Felice Matteucci in 1856;

– the two-stroke combustion engine built by Étienne Lenoir in 1859;

– the four-stroke combustion engine invented by Beau de Rochas in 1862, then developed by Nikolaus Otto in 1867 and perfected by Gottlieb Daimler and Wilhelm Maybach in 1887;

– the diesel engine in 1893.

The electric motor – the first developments of which can be traced back to 1881 – was first used in terms of transport for the electric tramway and then for use on an airship in 1884.

The deployment of the gasoline distribution network made a significant contribution to the development of the combustion-powered automobile [LOU 01]. Initially, it was small businesses that were very highly established in society, such as grocery stores and pharmacies, that sold the first quantities of gasoline for cars. Gasoline was initially packaged in 2-, 5- or 10-liter cans, then soon after wooden crates containing ten 5-liter cans were gradually placed at store fronts for public consumption. The American manufacturer Bowser developed the first pumps imported into Europe, which appeared in 1900, mainly among industrial owners of large car fleets.

The First World War marked the massive use of internal combustion engine means of transport, such as trucks, cars, aircraft and field artillery by tractors.

On the commercial side, France welcomed the first gas stations in the early 1920s, marking the beginning of automobile democracy. *L'Économique*, representing Standard Oil (an American oil refining and distribution company founded by John D. Rockefeller and his associates in 1870), was one of the first to create "service stations". In 1928, there were between 40,000 and 50,000 fuel sale points, a ratio of one outlet for every 20 cars.

The 1930s saw the arrival of the first premium fuels. With the help of marketing, "super" fuel became more and more widespread, and by 1934, a third of stations sold it.

Evidently (Figure I.2), the Second World War impeded the development of vehicles for private individuals since it was not until 1953 that the same number of vehicles returned to pre-war levels. From that date, the number only increased until 2017. From 1953 to the mid-1970s, this was mainly due to household equipment, followed by the rise in households with multiple cars, associated with urban sprawl, the increase in women's labor rates and the increase in home-to-work travel distances.

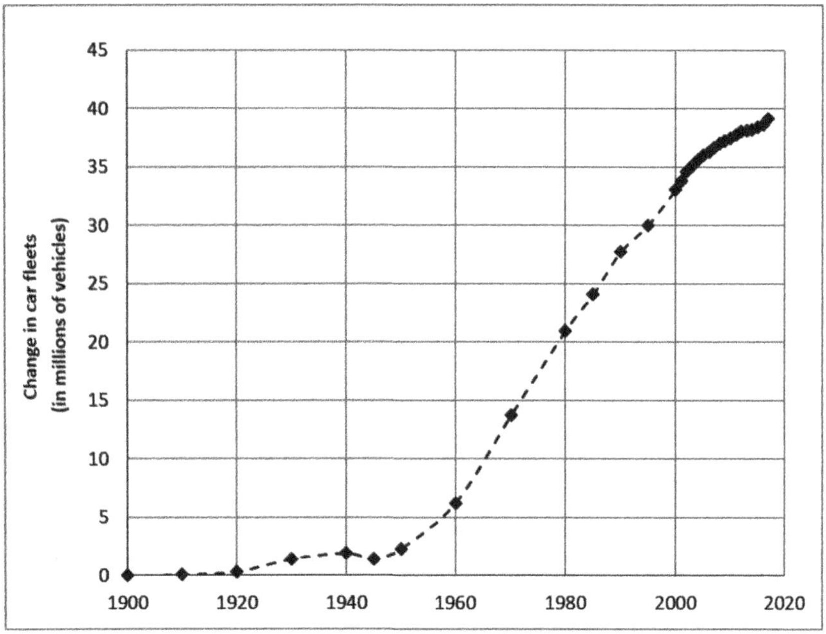

Figure I.2. *Change in car fleets from 1900 to 2020 (source:* Comité des constructeurs français d'automobiles*)*

This change in the number of cars contributed to the development of the road network. This was achieved thanks to the development of local roads (municipal roads) following the urbanization of society, and infrastructure for rapid long-distance transport. France had 963 km of highways in 1968 and 7,000 km by January 1, 1991; this network continued to grow between 1994 and 2014: 2.5% per year between 1994 and 2004 for conceded motorways, then 1% per year since 2006.

Since the 1960s, the number of private vehicles has increased sharply, and the distribution of diesel and gasoline energy has changed (Figure I.3), particularly in response to the policies and incentives offered by government institutions.

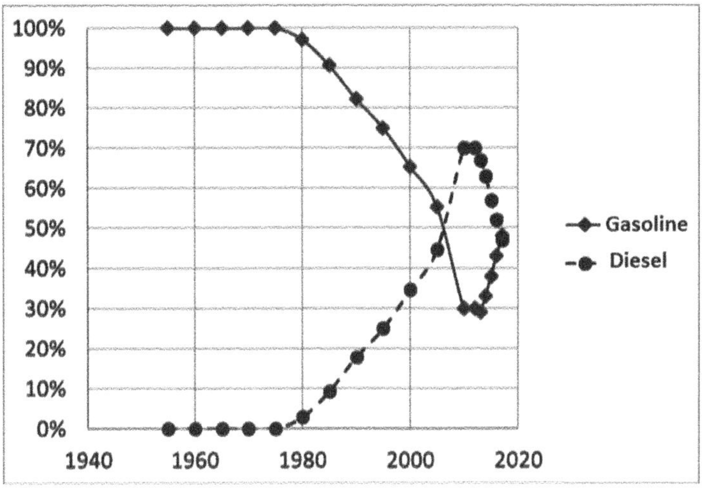

Figure I.3. *Evolution of the gasoline and diesel automobile fleet in France (source:* Comité des constructeurs français de l'automobile*). For a color version of this figure, see www.iste.co.uk/krysinski/automotive.zip*

Despite the implementation of a policy encouraging the use of electricity, among other alternative solutions, the market share of these new technologies remained marginal compared to fossil fuels until 2018.

	2013	2014	2015	2016	2017
Gasoline	29.70%	33.00%	38.60%	43.80%	47.50%
Diesel	67.00%	63.85%	57.20%	52.10%	47.30%
Hybrid	2.60%	2.40%	3.20%	2.90%	3.90%
100% electric	0.50%	0.60%	0.90%	1.10%	1.20%
Other	0.20%	0.15%	0.10%	0.10%	0.10%

Table I.1. *Development of hybridization and electric propulsion in vehicle fleets according to type of energy*

In France, the fuel distribution network was estimated at 11,000 stations in 2016, including 6,000 situated along traditional road networks and 5,000 in supermarkets and hypermarkets.

By comparison – and according to the same logic – the development of electric vehicles requires a major adaptation of the energy distribution system to become a viable alternative to piston-powered vehicles. It will be necessary to introduce a sufficient number of charging sites for individuals (in gas stations, mall car parks and other public spaces) and transport fleets (services, post offices, taxis, business parks, eco-neighborhoods, shared vehicle fleets) with their own charging sites to enable rapid charging. According to the *Association pour le développement de la mobilité électrique* (AVERE), at the end of 2017 France had 125,000 charging points, 51% were situated on company land, 36% in private individuals' garages or in collective housing, and 13% were accessible to the public, i.e. nearly 20,000 points spread across the country; the AVERE considers that the number of stations accessible to the public will rise to 100,000 by 2025.

The idea of using the electrical network for vehicle charging, taking into account the growing number of vehicles, the significant recharge time and periods of high vehicle fleet charging demand, leads to a redefinition of the overall strategy for electricity production. Furthermore, it leads, above all, to clever management of vehicle charging over a 24-hour period.

I.1.3. *Aeronautics – aircraft*

As far as the aeronautics sector is concerned, several key points demonstrate the historical development of techniques and propulsion systems in particular. After a period of trial and error for aerostation and gliding flight, the concept of "heavier than air" led to the first flights of motor vehicles able to take off on their own. First steam engines (Félix and Louis du Temple – 1877), then quickly followed by piston combustion engines [CHA 06].

The First World War marked the launch of the first mass production of aircraft. The inter-war period marked the launch of commercial air transport and, first and foremost, mail transport. This made it possible to develop aircraft and propulsion systems.

This period marked the first aeronautical revolution: on January 1, 1914, for the first time in the history of aviation, a passenger paid for a ticket on the first ever airline. The St. Petersburg-Tampa Airboat Line connected Tampa to St. Petersburg, Florida in 23 minutes. The price of the ticket was $400, the equivalent in 2017 would have been $9,000. This significant cost of the ticket did not prevent the airline from welcoming customers because the amount of time saved by the flight was considerable.

The Second World War was the height of aircraft using a piston engine and a propeller as a means of propulsion. Military aircraft were used intensively in different forms and for different missions: transport aircraft, bombers, reconnaissance aircraft, fighter aircraft, training aircraft, seaplanes, etc. Each of these missions led to different engine power requirements and therefore to improved performance of these technologies.

At the end of the Second World War, the jet engine arrived on the scene. This was the beginning of commercial air transport for regular "all-weather" flights; capable of flying in all weather conditions and achieving the practice of blind flight. It was the era of the jet and then supersonic flight. The first four-engine airliners appeared; the first supersonic civil flight was launched in 1976 (Concorde). Air transport became accessible to all. A century after the beginning of aviation, airlines now transport more than 3 billion passengers per year. In 2017, every minute 52 civilian flights took off from airports around the world

The world's aircraft fleet is expected to double by 2036. From a commercial point of view, since the late 1990s, the market has been dominated by Boeing and Airbus, which are competing in the field of civil aircraft and increasingly in the field of defense. The European manufacturer Airbus estimates that 35,000 new aircraft will have to be produced by then. Boeing is even more optimistic and estimates that 41,000 new aircraft will be needed within 20 years.

We can imagine that the second revolution in aeronautics will be achieved by seeking alternative solutions to the use of kerosene. Pollution and air traffic congestion remain the major concern. ADEME estimates that for a 500 km journey, aircraft can emit up to 241 kg of CO_2 per passenger while a car emits about 170 kg of CO_2 over the same distance. As there are many more cars than planes, road traffic is by far the most polluting source. In Europe, road traffic is responsible for 72% of greenhouse gas emissions compared to 12.3% for aviation. Despite the upcoming explosion in traffic and aircraft numbers, the air transport industry is nevertheless committed to halving its emissions by 2050 for clean and sustainable aviation. The International Air Transport Association (IATA) intends to stabilize CO_2 emissions from the sector by 2020, then reduce them by 50% by 2050 compared to 2005. To achieve this, all solutions must be considered, whether in terms of air traffic management and aircraft performance with more fuel-efficient engines and lighter equipment, not to mention the substitution of all or at least some of the kerosene by a less polluting fuel (biofuel) or energy (electricity).

Since the first oil shocks in 1974, various projects to develop electric motors have been carried out for small aircraft. These include the Solar One project in 1979, the development of HALE-type drones in the 1990s, the Solar Impulse project in 2009 and the Airbus Group E-Fan project, a small two-seater equipped with an electric motor combined with lithium-ion-polymer batteries that made its first flight in 2014. For large aircraft, everything will depend on technological progress in storage facilities, electrochemical or hydrogen, whose performance must increase significantly, particularly in terms of the amount of energy stored per kilogram, and on the development of new aircraft architectures to reduce energy requirements during the usage phase.

I.1.4. *Helicopters*

The first engine to equip a rotary wing was a steam engine that enabled flight at about 10 meters for about 30 seconds, invented in 1877 by Enrico Forlanini. Ten years later, Frenchman Gustave Trouvé launched a model equipped with an electric motor, connected from the ground by thin copper wires for power supply. In 1905, Maurice Léger built an aircraft with coaxial rotors equipped with a 6 hp engine, and brothers Henri and Armand Dufaux successfully achieved vertical take-off powered by an internal combustion engine [BOM 06, BOU 91].

The first "helicopter" flight is often considered to be that achieved by Louis Charles Breguet in August 1907. The vertical take-off aircraft could be converted to achieve horizontal flight using two rotating propellers (gyroplane) and thanks to being equipped with a piston engine [CHA 06]. Collective memory sometimes attributes the first free flight to Paul Cornu in November 1907. The helicopter was equipped with a 24 hp (18 kW) Antoinette piston engine that drove two twin-bladed rotors. The momentum was achieved by means of large belts. The conventional helicopter as it is designed today is unanimously attributed to the Ukrainian-born American engineer Igor Sikorski. His first flight on September 14, 1939, on the Vought-Sikorsky 300 (VS-300), was achieved via a cable end. A 75 hp motor (56 kW) was used to drive the single three-bladed main rotor. In this configuration, a secondary rotor called an "anti-torque" was placed vertically at the end of the tail to compensate for the horizontal rotor's reaction torque.

Helicopters were first equipped with internal combustion engines with cylinders positioned in a star-shaped or an in-line position, as in aircraft or in automobiles. The Alouette II built by Sud-Aviation was the first production aircraft, from 1955, to be powered by a gas turbine. The first commercial helicopter flight is considered to have been made in June 1949 when the United States Civil Aviation Administration first authorized a helicopter to conduct commercial operations. The first rotations connected partner airports such as Newark International Airport in New Jersey and Long Island International Airport in New York, or San Francisco International Airport and Oakland International Airport. In 1953, New York Airways became the first scheduled helicopter carrier to operate in the United States.

The 2010s saw the emergence of new projects aimed at reducing some of the limits of the classic conventional helicopter formula. The ambition was to increase the velocity and range of rotating wings while maintaining hover capabilities. The success of the Airbus helicopters X3 prototype in 2010 validated the concept of combining a main rotor with two propellers and two side wings. The RACER (Rapid and Cost-Effective Rotor Craft) project, which is the next step, plans to achieve more than 400 km/h in cruise flight while ensuring a compromise between respect for the environment through low noise emissions and lower fuel consumption, and mission performance. An alternative proposal is presented by Agusta's tilt rotor formula. The aircraft is made up of tilting rotors allowing the combination of vertical take-off and landing phases (the rotors having a vertical axis to ensure lift), and the forward flight phase (the rotors having a horizontal axis to ensure traction and lift being then provided by the wings). The velocity reached in cruise flight is significantly higher than that achieved in a conventional helicopter. For tilt wings formulas such as the Hiller X-18 (1959) or the Canadair CL-84 (1965), it is the entire wing that tilts with the rotors. They have a similar purpose to the previous formulas, but have not undergone any major developments, as this formula is less effective in hover flight.

In 2012, Pascal Chrétien and Solution F developed the first helicopter in the world to have an electric propulsion. A first hover flight without ground effect with an autonomy of 10 minutes was carried out using a prototype. Another electric helicopter project in the United States was able to fly in September 2016 with a Robinson R44 whose Lycoming Engines IO-540 internal combustion engine was replaced by two electric motors. The prototype was able to fly for 15 minutes with a velocity level of 80 kts. In October 2016, a third prototype of a single-seat electric helicopter with a payload of 115 kg, the Volta, built by a Toulouse-based SME, was able to make its first flights in France. At the same time, Volocopter produced several electric-powered prototypes. The VC2 has 18 propellers with a direct propeller drive motor suspended around an aluminum chassis with a central seat and batteries. The development of electric propulsion leads to the creation of hybrid series solutions (to improve range) combining piston or turbine thermal engines in series with electric propulsion. This type of system aims to increase aircraft autonomy to values that current battery technologies do not allow us to achieve, while having the qualities of electric propulsion (acoustic or consumption) in certain flight phases. Aurora's XV-24 project is composed of 24 fans driven by electric motors, powered by three generators driven by two Rolls-Royce turboshaft engines. The E-Fan X project by Airbus and Siemens takes this idea up with an aircraft by combining a Rolls-Royce turboshaft engine, a generator and electric motors that drive the propellers.

In conclusion, we can observe that the history of aeronautics is marked by technological leaps that have enabled the product to evolve over relatively long cycles. In the coming years, just as we have seen major transformations in the field of telecommunications and information technology, we can envision similar transformations in the field of transport. In the near future, urban mobility will be defined by a wide range of transport modes, including autonomous air taxis (urban air mobility), which are the new challenges facing aviation manufacturers.

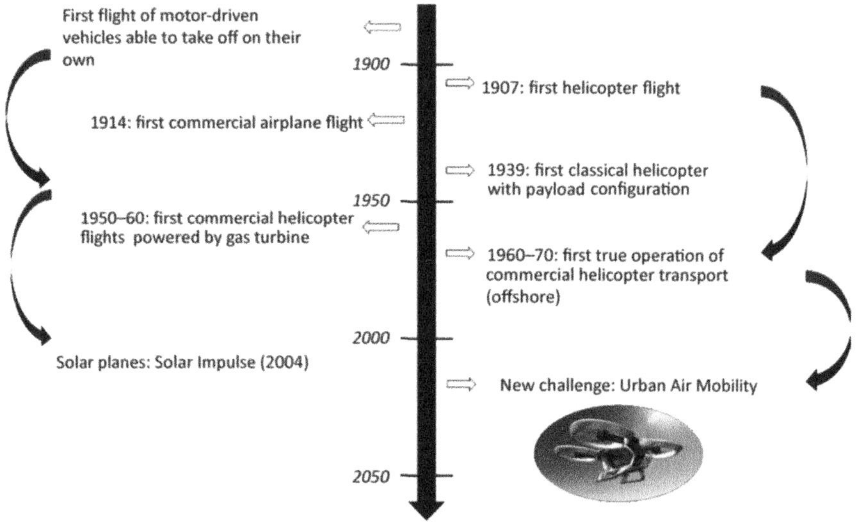

Figure I.4. *Evolution of aeronautics*

I.2. Functional considerations

The question arises as to how to link the distance covered or to be covered with the characteristics of a propulsion system and the qualities of the aircraft, such as its aerodynamic qualities, its own mass and the useful load to be transported.

The proposed approach enables a first approximation of the range that can be covered according to these parameters. It is based on several concepts defined in the following sections.

I.2.1. *L/D ratio*

Whatever the transport system, the system's mechanical balance can be studied by applying the fundamental principle of dynamics. The first approximation is the configuration of a straight horizontal trajectory at constant velocity, level flight for an aircraft, on a slope-free road for a vehicle.

The system being studied undergoes: forward resistance actions, weight, traction action related to the powertrain and lift action for an aircraft (aircraft or helicopter), or road reaction for a vehicle (Figure I.5). Initially, all these actions are assumed to apply to the structure's center of gravity.

The inertia effects are zero at constant velocity. The system's equilibrium equations are obtained by projecting the fundamental principle of dynamics (FPD) onto the horizontal and vertical axes.

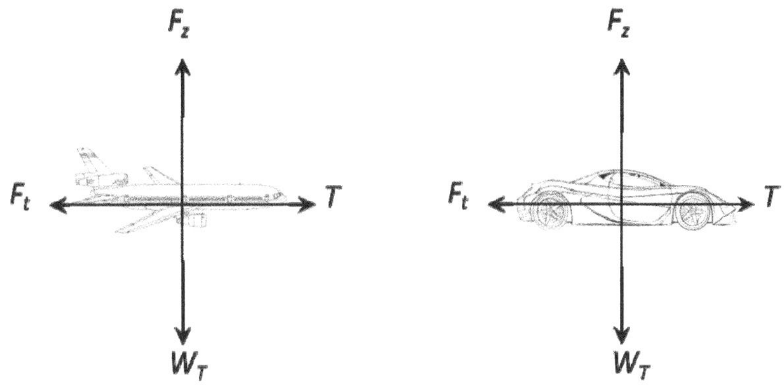

Figure I.5. *Simplified model of mechanical actions on an aircraft or a car*

The equations are then:

$$\begin{cases} 0 = T - F_t \\ 0 = F_z - W_T \end{cases} \text{hence} \begin{cases} T = F_t \\ F_z = W_T = M\ g \end{cases} \quad [I.1]$$

with:

– F_z: the lift force or vertical component of the ground reaction;

– F_t: the effort of resistance or drag;

– W_T: the system's total weight (M system mass);

– T: the tractive effort induced by powertrain.

It is common to introduce the L/D ratio *f*, the aerodynamic characteristic and the dimensionless structural quality of the structure [FIL 12]. For a profile, this translates the ratio between the lift P and the forward resistance force (drag T) (Figure I.6). This concept can be applied to the aircraft model, that is:

$$f = \frac{F_z}{F_t} \quad [I.2]$$

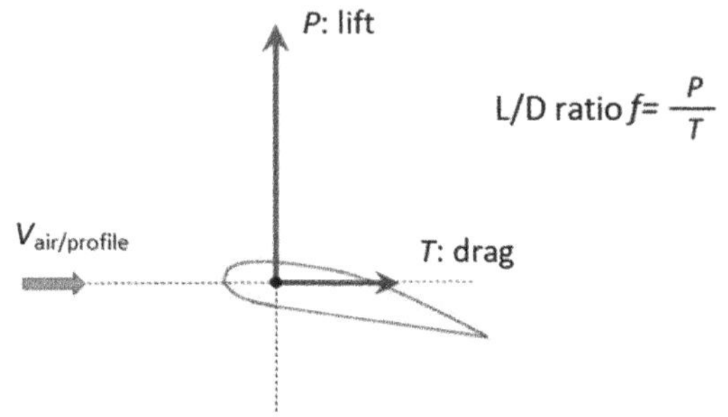

Figure I.6. *Definition of the aerodynamic L/D ratio of a profile. For a color version of this figure, see www.iste.co.uk/krysinski/automotive.zip*

The dimensionless parameter P_1 is introduced to reflect the drag ratio relative to the lift, which can also be seen using the FPD as the ratio of resistance actions F_t to weight W_T:

$$\Pi_1 = \frac{1}{f} = \frac{F_t}{W_T} \qquad [I.3]$$

By analogy, the same can be done for a land transport vehicle such as a car. The resistance is then the combination of the aerodynamic drag effects F_{ra} on the vehicle and the rolling resistance actions F_{rr}:

$$\begin{cases} F_t = F_{ra} + F_{rr} \\ F_z = W_T = Mg \end{cases} \qquad [I.4]$$

with:

– F_{rr}: the rolling resistance;

– F_{ra}: the aerodynamic resistance forces on the vehicle;

– F_z: the vertical component of ground reaction forces.

By analogy with aircraft, we obtain:

$$\Pi_1 = \frac{1}{f} = \frac{F_{rr}+F_{ra}}{W_T} \qquad [I.5]$$

In the case of systems with several lift and drag sources, such as for a combined formula (Figure I.7), it is possible to define by analogy:

$$\begin{cases} T = F_{t\ fuselage} + F_{t\ wing} + F_{t\ rotor} \\ F_z = F_{wing} + F_{rotor} = W_T = Mg \end{cases} \qquad [I.6]$$

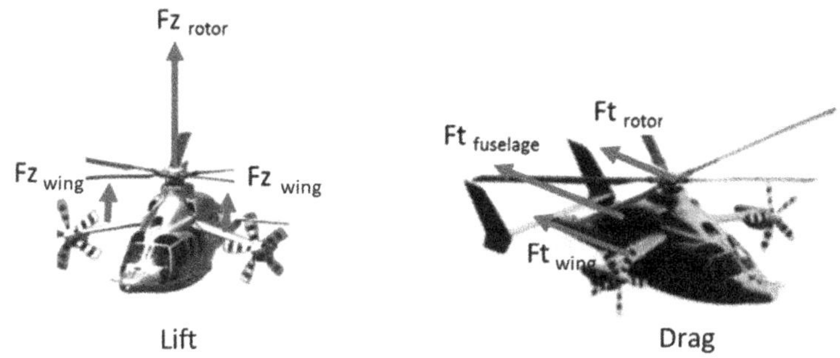

Figure I.7. *Actions involving a compound helicopter formula (source: Airbus). For a color version of this figure, see www.iste.co.uk/krysinski/automotive.zip*

Under these conditions, it is possible to define:

$$\Pi_1 = \frac{1}{f} = \frac{F_{t\ fuselage} + F_{t\ wing} + F_{t\ rotor}}{W_T} \qquad [I.7]$$

Parameter P1 is obtained by adding up each component's aerodynamic drag contribution: the fuselage, wing or rotor. We show that the inverse of the overall L/D ratio is the sum of each component's L/D ratio in the first approximation.

I.2.2. *Efficiency and energy losses from the engine to the propulsion system*

For all the systems proposed, a propulsion efficiency corresponding to the power ratio supplied by the power system, including the power chain, to the power from the storage system can be defined (Figure I.8) [LEP 91]. This propulsive efficiency includes engine efficiency and propulsive system efficiency.

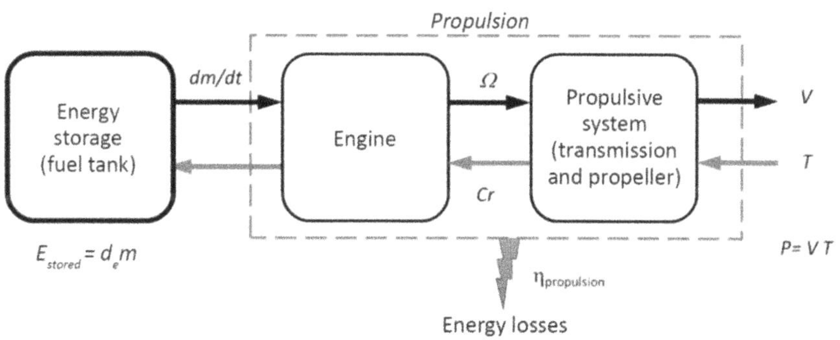

Figure I.8. *Energy consideration and propulsive efficiency*

By definition, the power supplied by the tank to the engine is given by:

$$P_{provided} = \frac{d\,E}{dt} = d_e\,\dot{m} \qquad [I.8]$$

with:

– d_e: the energy density [J/kg];

– $\dot{m}$: the mass flow rate [kg/s].

For the field of transport that is associated with energy storage due to the autonomy required, the energy density refers either to the mass energy density [J/kg] or to the volume energy density [J/L]. The higher the energy density, the more energy can be stored or transported for a given volume or mass.

This allows us to define another dimensionless quantity, corresponding to the propulsive efficiency associated with the efficiency of the engine and propulsion system including the power transmission and the system generating the traction (propeller or rotor) (Figure I.9). Let:

$$\Pi_2 = \eta_{pro} = \eta_{engine} \times \eta_{propulsive\ system} \qquad [I.9]$$

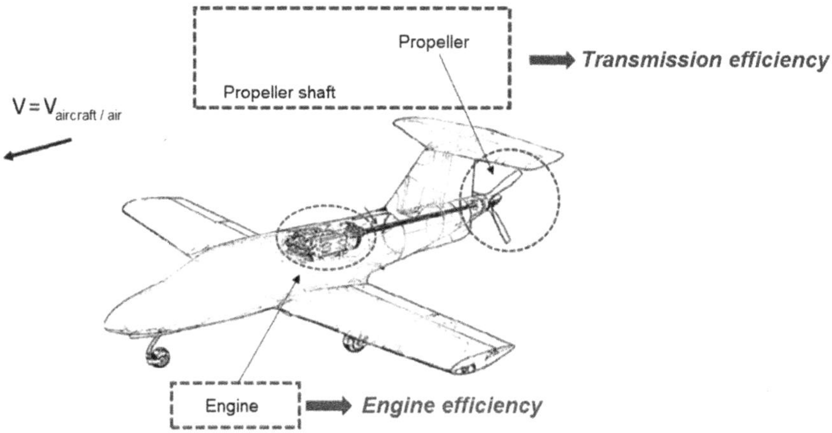

Figure I.9. *Energy loss and propulsive efficiency. For a color version of this figure, see www.iste.co.uk/krysinski/automotive.zip*

This results in:

$$\eta_{prop} = \frac{T\,V}{\dot{m}\,d_e} \Rightarrow \dot{m} = \frac{T\,V}{\eta_{prop} d_e} \qquad [I.10]$$

This relationship makes it possible to evaluate the mass flow required to create the tractive effort for a given distance to be covered. It is not generally exploited in this form.

I.2.3. *Concept of specific consumption: specific case of the aircraft*

Rather than directly assessing fuel flow, the concept of specific fuel consumption, known as SFC, is generally used to specify the mass of fuel required to provide power or thrust in a given time. Depending on the case, it is usually expressed in g/(kW h) or g/(kN s).

The SFC depends on the engine design. Differences in specific fuel consumption between engines using the same technology tend to be quite small. It allows the efficiency of different engines to be compared.

We can thus define:

$$SFC = \frac{\dot{m}}{P} \text{ or } SFC = \frac{\dot{m}}{T} \qquad [I.11]$$

or by combining [I.10] and [I.11]:

$$SFC = \frac{V}{\eta_{prop} d_e} \qquad [I.12]$$

At constant forward velocity *V*, the specific consumption is proportional to the velocity and is inversely proportional to the efficiency of the energy chain and the energy density of the fuel.

I.2.4. *Breguet-Leduc's formulation of the range to be covered*

The idea is to propose a sizing formula that leads to the analysis of an aircraft's range according to the system parameters [BOI 98].

A distinction is made between cases where fuel consumption results in a significant reduction in the system's mass – such as diesel, gasoline or kerosene – and cases where there is no change in mass, such as electric power units associated with batteries.

I.2.4.1. *System including variable mass energy storage: gasoline, diesel or kerosene*

The total W_T system weight is broken down into payload, fuel and structure:

$$W_T = W_{\text{payload}} + W_{\text{fuel}} + W_{\text{structure}} \tag{I.13}$$

with:

– $W_{payload}$: the payload's weight;

– W_{fuel}: the weight of the fuel to ensure the mission;

– $W_{structure}$: the structure's weight.

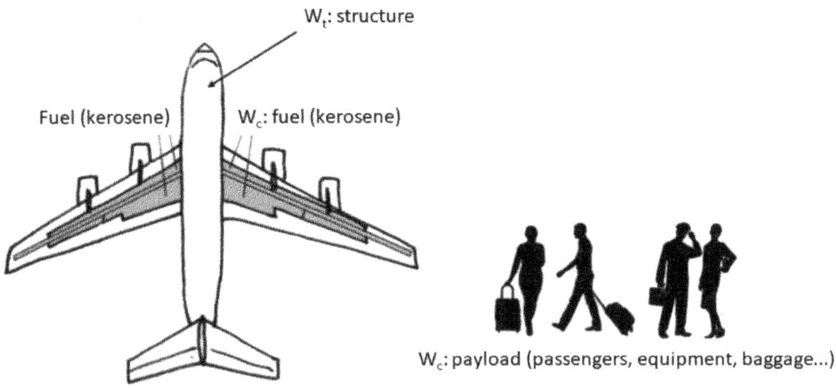

Figure I.10. *Definition and distribution of masses on an aircraft structure*

The change in weight is then due to the change in fuel weight during the flight or journey. This is negative. That is, gravity being constant:

$$\frac{d\,W_T}{dt} = \frac{d\,(W_{fuel})}{dt} = \dot{m}\;g \quad (\leq 0) \tag{I.14}$$

Taking into account the relationship expressing the conservation of power and introducing the L/D ratio f, we obtain:

$$\frac{d\,W_T}{dt} = \frac{T\,V}{\eta_{prop}\,\frac{d_e}{g}} = \frac{W_T\,V}{f\eta_{prop}\,\frac{d_e}{g}} \tag{I.15}$$

At constant velocity, we obtain by integration:

$$\frac{d\,W_T}{W_T} = \left(\frac{V}{f\eta_{prop}\frac{d_e}{g}}\right) dt \;\Rightarrow ln\left(\frac{W_{T\ initial}}{W_{T\ final}}\right) = \frac{V}{f\eta_{prop}\frac{d_e}{g}}\,\Delta t \qquad [I.16]$$

So:

$$\text{distance} = V\ \Delta t = f\eta_{prop}\ \frac{d_e}{g} ln\left(\frac{W_{T\ initial}}{W_{T\ final}}\right) \qquad [I.17]$$

The final mass at the end of the mission is defined as the structure mass plus the payload, the initial mass being the final mass plus the fuel mass.

By introducing the structure mass $M_{structure}$, the useful mass $M_{payload}$ and the fuel mass M_{fuel}, we obtain the following Breguet-Leduc formula:

$$\text{distance} = f\eta_{prop}\ \frac{d_e}{g} ln\left(1 + \frac{M_{\text{fuel}}}{M_{\text{structure}}+M_{\text{payload}}}\right) \qquad [I.18]$$

By introducing the SFC (see definition [I.12]), the following wording can also be used, which then introduces the velocity reduced to the specific consumption in the formulation:

$$\text{distance} = f\frac{V}{SFC\quad g} ln\left(1 + \frac{M_{\text{fuel}}}{M_{\text{structure}}+M_{\text{payload}}}\right) \qquad [I.19]$$

This formulation can also be written using dimensionless coefficients P_i. We note:

$$\Pi_3 = \frac{M_{\text{fuel}}}{M_{\text{structure}}+M_{\text{payload}}} \qquad [I.20]$$

In general terms, we could retain that the range, reduced to the energy density of the fuel and gravity, is such that:

$$\text{distance} = \left(\frac{d_e}{g}\right)\ \frac{\Pi_2}{\Pi_1}\ ln(1+\Pi_3) \qquad [I.21]$$

with dimensionless parameters:

– P_1: coefficient of resistance to forward flight based on weight or lift (similar to the opposite of L/D ratio *f*);

– P_2: engine power and propulsion system efficiency;

– P_3: ratio of fuel mass to system mass at the end of the mission (structure mass + payload).

The expression of the distance according to these dimensionless parameters shows that it is inversely proportional to the relative effect of the driving resistance brought back to the lift, proportional to the system's efficiency and above all as a function of the ratio of fuel mass to dry mass (at the end of the mission) through the function ln.

I.2.4.2. *A system including constant mass energy storage: batteries*

For electric motors, the energy is stored in batteries. Their mass is independent of the charge rate. There is therefore no mass variation during the flight or journey time. The power supplied by the engine is related to the power used to create the tractive effort to obtain a cruising velocity *V* by:

$$P_{useful} = \eta_{prop}\, P_{provided} = \eta_{prop} \frac{d\,E}{dt} \qquad [I.22]$$

by introducing the definition of velocity:

$$dt = \eta_{prop} \frac{dE}{P_{useful}} \quad \rightarrow dx = \eta_{prop} \frac{dE}{P_{useful}} V \qquad [I.23]$$

By integration, assuming that the velocity is constant, we obtain the range:

$$dx = \eta_{prop} \frac{f}{W_T} dE \quad \rightarrow d = \eta_{prop} \frac{f}{W_T} E_{stored} \qquad [I.24]$$

By introducing the definition of stored energy, the expression of distance takes the following form:

$$d = \eta_{prop} \frac{f}{W_T} [d_e\; M_{bat}] = \eta_{prop}\; f\; \frac{M_{bat}}{M_T} \left[\frac{d_e}{g}\right] \qquad [I.25]$$

We note:

$$\Pi_{3'} = \frac{M_{bat}}{M_T} \qquad [I.26]$$

By analogy with what has been done in the case of variable fuel mass, we obtain by introducing the dimensionless coefficients:

$$\text{distance} = \left[\frac{d_e}{g}\right] \frac{\Pi_2}{\Pi_1}\; \Pi_{3'} \qquad [I.27]$$

with dimensionless parameters:

– P_1: coefficient of resistance to forward flight based on weight or lift capacity (similar to the opposite of the L/D ratio *f*);

– P_2: engine power and propulsion system efficiency;

– $P_{3'}$: ratio of battery mass to system mass (total).

We thus obtained the two relationships making it possible to estimate the range for a system the mass of which varies (gasoline, diesel or kerosene) and for a system whose mass remains constant (battery):

$$\begin{cases} \text{variable mass} \rightarrow \text{distance} = \left(\frac{d_e}{g}\right) \frac{\Pi_2}{\Pi_1} \ ln(1+\Pi_3) \\ \text{fixed mass} \rightarrow \text{distance} = \left[\frac{d_e}{g}\right] \frac{\Pi_2}{\Pi_1} \ \Pi_{3'} \end{cases} \quad [I.28]$$

I.2.5. *Analysis and use of the Breguet-Leduc formula*

Using the previous formulations, it can be observed that there are several ways to increase the aircraft's distance. It is possible to work on the aerodynamics or on the propulsion system

I.2.5.1. *Influence of aerodynamics*

An aerodynamicist can increase the L/D ratio *f* (or decrease P_1) to improve this distance. For example, gliders have an L/D ratio in the range of 20 to 70 (70 for the Solar Impulse project); for a commercial aircraft the coefficient varies between 15 and 20, for a fighter aircraft from 5 to 10 and for a conventional helicopter from 2 to 3 [FIL 10].

A comparison can be made with the car. By analogy, the resistance effect on a vehicle is related to the aerodynamics and resistance effects of tires on the ground. The reaction of the road does not depend in the first order on velocity, the aerodynamic drag being (as it is for an aircraft) dependent on square velocity. We could define for a car an L/D ratio equivalent to different velocities reflecting the ratio between the weight since there is no lift, and the resistance effects (see the formulation [I.5] proposed at the beginning of the chapter). We then show that this equivalent L/D ratio is a function of velocity and the orders of magnitude are then:

$$\begin{vmatrix} \text{at 120 km/h} \quad f = \frac{Mg}{F_{rr}+F_{aer}(V)} \approx 20 \\ \text{at 200 km/h} \quad f = \frac{Mg}{F_{rr}+F_{aer}(V)} \approx 8 \end{vmatrix} \quad [I.29]$$

By comparison, the automobile has the L/D ratio of an airliner for operating velocities around regulatory velocities on highways or fast lanes.

I.2.5.2. *Influence of the propulsion system*

Powertrain characteristics are used in Breguet-Leduc's formula using three parameters: energy density, propulsive efficiency and engine mass included in the structure mass.

Fuel type has a direct influence on distance since it is proportional to d_e. For example, the orders of magnitude of this quantity are given in Table I.2 which shows that batteries have a much lower density than gasoline, diesel and kerosene fuels. The distance travelled can be increased through the energy efficiency (coefficient P_2) which includes the engine efficiency, transmission efficiency and system efficiency which generates the traction action T (rotor or propeller). From this point of view, not all systems are identical. Propulsive efficiencies will be detailed in the following chapters.

Compressed hydrogen	123 MJ/kg namely 34 kWh/kg	**Lithium battery**	1 MJ/kg namely 0.27 kWh/kg
Gasoline	47.2 MJ/kg namely 13 kWh/kg	**Super capacitor**	0.56 MJ/kg namely 0.15 kWh/kg
Diesel	45.4 MJ/kg namely 12.5 kWh/kg	**KERS**	0.5 MJ/kg namely 0.14 kWh/kg
Kerosene	43 MJ/kg namely 11.6 kWh/kg		

Table I.2. *Comparison of energy densities from different sources*

Similarly, the engines' mass powers are not identical. It can be seen (Figures I.12 and I.13) that the turbo engines associated with the helicopter are on a line close to 0.2 kg/kW, while the combustion engines associated with the automobile are close to 1.6 kg/kW and the fuel cell is about 2 kg/kW. The old electric motor and diesel engine technologies of the 1980s are on a line close to 4 kg/kW, the synchronous electric machine of the TGV system is about 1.33 kg/kW.

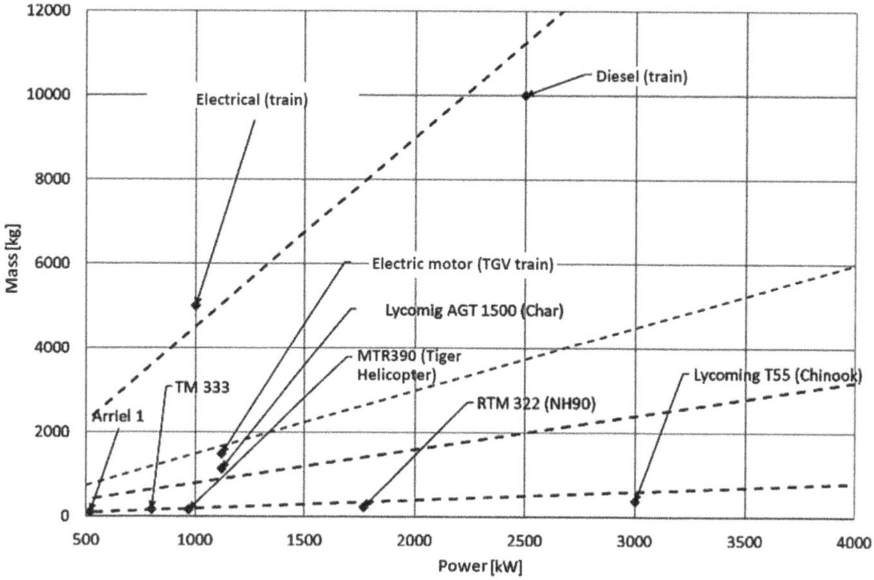

Figure I.11. *Specific power for different engines*

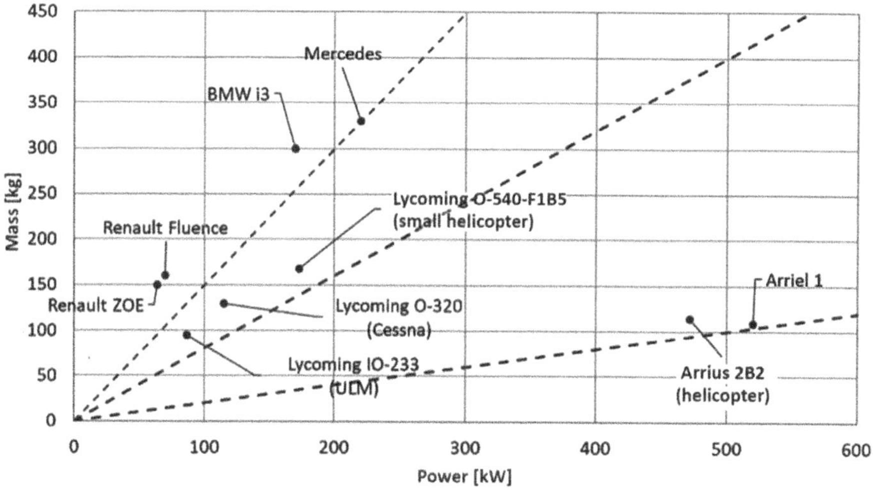

Figure I.12. *Comparison of specific powers: medium powers*

It is indeed the total power-to-weight ratio of the power source and the energy density – each associated with a type of energy source (hydrogen, electric, fuel, etc.) – that must be considered if we want to compare energy sources.

The ratio (P_3) of fuel mass to the sum of structural mass and payload mass is therefore important. Using Breguet-Leduc's two formulas:

$$\begin{cases} \text{variable} \rightarrow \text{distance} = \left(\frac{d_e}{g}\right) \frac{\Pi_2}{\Pi_1} \, ln\left(1 + \frac{M_{fuel}}{M_f}\right) \\ \text{fixed} \rightarrow \text{distance} = \left[\frac{d_e}{g}\right] \frac{\Pi_2}{\Pi_1} \frac{M_{bat}}{M_{bat}+M_f} = \left[\frac{d_e}{g}\right] \frac{\Pi_2}{\Pi_1} \frac{1}{1+\frac{M_f}{M_{bat}}} \end{cases} \qquad [I.30]$$

By comparing two systems, of the same L/D ratio and propulsive efficiency, assuming that the final mass, or dry mass, consisting of the payload and the structural mass is identical, the range can be compared as a function of the fuel mass or the battery mass.

Consideration should also be given to the impact on the structure mass. As the battery mass has increased, the structure mass will increase. It is possible to analyze this relationship using the constructive index i which translates the dry mass ratio (payload + structure) to total mass:

$$i = \frac{M_{Structure}+M_{\text{payload}}}{(M_{Structure}+M_{\text{payload}})+M_{fuel}} \qquad [I.31]$$

	M_T	M_{fuel}	$M_{payload}$	$M_{structure}$	i	d
Large airliner	575.00 T	254.00 T	102.00 T	168.00 T	0.52	15400 km
Airliner	77.00 T	200.00 T	101.00 T	42.40 T	0.17	6150 km
Heavy-lift helicopter	8.60 T	2.37 T	4.00 T	2.23 T	0.72	840 km
Average-lift helicopter	1.90 T	0.56 T	0.83 T	0.51 T	0.71	620 km
Small-city car	1.00 T	0.04 T	0.35 T	0.62 T	0.97	700 km

Table I.3. *Comparison of masses and constructive index*

It can be observed that, depending on the type of product or the task to be performed, the mass distribution is not identical (Figure I.13). For a car, the fuel mass is relatively low (4%) in terms of load and especially in terms of structure, while for an aircraft it is higher (40%).

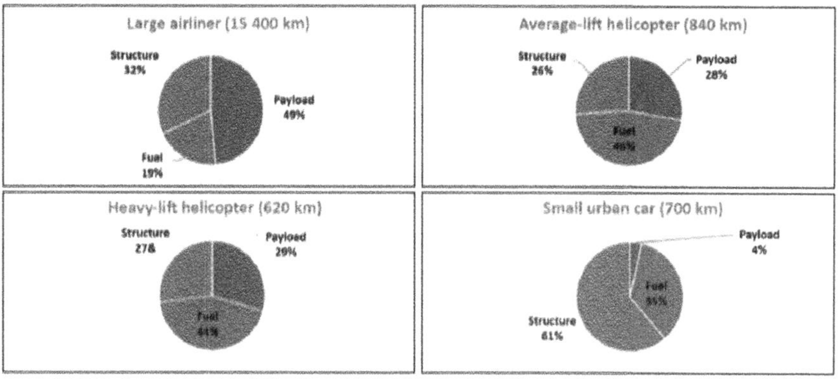

Figure I.13. *Weight distribution according to product type. For a color version of this figure, see www.iste.co.uk/krysinski/automotive.zip*

In fact, the concerns of car vehicle manufacturers, aircraft manufacturers or helicopter manufacturers are not the same.

The mass distribution between a gasoline vehicle and an electric vehicle can also be compared.

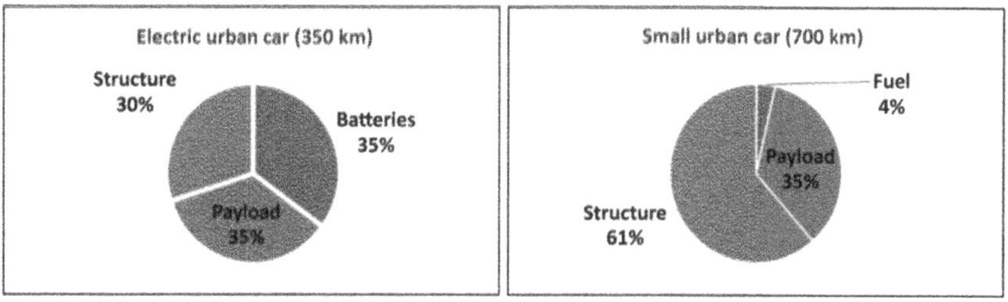

Figure I.14. *Comparison of the mass distribution of electric cars and gasoline cars. For a color version of this figure, see www.iste.co.uk/krysinski/automotive.zip*

On an aircraft, the amount of fuel is relatively large compared to the payload in order to carry out its transport mission. By tracing the change in range according to the battery or fuel mass, everything being identical, we observe (Figure I.15) that with batteries from 2017, it was impossible to offer the same range as for kerosene.

Assuming that the batteries have made sufficient progress to provide an energy density comparable to that of kerosene, it can be seen that the range is shorter. The battery mass would have to be increased by 10% to regain the same range.

The structure weight on a car is – compared to other masses – very significant. It is mainly related to the engine's power-to-weight ratio, in particular to ensure acceleration performance, safety and accessories.

It should be noted that historically, after a phase of decreasing vehicle mass, there has been an increase in vehicle mass since the 1980s, while much technological progress has been made. The increase in power consumption so that the consumer has more acceleration comfort, the development of safety standards and the increasing number of accessories lead to an increase in the vehicle's dry mass and thus mask the progress made to optimize consumption.

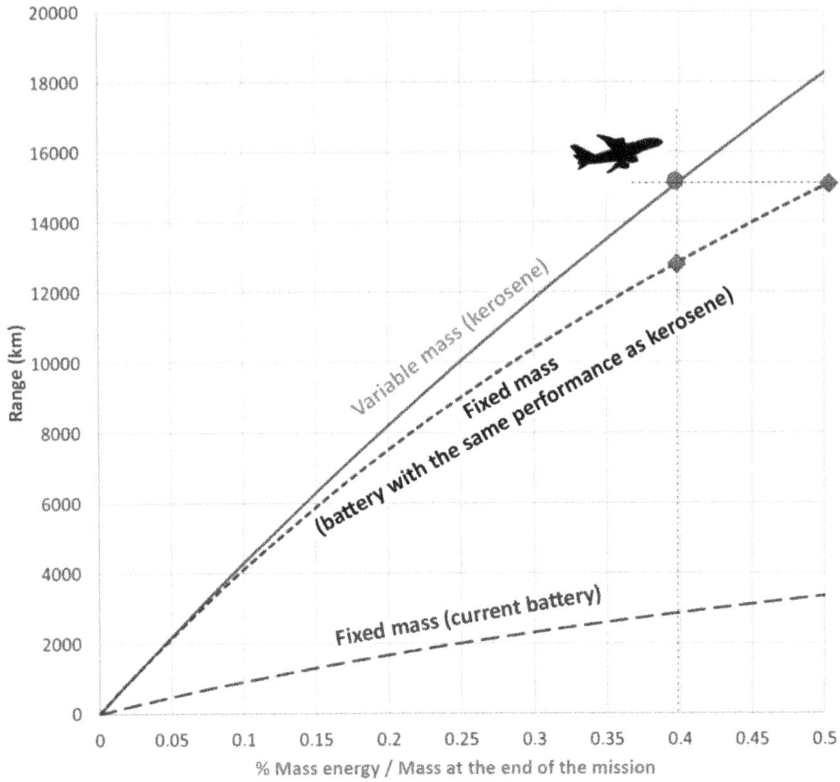

Figure I.15. *Influence of energy type on an aircraft's range. For a color version of this figure, see www.iste.co.uk/krysinski/automotive.zip*

When the proportion of fuel mass is compared with the dry mass, it is shown that the Breguet-Leduc formula can be written in the first order:

$$\begin{cases} \text{variable} \rightarrow \text{distance} = \left(\frac{d_e}{g}\right) \frac{\Pi_2}{\Pi_1} \ln\left(1 + \frac{M_{fuel}}{M_{final}}\right) \approx \left(\frac{d_e}{g}\right) \frac{\Pi_2}{\Pi_1} \frac{M_{fuel}}{M_{final}} \\ \text{fixed} \rightarrow \text{distance} = \left[\frac{d_e}{g}\right] \frac{\Pi_2}{\Pi_1} \frac{1}{1+\frac{M_{final}}{M_{bat}}} \approx \left[\frac{d_e}{g}\right] \frac{\Pi_2}{\Pi_1} \frac{M_{bat}}{M_{final}} \end{cases} \quad [I.32]$$

This shows that the two formulations are identical. In fact, the difference is directly related to energy density. It can be seen that battery technologies in 2017 still made a significant difference in range with gasoline, but that progress in battery performance had a direct impact on range.

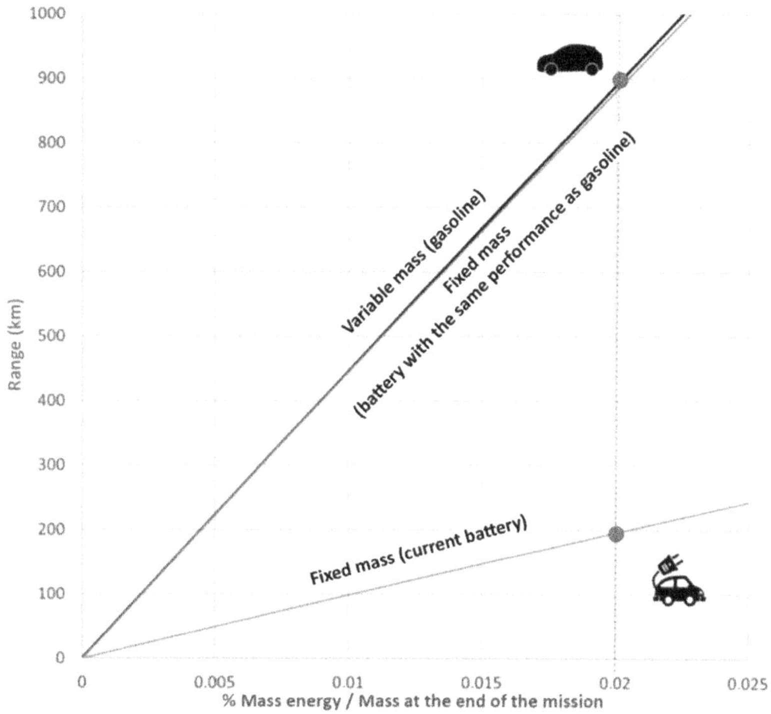

Figure I.16. *Influence of energy type on a car's range. For a color version of this figure, see www.iste.co.uk/krysinski/automotive.zip*

This Breguet-Leduc formula approach is interesting in terms of the trends it can reveal. It has also made it possible to show the importance of engine power through various characteristics (efficiency, power-to-weight ratio, type of energy) on the product's performance. The following chapters describe the propulsion systems and their characteristics in more detail.

I.3. Reflection on aircraft structure

In the field of aeronautics, a number of aircraft structures are put into competition to provide vertical take-offs or landings, a more or less long possibility of hover and cruise flight to cover a certain distance. It is interesting to compare the formulas according to the energy used to hover. This is done by comparing the ratio of the maximum aircraft weight to the hover power according to the ratio of the maximum weight to the wing surface (surface enabling lift) (Figure I.17).

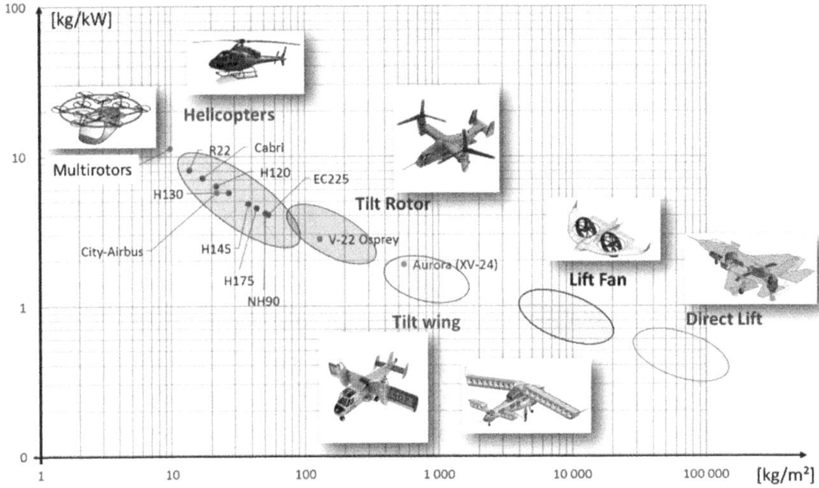

Figure I.17. *Comparison of formulas; hover efficiency. For a color version of this figure, see www.iste.co.uk/krysinski/automotive.zip*

It can be seen that for helicopters, the change in the two parameters follows a linear law in *log* scale which is linked to the classical mass law allowing us to size the rotors:

$$P = \sqrt{\frac{F^3}{2\rho S}} \Rightarrow \frac{P}{M} = \sqrt{\frac{g^3}{2\rho}\frac{M}{S}} \Rightarrow \frac{M}{P} = \sqrt{\frac{2\rho}{g^3}}\frac{1}{\sqrt{\frac{M}{S}}} \qquad [I.33]$$

Let:

$$log\left(\frac{M}{P}\right) = log\left(\sqrt{\frac{2\rho}{g^3}}\right) - \frac{1}{2}log\left(\frac{M}{S}\right)$$

$$\Rightarrow Y[kg/kW] = A - B\ X[kg/m^2] \qquad [I.34]$$

The other formulas, tilt rotor, tilt wing, lift fan and direct lift, require less and less power for a smaller load-bearing surface at the aircraft's maximum iso mass. Nevertheless, it can be observed that they are increasingly unable to hover for long periods of time (Figure I.18). Through typical missions, a helicopter is estimated to have hover flight times during a mission of the order of magnitude of 35–40 min. However, direct lift will only be seen in hover flight for take-off and landing stages.

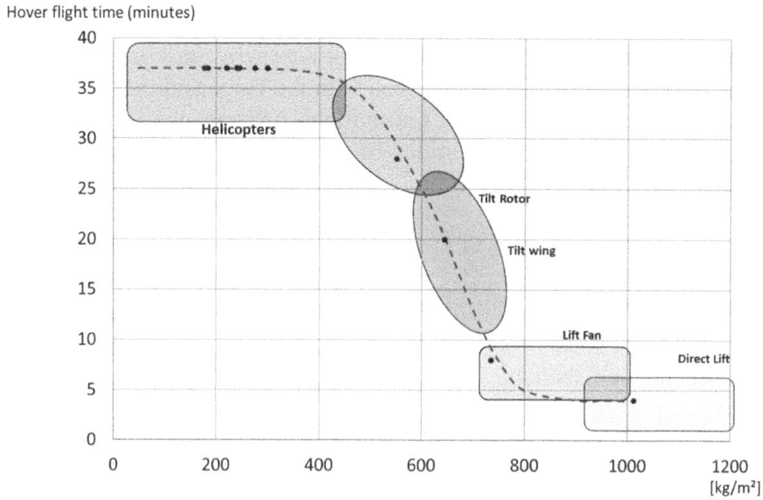

Figure I.18. *Comparison of hover flight time formulas. For a color version of this figure, see www.iste.co.uk/krysinski/automotive.zip*

In the case of these vertical take-off and landing aircraft (VTOL) formulas, we observe that each of these formulas can find its place in missions specific to it according to the time reserved for hover flight in the context of standard missions.

I.4. Reflection on the productivity of aircraft formulas

We are currently seeing a significant emergence of vertical take-off and landing formulas (VTOLs) to prepare the future market for electric urban taxis. As these formulas are very varied, it is important to be able to classify them according to the energy performance aspect. This is because the electrochemical storage capacity (batteries), as well as the storage in liquid or gaseous hydrogen, is limited.

To compare formulas, from a user or manufacturer's point of view, it is necessary to measure the product's productivity. In the context of aeronautics and for products, the mission of which is to achieve distance, such as an aircraft, productivity can be measured by the product of the payload, the range and the cruise velocities reduced to the cost:

$$\text{Productivity} = \frac{M_{PL} \times d \times V}{cost} \qquad [I.35]$$

Similarly, for VTOL that perform a cumulative hover mission with a distance to be flown, it is necessary to quantify their performance in these two flight configurations in order to compare them against the energy cost. In cruise flight, it is considered to be the product of the payload with the range and the cruising velocity reduced to the energy consumed. In hover flight, performance is measured by the product of the payload with the duration of the hover flight reduced to the energy consumed:

$$\left| \begin{array}{l} \text{productivity}_{\text{hover flight}} = \dfrac{M_{PL}\, T}{E_C} \\ \text{productivity}_{\text{cruise flight}} = \dfrac{M_{PL}\, d\, V}{E_C} \end{array} \right. \qquad [I.36]$$

The formulas can then be compared in Figure I.19 in terms of their hover performance versus their cruise performance. We observe that some formulas, such as Volocopter's VC200, achieve very good performance in hover flight but are less efficient compared to cruise flight. The tilt rotor formulas are effective when cruising. Conventional helicopters remain the ones that achieve the best compromise between hover and cruise flight missions.

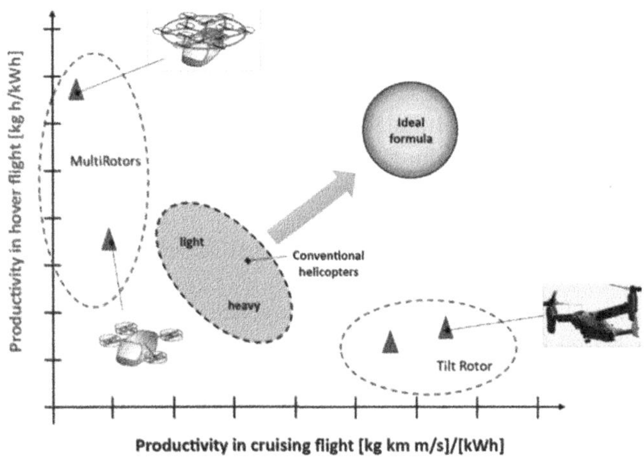

Figure I.19. *Comparison of the productivity of different formulas of vertical take-off and landing aircraft (VTOL). For a color version of this figure, see www.iste.co.uk/krysinski/automotive.zip*

The objective would be to have a VTOL structure that would allow a better compromise between hover and forward flight to compete with the conventional helicopter formula.

Other concepts, more difficult to quantify, must also be taken into account when comparing the relevance of each formula.

A first safety criterion involves analyzing the tolerance to possible component failures for hover or cruise flight functions: how many rotors can be lost without loss of control of the aircraft and so that the pilot remains able to ensure a relatively safe landing? What is the chance of survival if there is a loss of the tilting function of the rotors or wings?

A second criterion involves analyzing the complexity of the vehicle structure from a technological point of view, taking into account development cost, user cost and maintenance cost. This complexity is more difficult to quantify. It could be analyzed by a component count, a coupling element count and a nominal state count of the product.

The flight domain, which depends on the structure chosen, is a third criterion for comparison. Some flight areas may be prohibited for certain structures, giving the advantage to others depending on the desired missions.

Structure also has an impact on the choice of flight laws to be used to ensure the safety or flight qualities of an aircraft. The ability to fly without complex piloting laws is an important point of comparison between the various formulas.

I.5. Conclusion

We have seen in the various parts of this chapter that it is necessary – in order to define engine and propulsion system components – to take into account structure mass in relation to the payload using a constructive index.

The desire to introduce electric motors in transport, or even to substitute them for fossil fuels, cannot be taken into account without this consideration. We have seen that it is possible – in the current state of technology – to achieve this change in the short or medium term in relation to the car: there is the possibility of recovering energy during the braking phases and, as long as the desired range is not too high, the constructive index is relatively low. In recent years, we have seen the development of vehicles with 100% electric energy (Zero Emission Vehicles or ZEVs) or hybrid vehicles that combine fossil and electric energy.

For aeronautics, the energy recovery phases as on a land vehicle do not exist and the constructive index is not of the same order of magnitude. Nevertheless, in certain particular operating phases, the use of electricity in hybridization may be considered. This may be the case for electric motors working in case of a failure of the internal combustion engine for helicopter autorotation, during the transient modes of the internal combustion engine to exceed their pumping limits or use in a "range extender" such as a forward flight that would only use an internal combustion engine. The electric energy of the battery would then be complementary in the energy-intensive take-off phases.

As for achieving 100% electric propulsion, applications are only possible in the short or medium term for uses whose payload would not be very large (1 to 4 passengers or equivalent payload) and for a relatively short range (30 mins flight time). These performances could meet the challenges of Urban Mobility by developing flying taxis, whether autonomous or not.

The objective of this book is to show the technical challenges and the different existing solutions associated with energy management of automotive and aeronautical means of transport. We wish to demonstrate this in the context of the energy transition that marks the current transport world.

1

Motorization and Reflection on Ideal Engines

The objective of this chapter is to show what the ideal engine should be for a helicopter-type rotary wing, then for an aircraft and finally for a car, freeing itself from engine technology and the type of energy, based on loads and power requirements.

This analysis will make it possible to compare the type of response provided by internal combustion engines, turbines or electric motors and, consequently, the power chain structures required for each mode of transportation.

1.1. Motorization for an aircraft

1.1.1. *Helicopters*

The conventional helicopter, as defined below, is equipped with a motorized main rotor, ensuring its lift and propulsion, and an anti-torque rotor. This is the most traditional configuration, but there are many other configurations that are not described here, such as helicopters with counter-rotating rotors or single-rotor helicopters with gas ejection through the tip of the blades [POU 07].

1.1.1.1. *Determination of hover flight power*

The study of ideal engines is carried out by analyzing the power required for certain flight configurations. This is unlike the automobile, which requires an analysis of the engine torque that is required according to the vehicle's forward speed. The basis of this analysis for rotary wings comes from Froude's theory.

1.1.1.1.1. Froude's theory

British hydraulician, Raymond Edmund Froude, proposed a theory in 1889 providing an estimate of the power required to obtain the thrust of a ship's propeller. This theory is the same one used by aircraft propeller and helicopter rotor designers [ROU 07]. It is used to describe the power balance of the operation of a rotor in hover flight and to estimate the velocity of the air passing through the rotor, which is known as Froude velocity. The assumptions are as follows:

– the air is supposed to be inviscid and incompressible;

– the flow can be considered as one dimensional, that is the velocity vector of the air set in motion by the rotor is always assumed to be parallel to the rotor axis and the modulus of this vector is constant in any section of the flow perpendicular to the rotor axis;

– the static pressure in the rotor environment is assumed to be constant, that is the barometric change between the upstream and downstream of the disc is not taken into account. A variation of 1 hectopascal, or 1 mbar every 28 ft, is considered, or about every 8.54 m at sea level in standard atmosphere.

With Froude's theory and the above hypotheses, the rotor is considered as a system that modifies the kinetic energy of the fluid flowing through it by giving it velocity (Figure 1.1). It is assumed that the air passing through the rotor disc is located far upstream, at a zero V_0 velocity and far downstream, in a cylinder, at a uniform velocity noted $V\infty$.

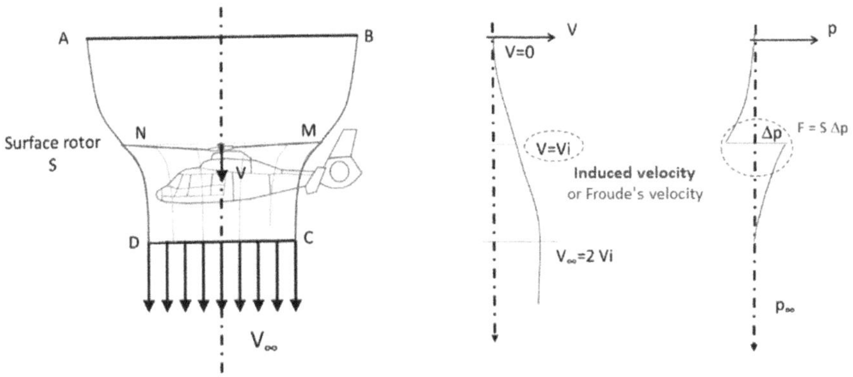

Figure 1.1. *Froude's theory. Diagram of flow and pressure. For a color version of this figure, see www.iste.co.uk/krysinski/automotive.zip*

The fundamental principle of dynamics can then be applied to all the air in the BMCDNA cylinder:

$$\sum_{ext \to fluid} \vec{F} = \iiint_V \rho \frac{d\vec{v}}{dt} dV \qquad [1.1]$$

Assuming that the regime is permanent – and therefore that there is flow retention – the fundamental principle of dynamics to a perfect fluid can be written as the Euler equation:

$$\sum_{ext \to fluid} \vec{F} = q_m(\vec{V}_\infty - \vec{V}_0) \qquad [1.2]$$

or in projection on the flow axis, and considering the zero V_0 velocity:

$$F = q_m V_\infty \qquad [1.3]$$

F corresponds to the force exerted by the propeller on the fluid. The contribution of the actions due to the pressure p is neglected. The air that passes through the rotor plane with velocity V then receives the power:

$$P = F\ V = q_m V_\infty V \qquad [1.4]$$

From an energy point of view, this power corresponds to the variation in kinetic energy of the air mass from AB to CD. To do this, we can write the force exerted by the propeller on the air as:

$$\Delta W = E_c(\infty) - E_c(0) = \frac{1}{2} m(V_\infty^2 - V_0^2) \qquad [1.5]$$

Assuming that the power is constant in a steady state, we have:

$$W = \int_{\Delta t} P\ dt \Rightarrow P_{average} = P = \frac{\Delta W}{\Delta t} \qquad [1.6]$$

Then we obtain:

$$P = \frac{\Delta E_c}{\Delta t} = \frac{1}{2}\frac{m}{\Delta t}(V_\infty^2 - V_0^2) = q_m \frac{V_\infty^2}{2} \qquad [1.7]$$

Comparing the two equations [1.4] and [1.7] demonstrates that:

$$V = \frac{V_\infty}{2} \qquad [1.8]$$

This velocity V is what the rotor "sees" and it is defined as the induced velocity noted V_i. The definition of the mass flow rate through the surface of the rotor disc is:

$$q_m = \rho\ S\ V \tag{1.9}$$

The expression of the force F is then:

$$F = \rho\ S\ VV_{\infty} = 2\ \rho\ S\ V^2\ \Rightarrow\ V = \sqrt{\frac{F}{2\rho\ S}} \tag{1.10}$$

Namely for the power:

$$P = F\ V = F\sqrt{\frac{F}{2\ \rho\ S}} = \frac{F^{\frac{3}{2}}}{\sqrt{2\rho S}} \tag{1.11}$$

If the rotor thrust force is attributed to the helicopter in hover flight, we can then write whilst introducing the mass of aircraft M:

$$P = \frac{(Mg)^{\frac{3}{2}}}{\sqrt{2\rho\ S}} = \frac{1}{R}\frac{(Mg)^{\frac{3}{2}}}{\sqrt{2\pi\rho}} \tag{1.12}$$

The analysis of the previous expression shows that:

– the power required for hover flight is proportional to the mass M of the aircraft at 3/2 power. We can thus observe that an increase of 10% in mass increases the power required by 15%;

– the power required is inversely proportional to the radius R. It is therefore necessary to have large rotor radii on helicopters. Other limiting constraints, presented elsewhere, limit their size;

– the power is a function of the density of air. This decreases with altitude: for a given mass, it will be necessary to have more power to ensure lift.

For these three points, the following elements can be specified:

1) Figure 1.2 shows the law linking the take-off weight and rotor radius for the same power. We can see that to go from 2,500 kg to 3,500 kg, the radius must go from about 10 m to 15 m.

2) One of the limits of excessive radii is due to the static deflection of the blades under their own weight when stopping or starting. This can be illustrated in a simplified approach, as in Figure 1.3.

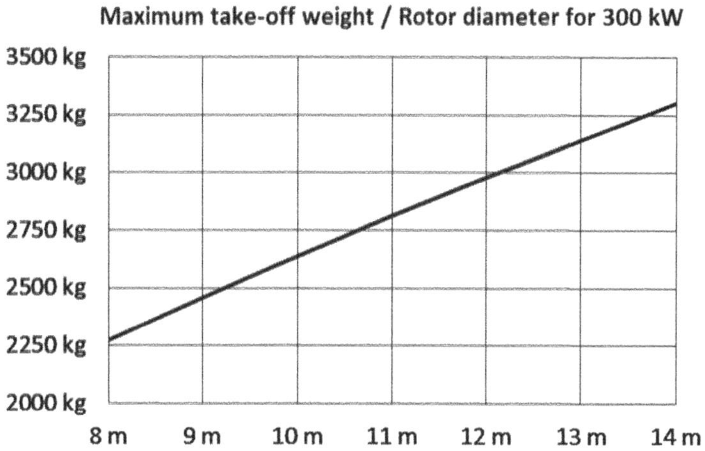

Figure 1.2. *Mass as a function of radius for constant power*

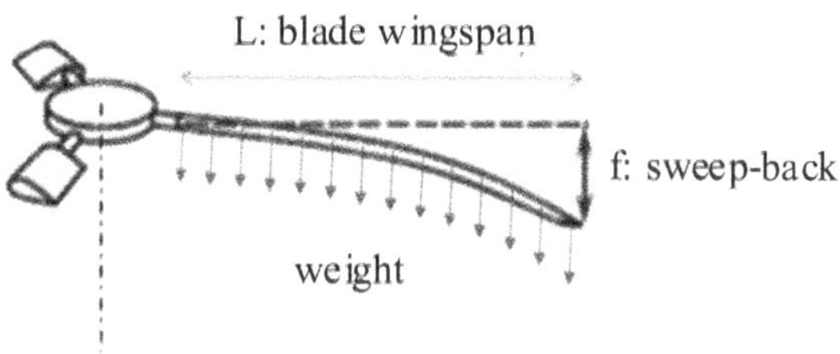

Figure 1.3. *The blade's bending deflection under its own weight*

It is assumed that the blade is embedded at its root, that the rotor hinge offset is neglected, and that the blade is assimilated to a homogeneous beam of length L. Using beam theory, under its own weight, uniformly distributed, we show that the deflection at its extremity is given by:

$$f = \frac{3\, \rho_l\, L^4}{24\, E\, I} \qquad [1.13]$$

with:

– ρ_l: the blade's linear mass;

– E: the blade's Young modulus;

– I: the blade profile's quadratic bending moment.

With identical blade characteristics, the same profile geometry, same material and same linear distribution, the deflection evolves like the length at power 4. A 10% increase in blade length leads to a 40% increase in sweep-back.

3) Another important element is to consider the variation in air density. This is a function of relative humidity, temperature and pressure. Indeed, altitude and meteorological conditions which influence the atmosphere's temperature and pressure are to be taken into account when analyzing the power requirement.

For example, under normal pressure and temperature conditions (1,013.25 hPa, 0 °C), the air density is 1.293 kg/m^3 while in the density table at an altitude of 0 m at 15 °C, the density is 1.225 kg/m^3. This can be modeled by considering the change in pressure using the hydrostatic law:

$$\frac{d\,p}{dz} + \rho\, g \,=\, 0 \qquad [1.14]$$

By combining with the ideal gas law, we obtain:

$$\frac{d\,p}{p} = -\,\frac{g}{RT}\,dz \qquad [1.15]$$

For a uniform temperature (rough but relevant simplification) T0 = 255 K, the barometric formula [MIL 10] is obtained:

$$p \,=\, p_0\, e^{-\frac{z}{H}} \qquad [1.16]$$

with:

$$H = \frac{R\,T_0}{g} \approx 7.4\,km;\;\; scale\ height$$

We observe that pressure is halved every 5 km (because H ln(2) ≈ 5 km), which has an impact on the variation in air density in relation to the altitude.

The example in Figure 1.4 demonstrates the evolution of density for dry air.

This change is often represented using Gil Denis' formula [DEN 99]:

$$\rho(h) = \rho_0\left(\frac{20000-h}{20000+h}\right) \qquad [1.17]$$

Through the evolution of air density according to altitude, it is shown that, when passing from 0 m to 7,000 m altitude, the density is halved. With the same aircraft mass, and the same rotor diameter, the power required for hover flight is 42% higher than at sea level.

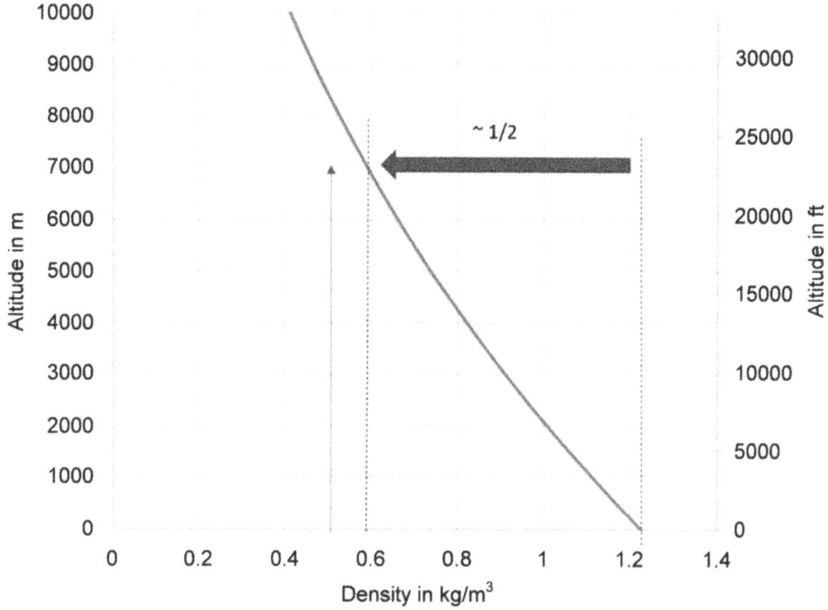

Figure 1.4. *Evolution of density as a function of altitude – dry air [DEN 99]. For a color version of this figure, see www.iste.co.uk/krysinski/automotive.zip*

On May 14, 2005, a helicopter was able to land on the summit of Everest (8,848 m) for more than two minutes. At this altitude, the density of the air was only about 0.480 kg/m^3, the power required was then increased by 60% due to the sea level. The helicopter used was a standard aircraft, a Squirrel AS350 B3, from which only unnecessary elements, such as passenger seats or comfort elements, had been removed, ultimately reducing the weight by 120 kg, with only one hour of fuel.

Froude's theory ideally allows us to estimate the power required for rotary wing lift. In reality, the power required for achieving lift by the rotor is more important: in particular, the fluid's viscosity effects on the blade must be taken into account and it is necessary to provide power to maintain the yaw of the helicopter in order to counterbalance the torque induced on the structure by the main rotor. However, this formulation has the advantage of being able to compare helicopters and associated rotors with each other.

Thus, through this formulation, it is possible to estimate whether a person would be able to maintain themselves in hover flight using the force of their own muscular forces. We consider a man weighing 80 kg, for an air density of 1.225 kg/m^3. It is assumed that the maximum muscle power for an accomplished athlete is 1 kWatt (by way of comparison, a professional cyclist reaches about 400 W on a climb). It is assumed that the rotor efficiency is 0.70.

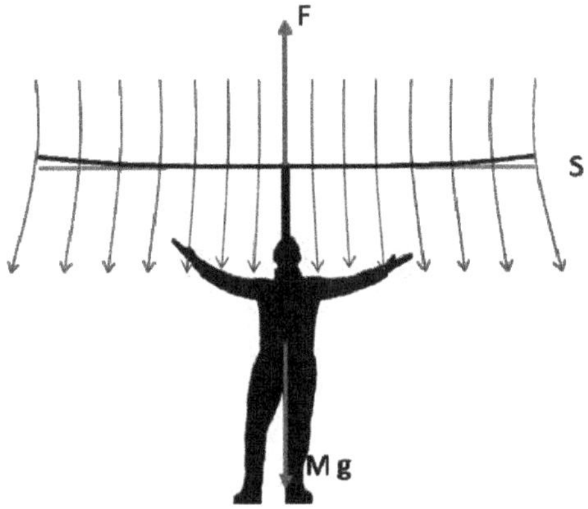

Figure 1.5. *Diagram of human-powered hover flight. For a color version of this figure, see www.iste.co.uk/krysinski/automotive.zip*

By means of Froude's formulation, it is possible to estimate the surface area S, and therefore the rotor diameter D required to ensure this lift according to power P:

$$S = \frac{(Mg)^3}{2\rho\ (\eta P)^2} D = \frac{\sqrt{2}}{\eta P}\sqrt{\frac{(Mg)^3}{\pi\ \rho}} \qquad [1.18]$$

In this case, it is estimated that it is necessary to have a surface area of 403 sqm, or a diameter of 23 m. This is enormous and poses technical problems in terms of realization.

For example, the Gamera II project, designed and built by students at the University of Maryland, succeeded in June 2012 in flying for 50 seconds with the sole strength of the pilot's muscles with four rotors 13 meters in diameter, that is a surface area of 531 m^2, for a structure weighing only 32 kg [BER 12].

This power, which is relatively high compared to the on-board mass, is specific to this mode of lift. This allows a comparison to be made with the case of a human-powered aircraft.

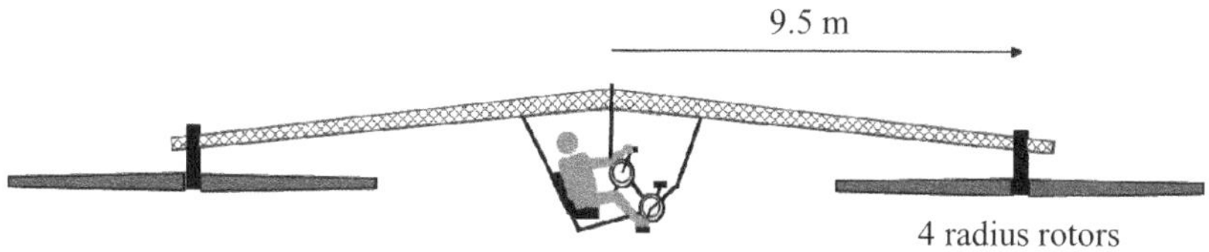

Figure 1.6. *Diagram of the Gamera II project [BER 2012]*

For an aircraft, it is shown that the power required to maintain a constant altitude and velocity follows the following law:

$$F_D = \frac{1}{2}\rho S C_D\, V^2 \text{ and } F_L = \frac{1}{2}\rho S C_L V^2 = Mg$$
$$\Rightarrow P_s = F_D V = \frac{1}{f}\sqrt{\frac{2\, g^3\, M^3}{\rho\, C_L S}} \quad [1.19]$$

with:

– M: the aircraft's mass;

– S: the aircraft's load-bearing surface;

– V: the forward velocity;

– g: gravity;

– ρ: the density of air;

– C_D: the aerodynamic drag coefficient;

– C_L: the aerodynamic lift coefficient;

– f: the aircraft's L/D ratio.

The Gossamer Albatros, a bicycle-type, human-powered aircraft, crossed the English Channel on June 12, 1979 in two hours and 49 minutes, traversing all 36 km, at a velocity of 13.669 km/h [GRO 04]. The mass of the machine alone was approximately 39 kg and the pilot 55 kg, for a lift surface of 79.4 m^2. The wingspan was almost 30 m, the L/D ratio about 17 and the aerodynamic coefficient of lift about 0.69.

By using the power equation formula, we can estimate:

$$P_s = \frac{1}{f}\sqrt{\frac{2\,g^3\,M^3}{\rho\,C_L\,S}} = \frac{1}{17}\sqrt{\frac{2\times 9.81^3\times 94^3}{1.225\times 0.69\times 79.24}} = 0.241\text{ kW} \quad [1.20]$$

Estimating the propeller's transmission efficiency of 0.80, the pilot power was in the order of 300 W. This meant a pilot-specific power of approximately 4.8 W/kg (pilot power/pilot weight).

We therefore have a specific power that is lower for a fixed wing compared to a rotating wing (4.8 W/kg versus 12.5 W/kg).

Figure 1.7. *The Glossamer Albatros (AVIONEWS – World Aeronautical Press Agency). For a color version of this figure, see www.iste.co.uk/krysinski/automotive.zip*

If the comparison is made for small-lift motorized aircraft, the same type of deviation is observed.

Thus, for a small Robinson R44 helicopter[1], a four-seater which has a 5.02 m rotor radius, with a maximum mass of 1,089 kg and a payload of 415 kg, the power required, based on Froude's theory with an efficiency of 70%, is:

$$P = \frac{1}{\eta}\frac{(Mg)^{\frac{3}{2}}}{\sqrt{2\rho\,S}} = 113\ kW \quad [1.21]$$

For a small Cessna 172[2], a four-seater aircraft with a wingspan of 11 m, a maximum mass of 1,200 kg (payload of 454 kg), a lift surface area of 16.2 m^2,

1 https://en.wikipedia.org/wiki/Robinson_R44.

2 https://en.wikipedia.org/wiki/Cessna_172.

an L/D ratio of about 17 and the wings' lift coefficient at 0.72 (at 130 km/h), the power required [1.19] is:

$$P_s = \frac{1}{f}\sqrt{\frac{2\,g^3\,M^3}{\rho\,C_L\,S}} = \frac{1}{17}\sqrt{\frac{2\times9.81^3\times\,1200^3}{1.225\times0.72\times16.2}} = 44\,kW \qquad [1.22]$$

We therefore have, once again, a specific power that is less for a fixed wing than for a rotating wing (104 W/kg versus 248 W/kg).

Propulsion systems are 145 hp and 180 hp respectively. The difference is due to the fact that there are design criteria other than useful power under the sole vision of lift (hover flight for the helicopter and constant velocity landing for an aircraft).

1.1.1.1.2. Profile power

The Froude model provides an ideal estimate of what can be expected, and thus compares the rotors with each other using an ideal model. To have a more predictive model and therefore one that is closer to reality, it is necessary to take into account the viscosity of air which induces an energy dissipation not taken into account by the Froude model. To do this, it is necessary to look at the blade and the power developed by the aerodynamic forces due to drag and lift:

$$P_T = \underbrace{F\,V_i}_{Induced} + \underbrace{P_d}_{profile} \qquad [1.23]$$

As shown in Figure 1.8, the aerodynamic forces on a profile come from lift and drag. Their projection into the rotation plane, via the radius, gives a resistant torque. This load develops a power that must be taken into account when estimating the power required for lift in addition to the power required to accelerate the air through the rotor disc.

On an entire blade, by integration, the torque C_r opposing the rotation created by the aerodynamic forces can be expressed by:

$$C_r = \int_0^R \frac{1}{2}\rho a C_x (\Omega r)^2 r dr \qquad [1.24]$$

with:

– ρ: air density;

– a: the rope;

– Ω: rotor angular speed;

– R: rotor radius;

– C_x: aerodynamic coefficient of the blade integrating the lift effects C_{x0} and C_z:

$$C_x = C_{x0} + k\ C_z^2 \quad [1.25]$$

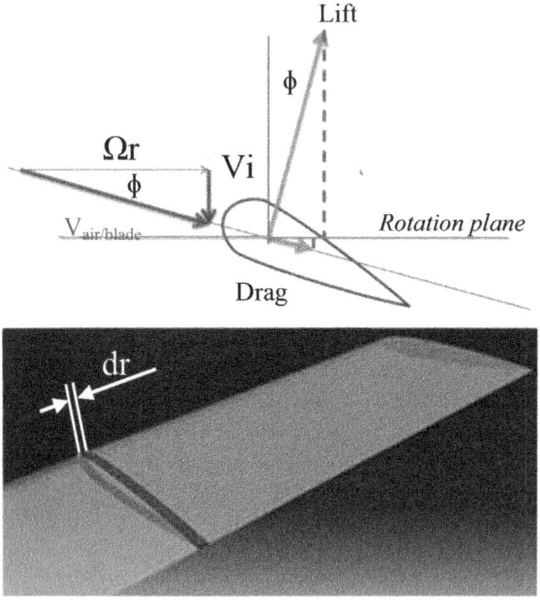

Figure 1.8. *A blade's aerodynamic lift and drag forces. For a color version of this figure, see www.iste.co.uk/krysinski/automotive.zip*

Then, for a rotor with N blades, the total power P_T required for lift becomes:

$$P_T = FV_i + \frac{N}{8}\rho a C_{x0} \Omega^3 R^4 + \frac{N}{8}\rho a k C_z^2 \Omega^3 R^4 \quad [1.26]$$

This formulation allows a more accurate estimate of the hover lift power taking into account the viscosity effects of the fluid. Contrary to Froude's formula, it is then necessary to have the design of the blades to use this formulation.

1.1.1.1.3. Effects of aircraft fairing

Some rotating wing boxes can be streamlined and this is the case with some drones or the rear rotor of a Fenestron helicopter. This may raise the question of the influence of this fairing on the evaluation of the power used for lift or thrust in terms of the rotor.

A fairing rotor is geometrically characterized by the diffusion ratio:

$$\delta = 2\frac{S_2}{S_1} \quad [1.27]$$

with:

– S_1: the rotor cross-section;

– S_2: the outlet section of the diffuser.

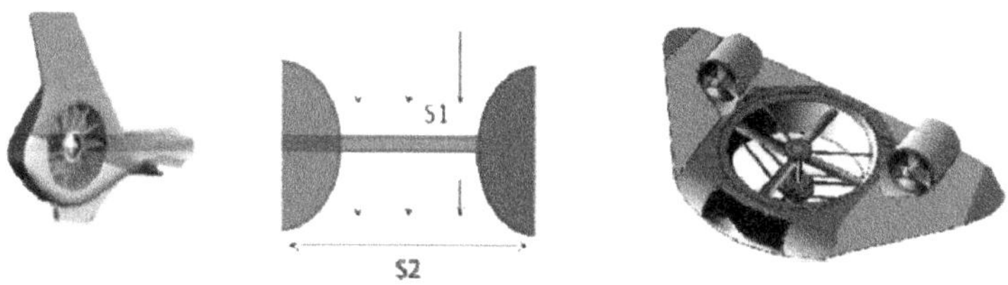

Figure 1.9. *Streamlined rotor. For a color version of this figure, see www.iste.co.uk/krysinski/automotive.zip*

Then, using Froude's theory, we can show that the rotor's power compared to non-powered:

$$P = \frac{P_{free}}{\sqrt{\delta}} \quad [1.28]$$

For a cylinder, the diffusion coefficient is 2, in which case the required power is only 70% of a free rotor's power. The main objectives are to reduce the velocity effects associated with drag, improve the safety of ground staff, reduce the risk of collisions with foreign objects such as stones, and to reduce noise levels. These are the advantages that enable the strength of the fairing on the rear rotors of Fenestron helicopters, and on some rotary wing drones claiming to go at high forward velocities.

1.1.1.2. *Determination of the forward flight power stage*

In order to define an ideal engine for a helicopter, it is useful to analyze the power required in forward level flight, at constant altitude and speed. This approach is done through the power balance approach.

This is mainly due to the induced power required to create lift, the drag related to blade rotation and the fuselage power induced by the aircraft's drag and rotor drag (Figure 1.10). Other power-consuming elements (mechanics, accessories, etc.) are neglected here.

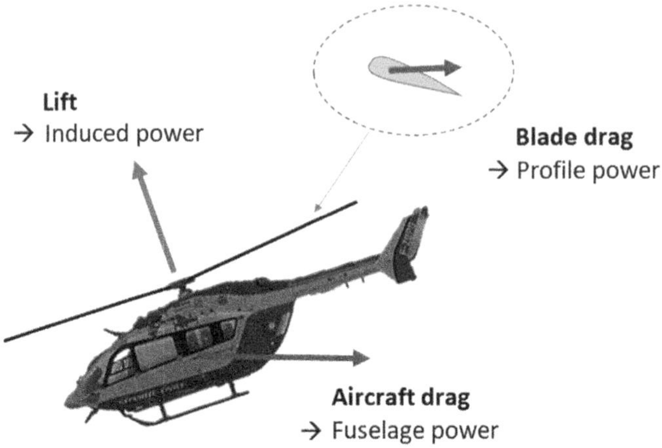

Figure 1.10. *Power balance of a helicopter in forward flight. For a color version of this figure, see www.iste.co.uk/krysinski/automotive.zip*

The induced power, or Froude power, is modified in forward flight. Figure 1.11 demonstrates the air flow induced by drag and the impact on the induced velocity.

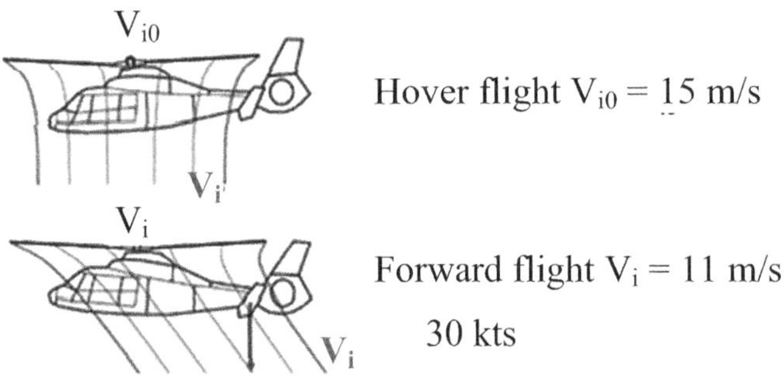

Figure 1.11. *Evolution of the induced flight in forward flight. For a color version of this figure, see www.iste.co.uk/krysinski/automotive.zip*

Figure 1.12 shows the evolution of induced power as a function of formal speed.

The power consumed by the general drag of the fuselage including the fuselage itself, accessories and interactions between subsystems is generally represented as follows:

$$P_{fus} = \frac{1}{2}\rho\left(C_x\, S_f\right)\, V^3 \qquad [1.29]$$

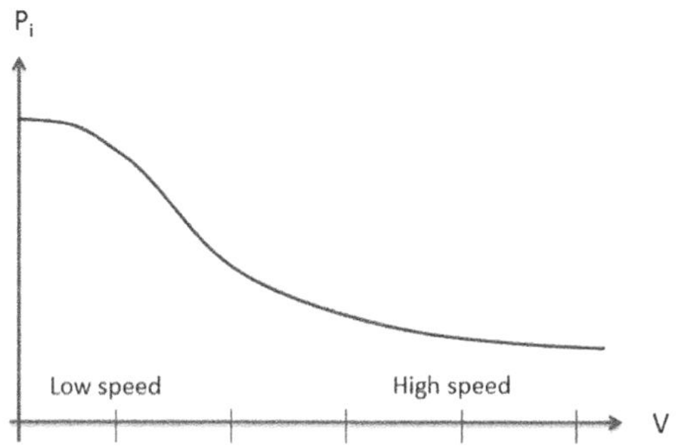

Figure 1.12. *Variation of induced power as a function of speed*

We observe that it increases proportionally to the cube velocity. The total of the induced power and the fuselage power is shown in Figure 1.13. We note that at low speeds and hover flight, it is the induced power that is predominant.

However, it is the fuselage's drag power that is predominant at high speeds, with the induced power becoming low (less than 10% of the total power). In forward flight, several specific needs are observed: V1 the minimum power required for flight, V2 the maximum L/D speed (maximum range speed) and V3 the maximum possible speed for an available power.

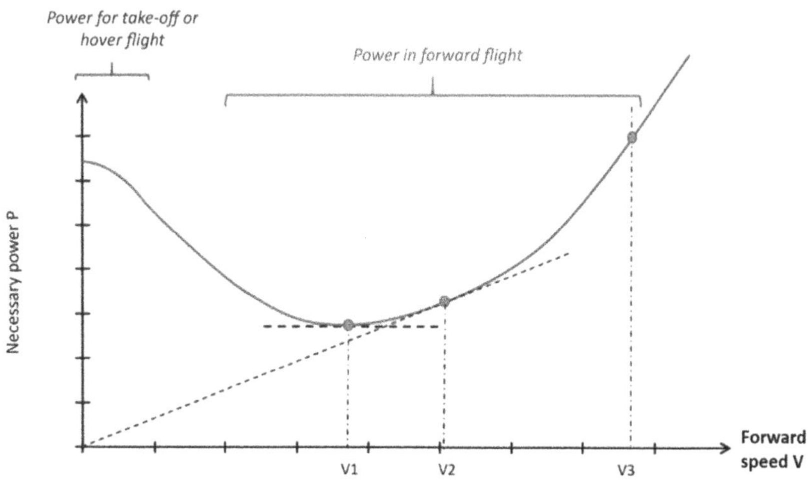

Figure 1.13. *Evolution of the power required in relation to forward velocity. For a color version of this figure, see www.iste.co.uk/krysinski/automotive.zip*

1.1.1.3. *Particularity of the ground effect*

Generally speaking, ground effect refers to the evolution of aerodynamic forces applied to a plane or helicopter when close to the ground. For helicopters, ground effect is defined as an increase in performance when the helicopter is approximately at a height of one rotor diameter from the ground. The helicopter needs less power to take off due to this ground effect [ROU 07].

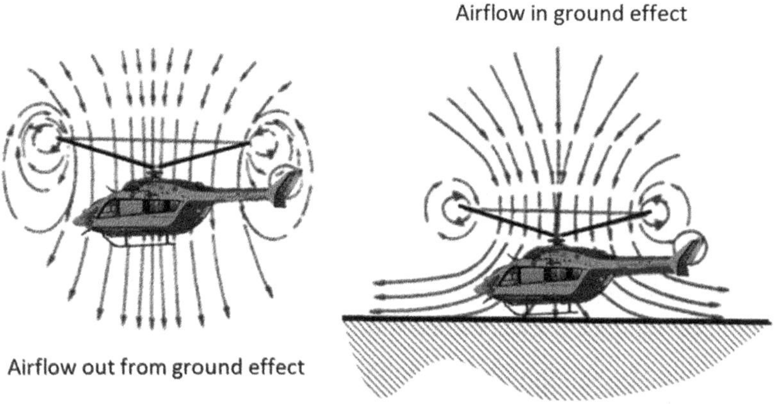

Figure 1.14. *Illustration of the ground effect. For a color version of this figure, see www.iste.co.uk/krysinski/automotive.zip*

1.1.1.4. *Notion of autorotation – problematizing engine failure*

Helicopter manufacturers must provide a scenario for occupant protection in case of engine failure. As a helicopter does not have wings, it cannot, like some aircraft, glide to land. In the event of engine failure, motor reversibility is used. When falling, the blade is rotated and that creates overall lift to slow the fall, and especially to store energy in the rotor, which stabilizes at a given angular velocity. The pilot, when close to the ground, uses this energy through a specific maneuver called flare, in order to land safely.

It should be noted that to perform this maneuver in good conditions, the rotor must be able to store as much energy as possible. This is taken into account in the blade design and in particular when defining the inertial and geometric characteristics of the blade. A second point is that the drop height must be sufficient to achieve an appropriate rate of descent (Figure 1.16).

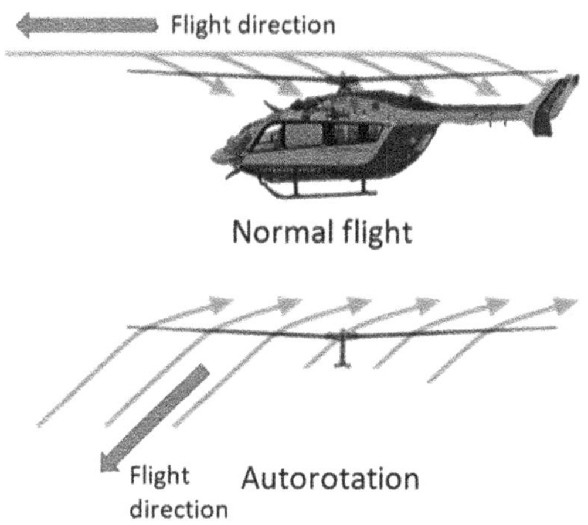

Figure 1.15. *Air flow from helicopter autorotation. For a color version of this figure, see www.iste.co.uk/krysinski/automotive.zip*

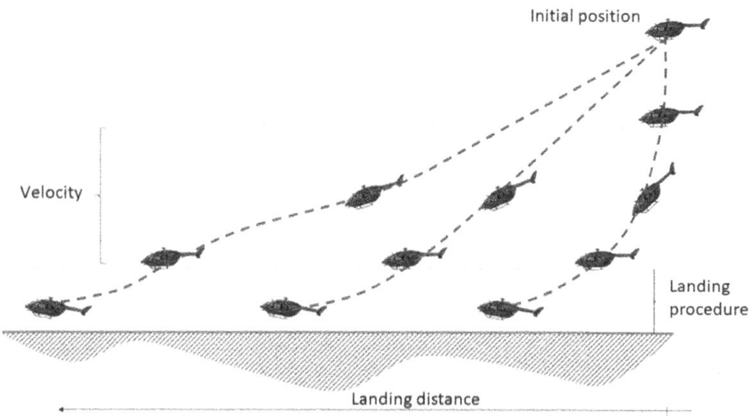

Figure 1.16. *Influence of velocity on landing success*

If the helicopter is in forward flight, this minimum drop height is reduced, allowing the pilot to use this speed to rotate the rotor. In the helicopter's height/velocity diagram, there is a danger zone where the pilot will not be able to land in good conditions. In the case of a single-engine helicopter, it is shown that this area is of the shape shown in Figure 1.17.

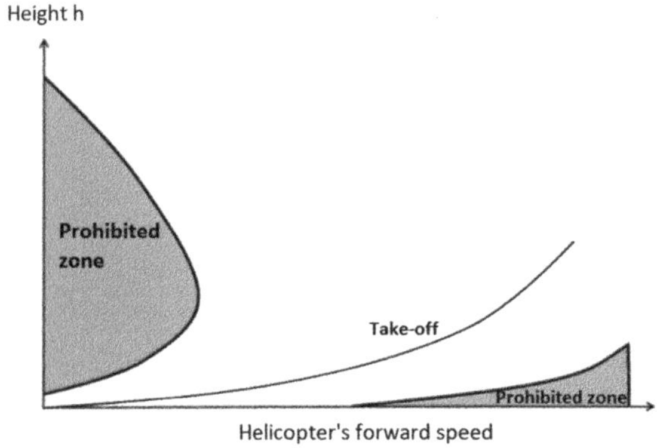

Figure 1.17. *Definition in autorotation (single-engine) of prohibited areas*

The ability to land safely therefore depends on the height and speed at the time of engine failure.

For multi-engine helicopters it is statistically accepted that both engines cannot fail at the same time. However, this does not mean that the helicopter can maintain hover flight with a single engine. Nevertheless, the failure procedure for a multi-engine helicopter takes into account that the helicopter still has power available. The idea is that the device can reach the flight speed where its power requirement is minimal (Figure 1.19). This V1 speed can be achieved if the pilot, in the event of engine failure, can nose-dive until this speed is attained, with enough power remaining for the pilot to be able to maintain level flight. Again, there is a restricted area in the height/velocity diagram to ensure that the pilot can achieve this speed but it is, by definition, smaller than for single-engine aircraft.

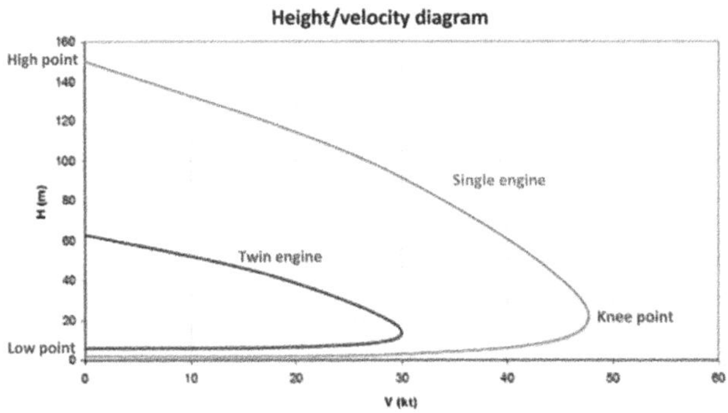

Figure 1.18. *Height/speed diagram for twin- and single-engine helicopters. For a color version of this figure, see www.iste.co.uk/krysinski/automotive.zip*

It can be seen that the ideal choice of engine power has an impact on the definition of this area.

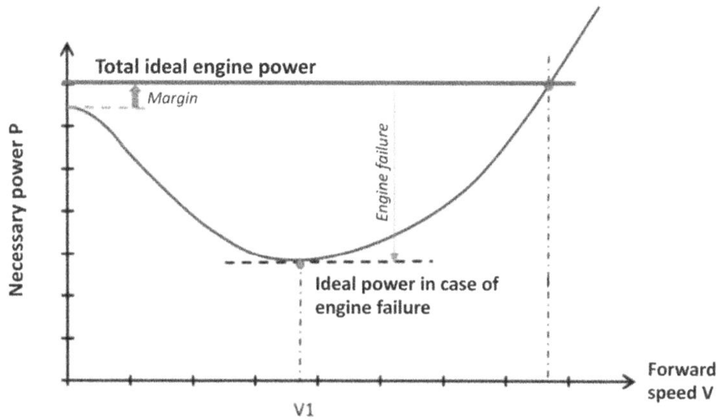

Figure 1.19. *Transportation powertrain principle for a twin engine*

It is shown that an ideal helicopter power system should cover twin-engine failures with the possibility of continuing the mission at an economical speed.

1.1.2. *Aircraft*

If we want to analyze the ideal motorization for an aircraft, it is also necessary to analyze the lift and drag forces during the different flight regimes.

To understand what is happening in different flight configurations, it is necessary to rely on the aircraft's load balance. The aircraft is considered to be an unalterable solid mass M with a center of gravity G. It is meant to fly in a calm and windless atmosphere. The mechanical actions that apply to the aircraft are weight, aerodynamic effects and propulsion forces. The aircraft's balance leads to six equations whose resolution makes it possible to evaluate the propulsion forces required in different flight regimes.

Let's examine the case of straight climb flight. It is assumed that the incidence is low enough to make the aircraft itself correspond to the aerodynamic unit and that the forces are all applied at the same point G (Figure 1.20).

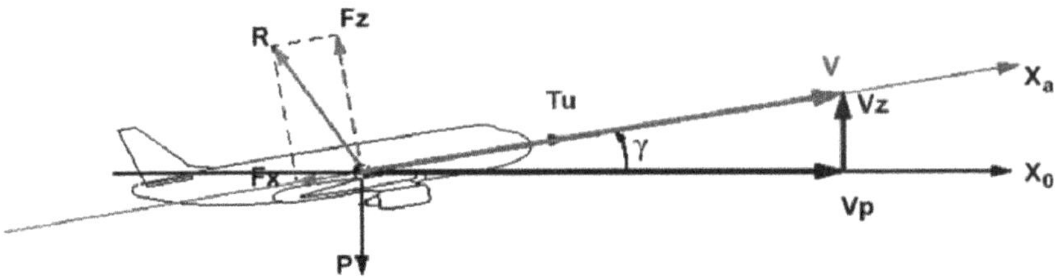

Figure 1.20. *Forces applied to a climbing. For a color version of this figure, see www.iste.co.uk/krysinski/automotive.zip*

With:

– Tu: engine thrust;

– Fz: the aerodynamic forces;

– Fx: the drag induced by the entire aircraft;

– P: the weight;

– Vp: horizontal velocity;

– Vz: vertical velocity;

– V: the resultant of the velocity.

The projection of the equilibrium then gives the following two relations:

$$\begin{cases} T_u = F_x + Mg \ \ sin(\gamma) \\ 0 = F_z - Mg \ \ cos(\gamma) \end{cases} \quad [1.30]$$

From these equations, we can see that the angle of climb is given by:

$$\gamma \approx sin(\gamma) \approx \frac{F_u - F_x}{Mg} \quad [1.31]$$

The minimum thrust required for level flight is given for:

$$(F_u)_{\text{mini}} = F_x \quad [1.32]$$

The angle of climb, reduced in percentage points, is then defined by:

$$\gamma = 100\left(\frac{F_u - (F_u)_{mini}}{Mg}\right) \quad [1.33]$$

A take-off without engine failure is the normal and most frequent take-off configuration. Nevertheless, it is necessary to consider the event of engine failure and the possibility of having to operate with the remaining N-1 engines; for example, when an engine fails in the critical phase, when the aircraft has exceeded a minimum engine failure velocity (V_{EF}). This corresponds to the time when the distance required to stop the aircraft is equal to the length of the remaining runway. Critical situations are located for this take-off phase with One Engine Inoperative (OEI) engine failure. Several critical segments (1, 2, 3 and 4) can then be defined (Figure 1.21).

In the event of engine failure, from the V_{EF} point onwards, minimum engine performance is required to ensure continued take-off to ensure climb gradients for the different phases. For example, the minimum stabilized climb gradient after the landing gear has been fully retracted and 400 ft is required at 2.4%. For an engine, it is 2.7% and for a four-engine, 3%.

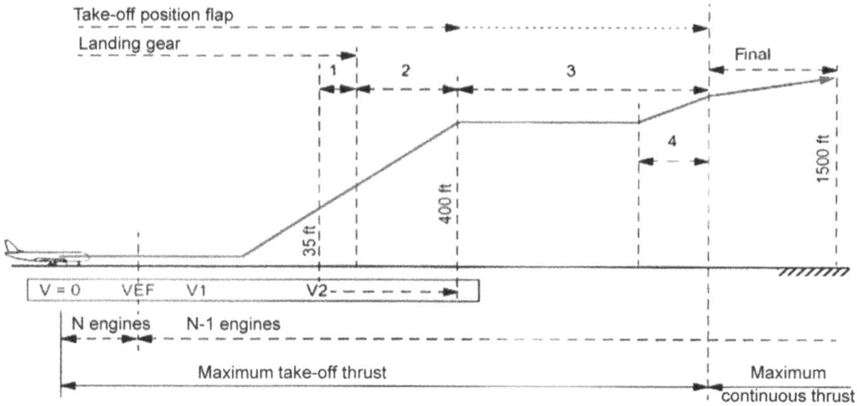

Figure 1.21. *Different phases of an aircraft's take-off (L'avionnaire)*

Thus, in segment 2, for a twin-engine aircraft with one engine failing, the remaining one must provide take-off at take-off power, with the landing gear retracted, the curvature flaps in the take-off position and at a climb velocity equal to V2:

$$\begin{cases} F_u = (F_u)_{mini} + 2.4\%\ Mgtwin\ engine \\ F_u = (F_u)_{mini} + 2.7\%\ Mgengine \\ F_u = (F_u)_{mini} + 3.0\%\ Mgfour - engine \end{cases} \quad [1.34]$$

1.1.3. *Compound formulas*

Rotary wings such as helicopters enable hover flight unlike fixed wings. However, due to the drag generated by the rotor, the rotary wings are limited in forward speed.

The manufacturers' idea is to propose new formulas, called "compound" formulas, in the sense that specific modules are added to the structure: either wings that generate lift, or thrusters such as propellers or additional engines to provide additional propulsion (Figure 1.22).

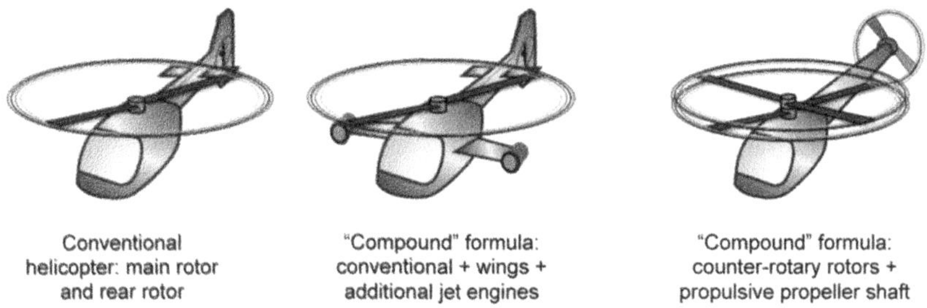

Figure 1.22. *Examples of derived "compound" formulas of the conventional helicopter*

This rather old idea is found on the Bell 533, an aircraft derived from the UH-1 Iroquois and modified accordingly. Work was carried out on the blades to reduce drag and two turbojets were added on fixed wings. The Bell also had its flight controls modified to use the rotor pitch control during low velocity phases and to switch to the principle of aircraft flight controls in high velocity flight phases. This prototype, qualified as a High-Performance Helicopter (HPH), reached 275 kts (509 km/h) in 1969.

Between 2008 and 2011, Sikorsky, with the X2 prototype, proposed an experimental prototype consisting of two counter-rotating rotors and an auxiliary propulsion system based on a propeller. A particular effort was made to reduce the drag of the counter-rotating rotors by using fairing elements on the rotor and by reducing the motor's rotational speed at high velocity flight. Its record flight was set in 2010 at 250 kts (460 km/h) in horizontal flight.

The manufacturer Eurocopter, now Airbus Helicopters, proposed the prototype for a faster helicopter in 2010. This so-called "compound" hybrid formula was composed of a main rotor and two propellers installed on the wings. In 2013, the aircraft was able to fly at 255 kts (472 km/h) in level flight (Figure 1.23).

Figure 1.23. *Airbus Helicopters X3 prototype (source: Airbus Helicopters)*

For all configurations, it can be observed that reducing the rotor drag power is an important step. It is conventional to use the advancement parameter μ which translates the ratio of the aircraft's velocity to the relative velocity of the blade tip in relation to the fuselage:

$$\mu = \frac{V}{\Omega\ R} \quad [1.35]$$

For example, for an 8 m blade and a rotor that rotates at 260 rpm, at a velocity of 20 kts (37 km/h), the parameter is 0.047 and for a velocity of 150 kts (278 km/h) the travel parameter is 0.354. Generally, for a conventional helicopter, it does not exceed 0.5. At high velocity, we show [POU 07] that the profile power in comparison with hover flight increases significantly.

$$\begin{cases} P = (1 + 5\ \mu^2)\ P_0 \\ P_0 \text{ power profile in hover flight} \end{cases} \quad [1.36]$$

With a hybrid system, lift is created by the rotor and wings. The propulsion effort can be achieved by the propellers and the main rotor. The drag forces are produced by the fuselage, propellers and rotor. Regarding a helicopter, the more the main rotor's propulsion effect is increased, the more it will tilt. The resistance force is then greatly increased. Depending on the power distribution required from the rotor or propellers, the ratio of the lift forces created by the different elements to the different drag effects can vary.

The L/D ratio of the compound can be expressed by:

$$f_{aircraft} = \frac{L}{D} = \left(\frac{1/2\ \rho\ \text{CxS}\ V^2}{m\ g\ \eta_p} + \frac{k}{f_{rotor}} + \frac{1 - \text{k}}{\eta_p\ f_{wing}} \right)^{-1} \quad [1.37]$$

with:

– η_p: the propellers' propulsive efficiency;

– k: the lift distribution assigned to the rotor (between 0 and 1);

– CxS: the fuselage's drag parameter.

It can be shown that the L/D ratio of the rotor, f_{rotor}, is strongly related to the advancement μ and the rotor's angle of incidence α. If the angle increases, the rotor's L/D ratio deteriorates. There is an attempt to fly flat, almost with no angle.

Similarly, too high CxS can affect the aircraft's overall L/D ratio. There is therefore a major interest in reducing the fuselage drag, particularly by streamlining the rotor head. It is also preferable to fly at higher altitudes where the density is lower.

All these considerations together can lead to a hybrid helicopter that has an L/D ratio that is twice as effective as a conventional helicopter. The impact of this high-velocity configuration allows one, for the same nominal power installed, to fly at higher velocities than a conventional helicopter, since the flattening of the rotor is possible thanks to the compensation given by wings and propulsions supplied by propellers.

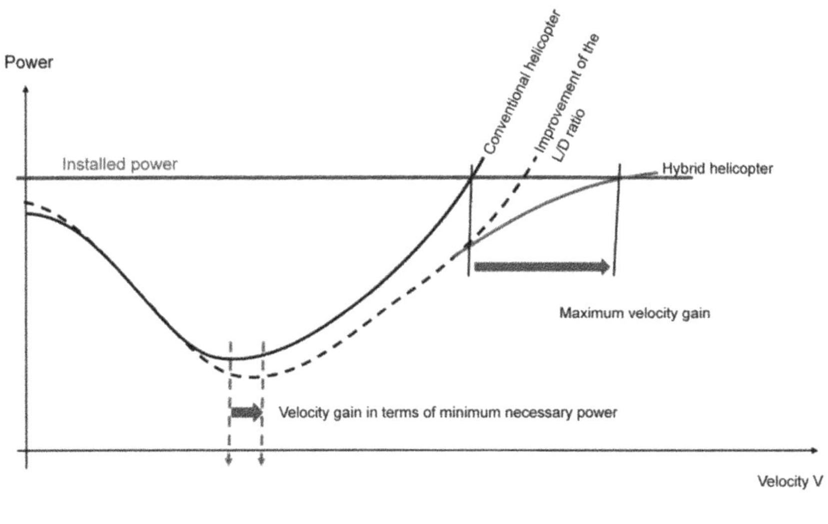

Figure 1.24. *Power required for a hybrid formula. For a color version of this figure, see www.iste.co.uk/krysinski/automotive.zip*

1.2. Motorization for an automobile

The objective of this chapter is to show, based on effort and power requirements, what the ideal engine for a passenger car could be, by freeing itself from engine technology and the type of energy. This analysis will make it possible to compare the type of response provided by thermal and electric engines, with the necessary drive train architectures specific to the transportation powertrain.

A vehicle is subjected to a combination of forces opposing motion (aerodynamics, rolling resistance, gravity and inertial loads due to acceleration). In fact, it is useful to define the vehicle's performance by focusing on the resistance loads at different vehicle velocities and for a given road based on energy requirements. In relation to automobile motorization, this means qualifying the engine in regards to its torque and power characteristics.

1.2.1. *Determining tractive force and useful power*

There are two factors to consider when defining the maximum tractive force that an engine should provide:

– the forces produced by the cumulative resistance effects, in the worst-case scenario: slope, acceleration, velocity (aerodynamics and rolling resistance);

– the maximum force transmissible by the wheel/ground contact induced by the possible slippage. It is useless to make an engine oversized when the wheel is not able to transmit as much torque.

For automobiles, performance is generally limited by the first factor, it being understood that the manufacturer will tend, under normal conditions of use, to size the tires and drive chain (reduction ratios) in accordance with the power requirement.

The resistant forces are mainly due to tire friction, aerodynamic forces and gravity effects during a slope to be climbed, as shown in Figure 1.25. In the case of a descending vehicle, the gravity forces are then the driving force.

We note:

– V: the vehicle velocity in [m/s];

– M: the vehicle mass [kg];

– g: the gravitational constant [m/s^{-2}];

– α: the road's gradient [rad];

– C_{rr}: the rolling resistance coefficient (no unit);

– C_x: the aerodynamic drag coefficient (no unit);

– ρ: the density of the air [kg/m^3];

– S: the projected surface (master torque surface) [m^2].

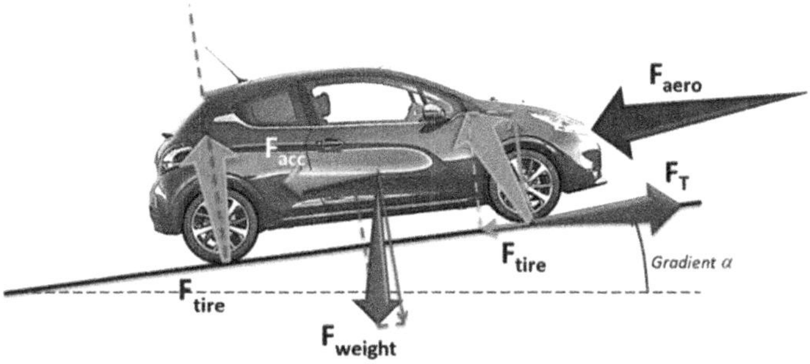

Figure 1.25. *External forces on a front-wheel drive vehicle (source: PSA). For a color version of this figure, see www.iste.co.uk/krysinski/automotive.zip*

The expressions of the different efforts are usually given by [BAD 13]:

$$\begin{cases} F_{tire} = C_{rr}(V)\ Mg\ \cos(\alpha) \\ F_{aero} = \frac{1}{2}\rho\ S\ C_x V^2 \\ F_{weight} = Mg\ \sin(\alpha) \end{cases} \quad [1.38]$$

If the vehicle is isolated, the fundamental principle of dynamics makes it possible to evaluate the traction force required:

$$F_T = M\frac{dV}{dt} + F_{tire} + F_{aero} + F_{weight} \quad [1.39]$$

This results in:

$$F_T = M\frac{dV}{dt} + C_{rr}(V)\ Mg\ \cos(\alpha) + \frac{1}{2}\rho\ S\ C_x V^2 + Mg\ \sin(\alpha) \quad [1.40]$$

The useful power at each moment is then given by:

$$P_u = \underbrace{MV\frac{dV}{dt}}_{P_{acc}} + \underbrace{C_{rr}(V)\ Mg\ \cos(\alpha)V + \frac{1}{2}\rho\ S\ C_x V^3 + Mg\ \sin(\alpha)V}_{P_{res}} \quad [1.41]$$

The power used for P_{acc} acceleration and the power lost due to the P_{res} resistance can be dissociated. The solution of providing an acceleration capability is dependent on speed and would lead to oversizing the engine power and thus, regardless of the engine technology, to increasing the engine mass. For passenger cars, it was ideally chosen to be able to provide constant power independent of speed, giving a higher acceleration capacity at low speed than at high speed and a sufficient maximum speed without oversizing the engine.

When calculating a consumption cycle during operation, it is considered that there is no slope. Slope consideration is important when sizing vehicle performance. Figure 1.26 compares the different powers involved as a function of velocity slope, assuming that the vehicle is on the road with a constant slope. The rolling resistance coefficient C_{rr} is considered constant. This coefficient can then be considered as the resistance force due to the tire reduced to the vertical load exerted on it. It is then expressed in kg/t. Thus, a rolling resistance coefficient of 10 kg/t corresponds to a C_{rr} coefficient of 0.01. Depending on the design of the tires, the inflation and ground type (bitumen, concrete, sand, etc.), this coefficient can change significantly (from 1 to 30 kg/t):

$$\begin{cases} M = 1000\ kg \quad g = 9.81\ ms^{-2} \quad C_{rr} = 0.01 \\ \rho = 1{,}225\ kgm^{-3} \quad S = 2.5\ m^2 \end{cases} \qquad [1.42]$$

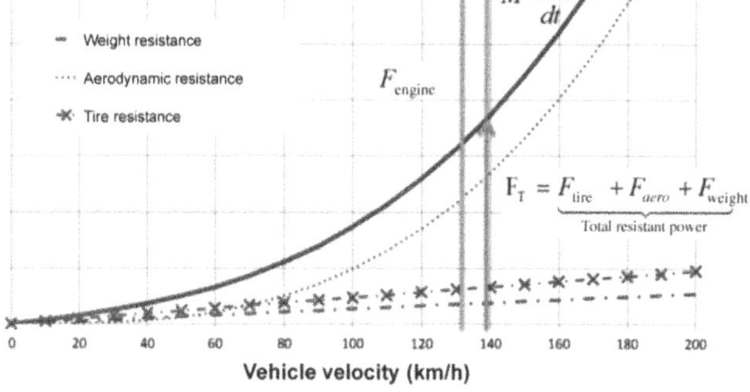

Figure 1.26. *Influence of velocity on the power required at constant speed for a constant slope. For a color version of this figure, see www.iste.co.uk/krysinski/automotive.zip*

The curve in Figure 1.27 illustrates the additional power required to achieve an acceleration of 1 m/s^2, or 0.1 g, or for a rolling resistance variation C_{rr} of 0.02 for a deflated tire, at isovelocity compared with the reference solution C_{rr} of 0.01 without acceleration.

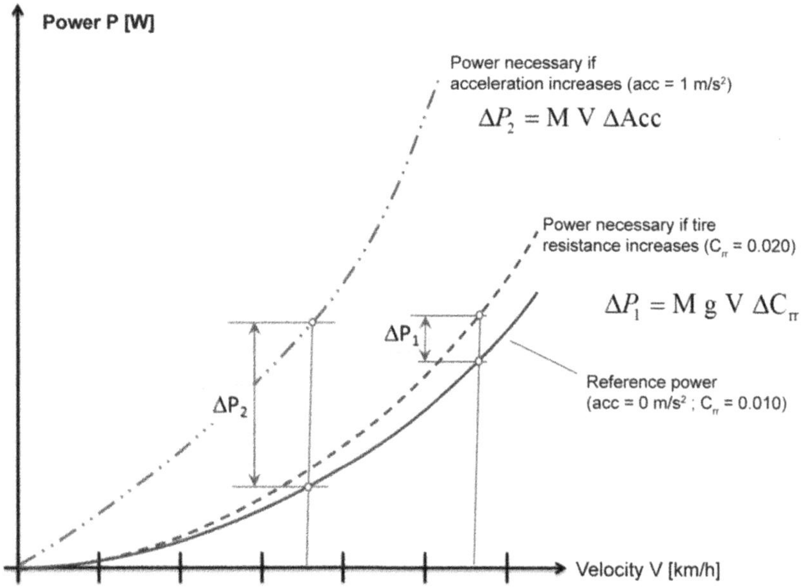

Figure 1.27. *Variation in a vehicle's power requirement at a given velocity. For a color version of this figure, see www.iste.co.uk/krysinski/automotive.zip*

To take road behavior more precisely into account, a road feature, known as LDR, was introduced, representing aerodynamic force, rolling resistance and transmission losses. It is in the form of a second order polynomial as a function of velocity with three coefficients F_0, F_1 and F_2 that depend on the vehicle, its configuration and the test procedure[3].

At first order, F_0 represents rolling resistance, F_1 represents transmission losses and F_2 represents aerodynamic resistance. We thus have:

$$F_{LDR} = F_0 + F_1\ V + F_2\ V^2 \qquad [1.43]$$

with V the vehicle's velocity in km/h.

3 A set of test procedures, including driving cycles, or approval cycles, harmonized worldwide for passenger cars and light commercial vehicles.

This gives for the expression of power:

$$P_{traction} = M\ V\ \frac{dV}{dt} + F_0\ V + F_1\ V^2 + F_2\ V^3 \quad [1.44]$$

The examples of LDRs for two passenger vehicles are given in Table 1.1.

Road laws (LDR)	Vehicle 1	Vehicle 2
Wheel radius (m)	0.3069	0.3069
Reference mass (kg)	1,495	1,598
Gross vehicle mass (GVM) (kg)	1,743	1,877
Fo (N)	123	129
F1 (N/km.h)	0.554	0.554
F2 (N/km.h^2)	0.03225	0.0309

Table 1.1. *Example of a road law for two touring cars*

To be more precise, it would also be necessary to take the inertia of the rotating elements into account, using an energy equivalence approach:

$$\begin{cases} E_c(equivalent\ system) = \frac{1}{2} M_{equ} V^2 \\ E_c(vehicle) = \frac{1}{2} \mathrm{M\ V}^2 + \sum_{no.\ \text{of wheels}} \frac{1}{2} I_{wheel} \Omega_{wheel}^2 + \frac{1}{2} \sum_i I_i \omega_i^2 \end{cases} \quad [1.45]$$

Kinematic relationships are established:

$$\begin{cases} V = R_{wheel} \Omega_{wheels} \\ \omega_i = k_i\ \Omega_{wheels} \end{cases} \quad [1.46]$$

This results in the following definition of the vehicle's equivalent mass:

$$M_{éq} = M + \sum_{no.\ of\ wheels} \frac{I_{wheel}}{R_{wheel}^2} + \sum_i \frac{I_i}{k_i^2 R_{wheel}^2} \quad [1.47]$$

with k_i the reduction ratio between the rotating element no. i of the transmission chain and the wheels.

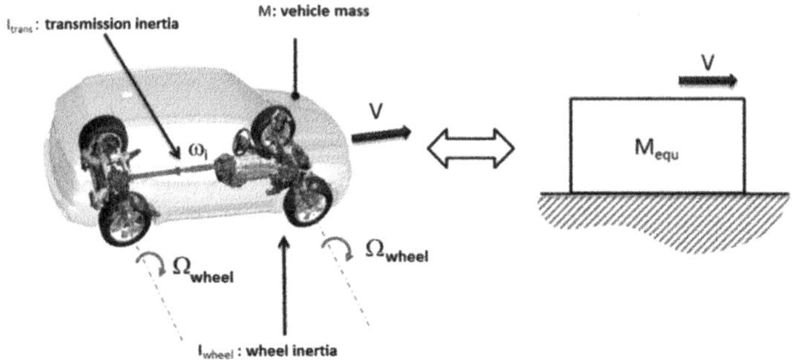

Figure 1.28. *Definition of the masses and inertias used to calculate the equivalent mass*

1.2.2. *Definition of ideal transportation powertrain*

The relationship between power and angular velocity makes it possible to obtain, for a constant transmission ratio, the evolution of the torque according to angular velocity that the constant engine power would provide (Figure 1.30).

$$P = C \times \omega$$
$$\Rightarrow \quad C = \frac{P}{\omega} \qquad [1.48]$$

However, we can see that in this case, the torque at low rotation speed becomes very significant. However, one of the factors mentioned above imposes a torque limit so that the wheel does not slip.

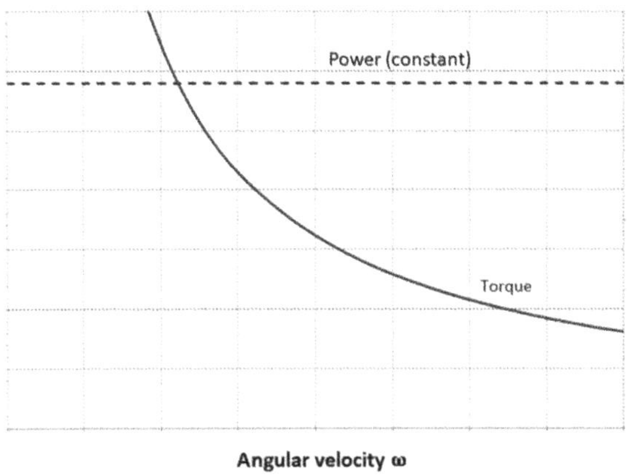

Figure 1.29. *Evolution of torque as a function of rotational speed for constant power*

This torque, when placed at the adhesion limit, can be evaluated using the model presented in Figure 1.30.

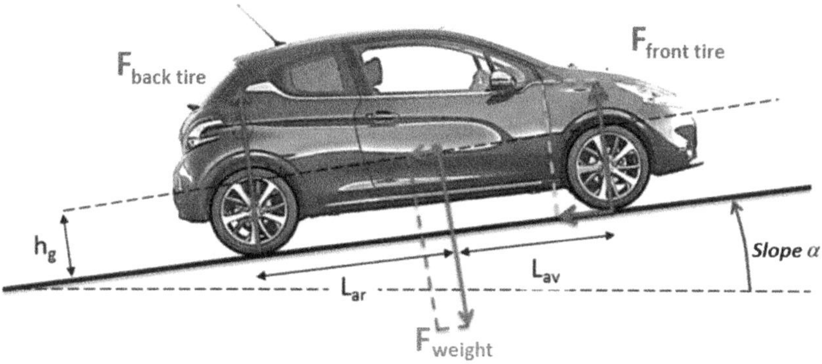

Figure 1.30. *Simplified model of the forces applied to a vehicle. For a color version of this figure, see www.iste.co.uk/krysinski/automotive.zip*

The fundamental principle of dynamics applied to the vehicle in translation on a slope (Figure 1.30) gives:

$$M\frac{dV}{dt} = F_t - \left(F_{ar} + F_{av} + F_{aero} + F_g\right) \qquad [1.49]$$

We are interested in the ground reaction forces on the wheels N_{ar} and N_{av} in order to determine the maximum tractive effort that wheel/ground contact can support (Figure 1.31).

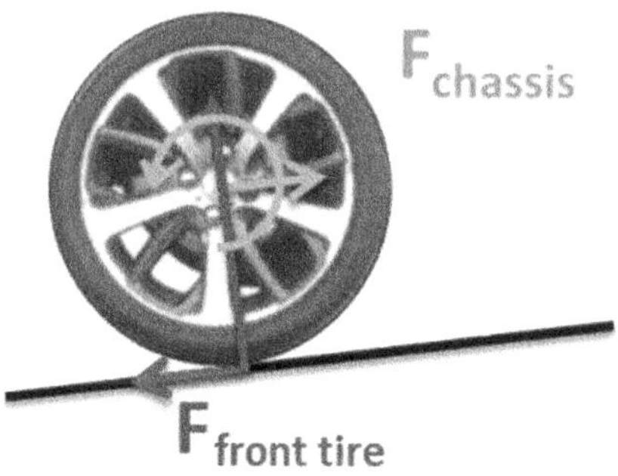

Figure 1.31. *Breakdown of actions on the wheel. For a color version of this figure, see www.iste.co.uk/krysinski/automotive.zip*

By applying the theorem of the resultant, we obtain:

$$N_{av} = \frac{L_{ar}}{L} Mg \cos\alpha - \frac{h_g}{L}\left(F_{aero} + F_g + Mg\frac{r_d}{h_g}\cos\alpha + M\frac{dV}{dt}\right) \qquad [1.50]$$

and:

$$N_{ar} = \frac{L_{av}}{L} Mg \cos\alpha - \frac{h_g}{L}\left(F_{aero} + F_g + Mg\frac{r_d}{h_g}\cos\alpha + M\frac{dV}{dt}\right) \qquad [1.51]$$

However, according to Coulomb's law, the maximum tractive effort that can be applied to prevent slippage must be less than or equal to μN. The maximum tractive effort can be rewritten in the case of a front-wheel drive vehicle by introducing equation [1.49] and applying the equation that corresponds to the total rolling resistance:

$$F_t = F_{front} + F_{back} = Mg \cos\alpha \, C_{rr} \qquad [1.52]$$

So:

$$F_{tmax} = \frac{\mu Mg \cos\alpha \frac{(L_{ar} - C_{rr}(h_g - r_d))}{L}}{1 - \frac{\mu h_g}{L}} \qquad [1.53]$$

For example, for a segment B vehicle on flat roads, with the following data: $\mu = 0.75$ (asphalt), M = 1,080 kg, L = 2.538 m, $L_{ar} = 1.5$ m, $r_d = 0.3$ m, $Cr_r = 0.008$, $h_g = 0.9$ m, a force of 6,377 N is obtained, corresponding to a maximum engine torque (overall ratio of 12) of about 160 Nm. Beyond this torque, the wheel slips.

We can therefore point out the characteristic curve of an ideal engine for a land vehicle, which is characterized by the following phenomena:

1) adherence limit (torque limitation by wheel slip);

2) motor characteristic at constant maximum power;

3) limitation in engine velocity (depending on engine technology).

The theoretical curve of an ideal motor taking into account these different limitations is given in Figure 1.32.

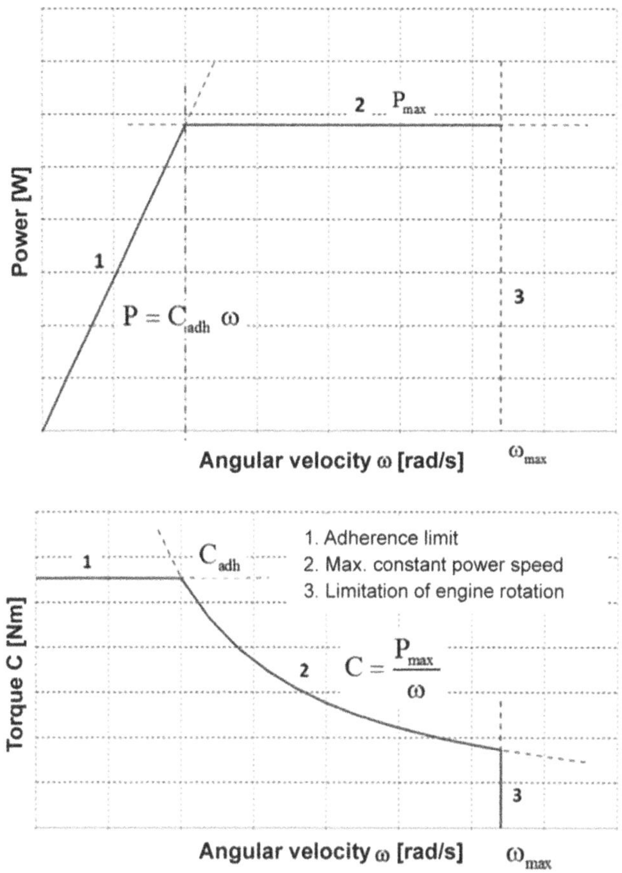

Figure 1.32. *Performance characteristics of a car engine, taking into account the limitations*

1.3. Conclusion

It has been seen that for helicopters, the choice of installed power is made according to the power required for hover flight or the maximum velocity (small aircraft). This can also be so by integrating the concept of safety for twin-engine helicopters (large aircraft), to give the possibility of providing cruise flight in economy mode in the event of single engine operation. For rotating wings, it is deemed preferable to operate the rotor at a constant angular velocity in order to minimize the effects of inertia that occur during the change of rotational speed. These could pose problems during the sizing of different components. Control is then achieved by varying the pitch of the blades.

For aircraft, sizing is based on the risk of losing an engine during take-off. Even if the risk of failure is very low, the remaining engine must allow take-off with a lower climb gradient and ensure flight until landing at the airport chosen by the pilot.

For cars, the choice of motorization is made on the basis of maximum torque at low velocity to allow high accelerations or to have the ability to operate or start on a hillside, and on the choice of maximum power at high velocities to allow for these velocities to have acceleration capabilities.

2

Engine Technologies

2.1. Introduction

The oldest means of land transport is said to be porterage. Later, humans used animals to assist in this activity. Cattle, domesticated since the fifth millennium BCE, have been used for centuries to pull loads. The donkey, domesticated in the first millennium BCE, has been used to pull and carry loads or people. The horse, a luxury animal when it was domesticated in the second millennium BCE, came into more common use later. Once the wheel was invented, transporting heavier goods became possible.

A major revolution came with the invention of the steam engine by Denis Papin, which made movement of land vehicles possible. A large number of studies and inventions created the possibility of developing different types of fuels which improved transportation powertrains and efficiency:

– gas, gasoline and oil in internal combustion engines or gas turbines;

– electricity from electric batteries;

– nuclear batteries in some submarines;

– fuel cells to supply electricity through turbines or directly to a winding (in an electric motor);

– hydrogen.

Today, some transport systems combine different energy sources. The notion of a "hybrid vehicle" often refers to a combination of a combustion engine and electric motor.

The objective of this chapter is to present different transportation powertrains through their basic principles of operation and improvements used to reduce their consumption and pollutant emissions. Gas turbines, internal combustion engines and electric engines are presented in this chapter.

It is important to mention in this introduction that the needs of propulsion systems are not the same according to two industrial sectors considered in this chapter: the aeronautics sector and the automotive sector. Some comparative elements are presented as examples in Table 2.1.

	Aeronautics	**Automobile**
Power consumption	~ 10^5 hp	~ 10^2 hp
Lifespan	~ 10^5 h	~ 10^3 h
Revision interval	~ 10^4 h	~ 10^2 h
Maintenance costs	~ 10^2 \$/h	~ 10^0 \$/h
Catalog price	~ 10^7 \$	~ 10^3 \$

Table 2.1. *Comparison of data from aeronautical/automotive specifications [AOU 15]*

2.2. Gas turbines

2.2.1. *General operating principles*

A *gas turbine*, abbreviated GT, is an internal combustion engine that produces two types of energy in the context of aeronautical applications:

– mechanical energy used for driving an aircraft propeller or helicopter rotor via a gearbox;

– kinetic energy given to the gases at the turbine outlet and expanded in the nozzle to produce thrust on an aircraft. A turbojet is a typical application of this type of propulsion.

A gas turbine is shown in Figure 2.1. Atmospheric air (point 1) is sucked in by a gas compressor (GC). The input parameters are pressure p_1 and temperature T_1. The air is then compressed to pressure p_2 at temperature T_2 (point 2). It then passes through the combustion chamber (CCh), where the liquid or gaseous fuel is supplied. The air mass flow rate is noted as q_{mc}.

Theoretically, combustion is carried out at constant pressure, but in practice, because of the drop in pressure, the pressure p_3 at the turbine inlet at T_3 is slightly lower than p_2 (it is approximately 95% to 98% of p_2).

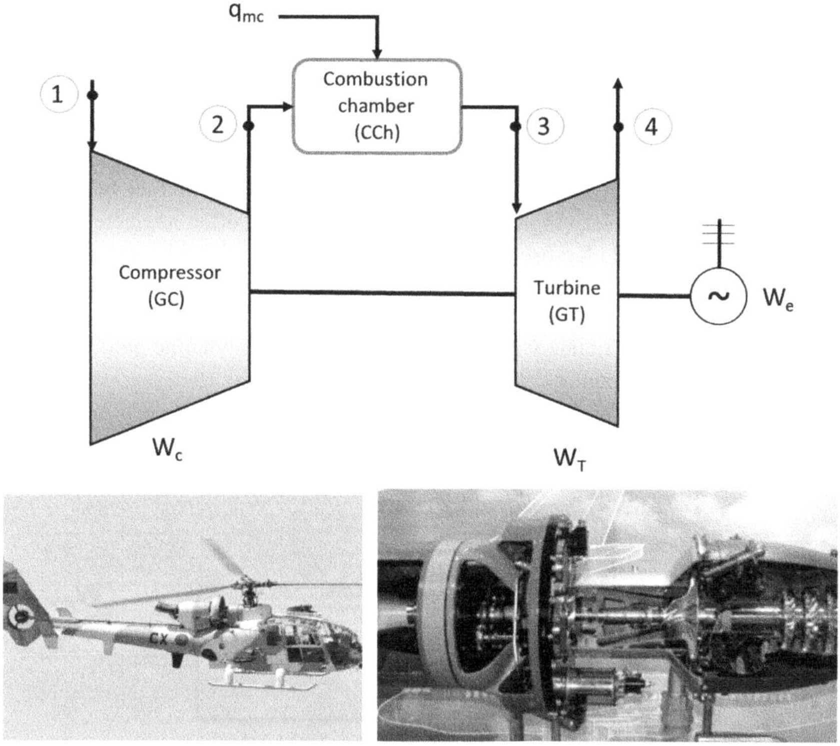

Figure 2.1. *Diagram of a single-shaft gas turbine. ASTAZOU single-shaft gas turbine in Gazelle (source: Airbus). For a color version of this figure, see www.iste.co.uk/krysinski/automotive.zip*

The waste gases, the parameters of which are p_3 and T_3, expand in the GT to point 4 (p_4 and T_4). The rotor is mounted on the same shaft as the compressor.

The power supplied by the W_T turbine is greater than the power expended by the W_C compressor, so the effective power of the W_e engine is as follows:

$$W_e = W_T - W_C \quad [2.1]$$

The overall efficiency of the gas turbine is thus defined as the ratio of the recovered power to the power supplied by the W_{CCh} fuel:

$$\eta = \frac{W_e}{W_{CCh}} = \frac{W_e}{q_{mc}\ PCI\ \eta_c} \quad [2.2]$$

with LCV (lower calorific value) of the fuel, which is a property of the fuel. This is the amount of heat released by the complete combustion of a fuel unit if steam is assumed to be non-condensed and the heat is not recovered.

As combustion is not ideal, there is still unburned fuel in the exhaust gases, and only the flow of burned fuel provides energy to the active fluid. Efficiency, or combustion efficiency η_c, is defined by the ratio of the flow of fuel burned to the total fuel flow. As it is very high (96% to 98%), losses are often neglected by taking an efficiency equal to 1. Kerosene used in aeronautics is a liquid fuel derived from crude oil with an LCV of about 44.5 MJ/kg.

The gas turbine's efficiency is then mainly of a thermodynamic nature, noted η_{th} and called thermal efficiency .

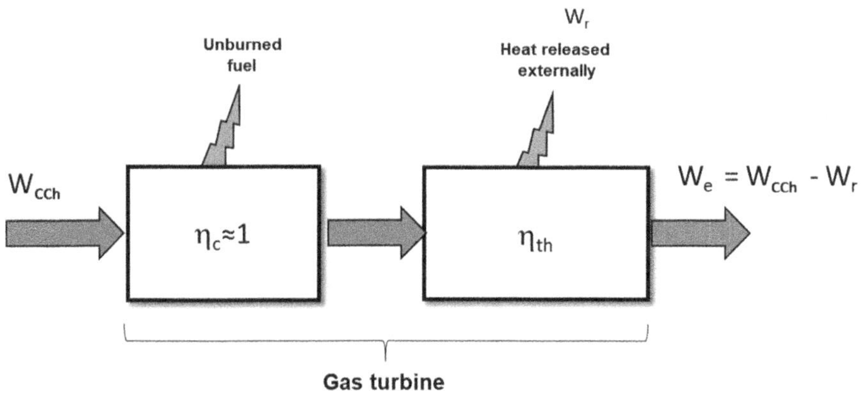

Figure 2.2. *Energy balance of a gas turbine*

2.2.1.1. *Single-shaft turbines and twin-shaft turbines*

The internal combustion turbine, with both the compressor and turbine rotors fixed on the same shaft, is called a "single-shaft turbine" (Figure 2.1). It ensures a good overall efficiency only at a velocity near to the nominal velocity. This is related to a mechanical connection of the compressor and turbine rotors. As a consequence, the decrease in the angular velocity N causes a decrease in the air flow rate q_m and the compression ratio $\Pi = p_2/p_1$ and, as a result, the effective efficiency decreases [KRY 94].

A single-shaft turbine does not generate a starting torque C_o (torque at zero velocity). A turbine cannot therefore start using machines that require torque in this phase, as is the case for motor vehicles[1], as well as machines that require high torque at low angular velocity (e.g. a ship's propeller in maneuvering conditions). Single-shaft turbines can drive machines with power characteristics for which power can be expressed as $P = k N^3$, and machines that can be loaded at rated velocity (electric alternators, rotary pumps and aircraft propellers). The first single-shaft turbine helicopters, such as the Gazelle, required a clutch to ensure start-up. At a certain speed, the rotor was then started by a centrifugal clutch.

During start-up, the turbocharger (the compressor's rotor and turbine mounted on the same shaft) is started by an electric starter or by another engine (a small turbine or piston engine). When a certain angular velocity is reached, the fuel is injected into the combustion chamber. The fuel flow is ignited by a spark of the starting spark plug. The waste gases begin to supply the turbine and, at a certain angular velocity, the turbine power becomes greater than the compressor power. The turbocharger starts to accelerate to nominal velocity at which the power can be adjusted from a zero value (no-load, where $W_e = 0$ and $\eta_e = 0$) to maximum power, which is limited by the T_3 temperature. The power variation is obtained by the variation in the flow rate of the fuel supplied to the combustion chamber (i.e. the T_3 temperature variation).

Since the single-shaft turbine does not withstand load fluctuations very well, a possible solution is to separate two functions of the turbine by splitting it into two parts: one, the CT compressor turbine, drives the compressor, the other, the usable power turbine (free turbine) FT, drives a receiving machine. Two groups are thus formed: a GG gas generator and an FT mechanical energy generator which are mechanically independent. The control is carried out by altering the gas generator's velocity, and therefore the compressor's flow rate driven by the compressor turbine.

For a constant velocity of the turbocharger (or gas generator), the power and torque of the free turbine can vary in a wide range. In the case of the free turbine, when it is stopped, its torque reaches the maximum value which ensures the traction of vehicles requiring a high starting torque. A twin-shaft turbine can withstand power fluctuations well: at half load, the efficiency is still 80–90% of the full load efficiency.

Figure 2.4 compares the torque curves for a single-shaft turbine and a twin-shaft turbine.

1 Except for special constructions, such as single-shaft turbines with torque converters or electric generators.

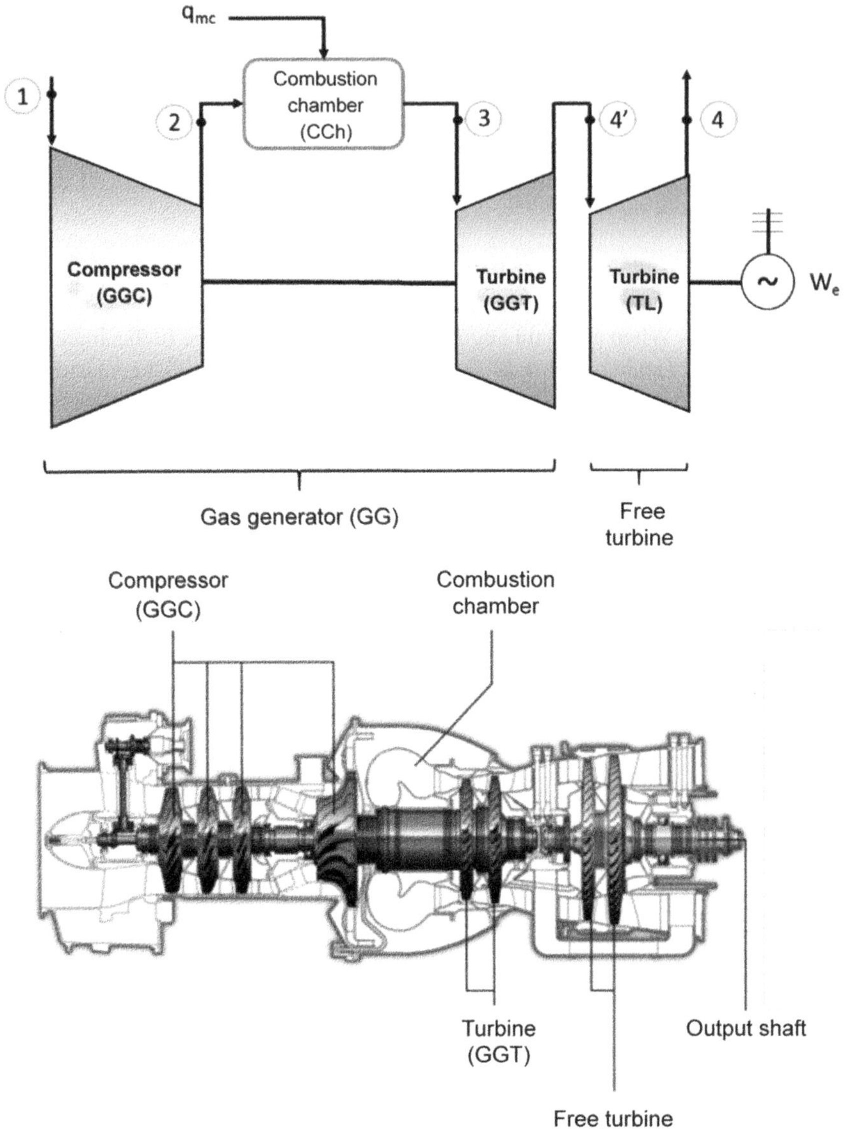

Figure 2.3. *Diagram of a twin-shaft gas turbine. Makila turboshaft on Airbus Super Puma Helicopters (source: Safran). For a color version of this figure, see www.iste.co.uk/krysinski/automotive.zip*

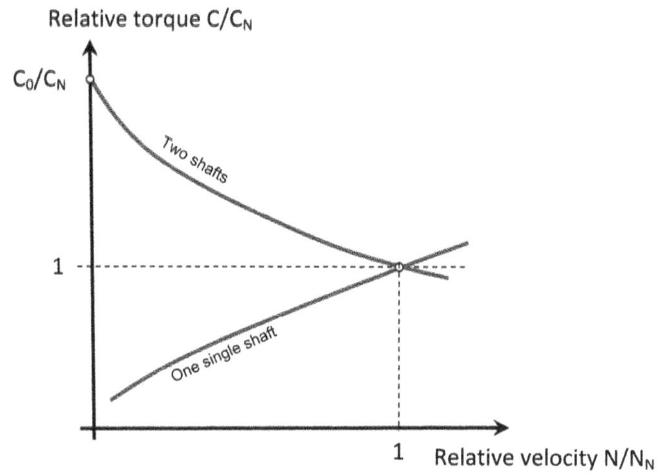

Figure 2.4. *Torque comparison for single-shaft and twin-shaft gas turbines and number*

2.2.1.2. *Representation of the thermodynamic cycle of gas turbines*

A representation of the thermodynamic cycle can be made using the h-s diagram (Figure 2.5), where h is the enthalpy and s is the entropy. The theoretical Joule cycle is represented by the points 1j – 2j – 3j – 4j. The real cycle of a single-shaft turbine is represented by the points 1-1′-2-3-4-4′.

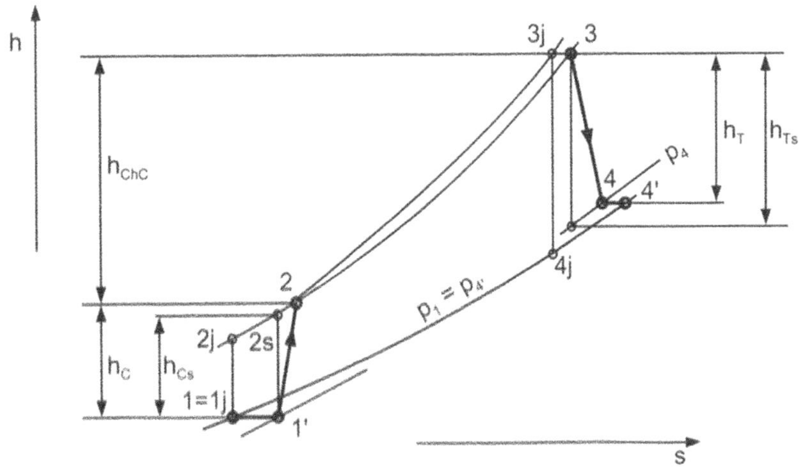

Figure 2.5. *(h-s) diagram of a single-shaft gas generator*

Figure 2.6 shows the same type of diagram for a twin-shaft and a free turbine. The index s indicates the case of an isentropic transformation (an ideal transformation with no losses).

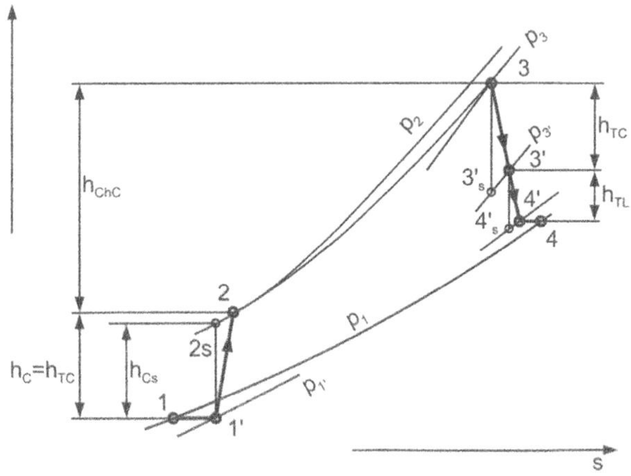

Figure 2.6. *(h-s) diagram of a twin-shaft gas generator*

The thermodynamic transformations represented in the h-s diagram and the enthalpy variations are as follows:

– 1-2 compression, $h_c = h_2 - h_1$ [J/kg];

– 2-3 combustion, $h_{CCh} = h_3 - h_2$;

– 3-4 expansion in a single-shaft turbine, $h_T = h_3 - h_4$.

For the free turbine system, the same can be defined:

– 3-3′ expansion in a gas generator turbine, h_{Tc};

– 3′-4 expansion in a free turbine FT, $h_{TL} = h_{3'} - h_4$.

It is easy to notice that the expansion in a twin-shaft turbine passes through the intermediate pressure $p_{4'}$. The fall in enthalpy (h_3 – $h_{3'}$) results from power absorption by the compressor. The fall ($h_{3'}$ – h_4) passes through the FT, which provides usable power.

The thermal efficiency of gas turbines is given by:

$$\eta_{th} = \frac{h_T - h_C}{h_{ChC}} = \frac{h_{TL}}{h_{ChC}} \quad [2.3]$$

The power W_C is expressed as the product of the enthalpy variation and the mass flow rate of the fluid. Thus:

$$W_C = h_C\, q_{mC}; W_{CCh} = q_{m\,car}\ PCI\ \eta_{CCh} \qquad [2.4]$$

with η_{CCh} the efficiency of the combustion chamber.

NOTE.– Points 1, 2, 3, 4′ and 4 represent a shutdown state of the fluid. Pressure drops have been neglected.

2.2.1.3. *Thermal efficiency of gas turbines*

The thermal efficiency η_{th} of a single cycle internal combustion turbine depends on the ratio $\tau_{th}=T_3/T_1$ (from the T_3 maximum fluid temperature to the minimum temperature of T_1), the compression ratio $\Pi = p_2/p_1$, and the efficiency of the compressor, combustion chamber and turbine.

The β coefficient is defined as:

$$\beta = \frac{\gamma - 1}{\gamma} \qquad [2.5]$$

The η_{st} isentropic expansion efficiency in the turbine and the η_{sc} isentropic compression efficiency are expressed by the following equations, assuming that the gas is a perfect gas:

$$\eta_{sT} = \frac{T_{4's} - T_3}{T_4 - T_3}$$
$$\eta_{sC} = \frac{T_{2s} - T_1}{T_{2'} - T_1} \qquad [2.6]$$

Single cycle efficiency, or thermal efficiency, is defined by [BAI 71]:

$$\eta_{th} = \frac{\left(\Pi^{\beta} - 1\right)\left(\eta_{sC}\eta_{sT}\tau - \Pi^{\beta}\right)}{\Pi^{\beta}\left(\eta_{sC}\left(\tau - 1\right) + 1 - \Pi^{\beta}\right)} \qquad [2.7]$$

The improvement of this thermal efficiency can be obtained mainly by increasing the cycle parameters (the T_3/T_1 temperature ratio and the p_2/p_1 pressure ratio). Isentropic efficiencies can also be improved. In particular, we can try to reduce the losses due to viscous friction during the flow at the blades by moving closer to an ideal cycle. Optimizing blade shape affects the compression and expansion efficiency. Nevertheless, it must be noted that isentropic efficiencies are already very high (close to 100%) and that their increase is limited.

For the case of the ideal Brayton cycle, efficiency η_{sc} and η_{st} of 1, we obtain:

$$\eta_{th} = \frac{\left(\Pi^{\beta} - 1\right)}{\Pi^{\beta}} = 1 - \frac{T_1}{T_3} \qquad [2.8]$$

For example, for a temperature T_1 of 15 °C and a temperature T_3 of 1,200 °C, we obtain an efficiency for an ideal Brayton cycle amounting to 80%. This consideration overestimates the thermal efficiency, which in reality is closer to the order of 40–50%.

Thus, in Figure 2.7, the variation in thermal efficiency η_{th} is shown as a function of Π for different T_3 values, for a single cycle. The considered values are such that the compressor inlet temperature T_1 is 15 °C, the isentropic exponent γ is 1.4, the isentropic efficiency of the compressor is 0.785 and the isentropic efficiency of the is turbine about 0.9.

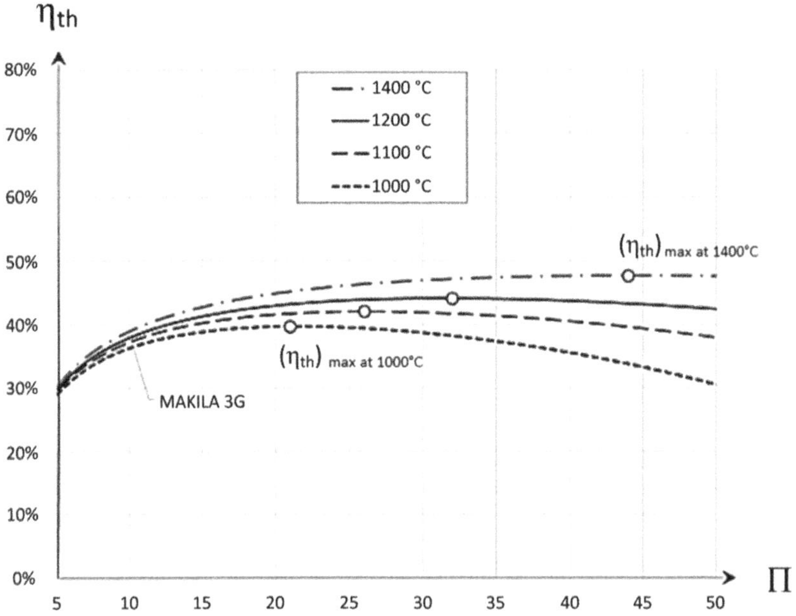

Figure 2.7. *Thermal efficiency of a single cycle internal combustion turbine depending on the compression ratio Π for different temperature values T_3*

It can be seen that, for the cycles considered, the thermal efficiency increases if the T_3 temperature rises. For a given temperature T_3, the efficiency reaches its maximum value for a given compression ratio $(\eta_{th})_{max}$.

The criterion of maximum thermal efficiency is not singular. It is also necessary to consider the dimensions, weight and manufacturing costs of an engine. To do this, we use a parameter called specific power and noted w_s, which is a ratio of the usable power to the air mass flow rate:

$$w_s = \frac{W_u}{q_m} \; [J/kg] \qquad [2.9]$$

with q_m being the mass flow rate of air drawn in by the compressor.

The mass power w_s indicates the power obtained for 1 kg of the compressor air flow rate. We can thus derive the following [BAI 71]:

$$w_s = C_p \; T_1 \; \frac{\Pi^{\beta} - 1}{\Pi^{\beta}} \left[\tau \, \eta_{sT} - \frac{\Pi^{\beta}}{\eta_{sC}} \right] \qquad [2.10]$$

Figure 2.8 shows the change in the specific power w_s according to the compression ratio for different T_3 temperature values. The conditions are the same as before. Figure 2.8 represents the value of C_p amounting to 1.0413 kJ/(kg·K).

Another representation of the Brayton cycle is shown on a T-s diagram in Figure 2.9:

– process 1-2 represents isentropic compression;

– process 2-3 represents isobaric combustion;

– process 3-4 represents isentropic expansion;

– process 1-2 represents isobaric exhaust.

On a T-s diagram, the surface of the cycle is representative of specific power. Figure 2.10 shows that if the compression ratio is too low or too high, specific power will be low and will be limited by the T_1 atmospheric temperature and by the T_3 temperature (the limits are permissible by the materials).

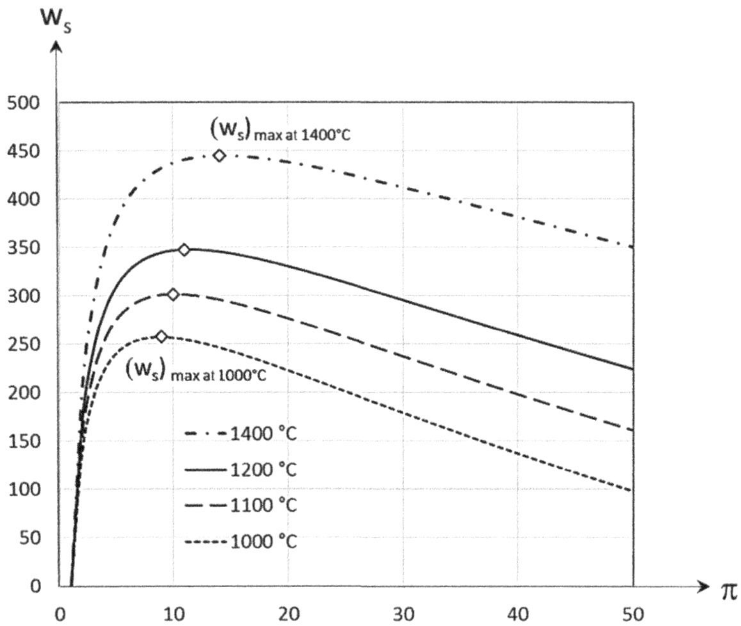

Figure 2.8. *Specific power of a single cycle internal combustion turbine according to the compression ratio for different T3 temperature values*

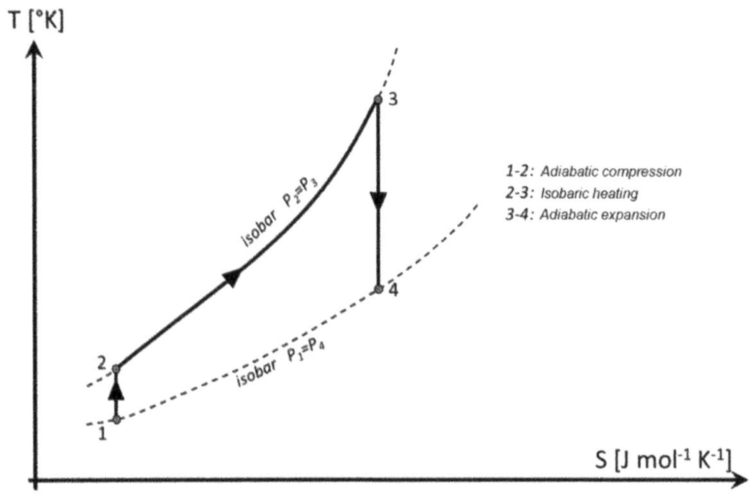

Figure 2.9. *T-s diagram. Brayton cycle of a gas turbine engine*

Indeed, if the compression ratio is too high (e.g. a Π of 20), the obtained T_2 temperature is high. The energy supply during combustion can only be reduced as it is limited by temperature T_3. Similarly, if the compression ratio is low (e.g. a Π of 5), the energy input from combustion is high, but the energy recovered from the shaft is low.

There is thus an optimal compression ratio, at a fixed T_1 and T_3, which allows achievement of an optimal specific power. For a perfect cycle (Brayton cycle), we would have:

$$w_s = C_p\, T_1 \frac{\Pi^{\beta} - 1}{\Pi^{\beta}}\left[\tau - \Pi^{\beta}\right] \quad [2.11]$$

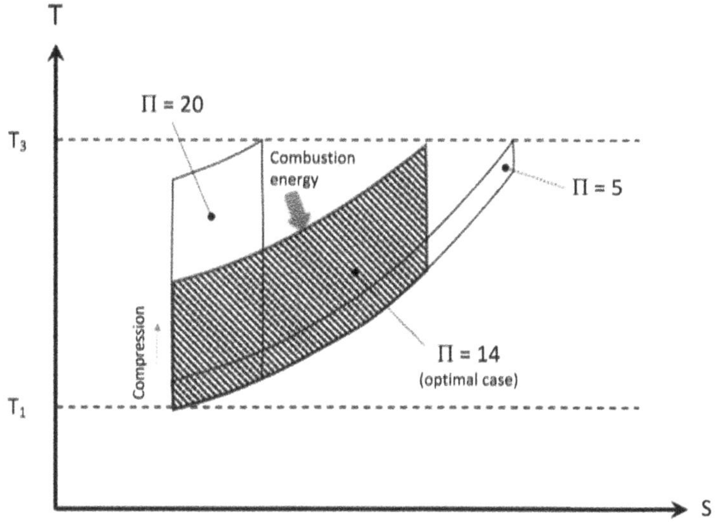

Figure 2.10. *Influence of the compression ratio on a gas turbine with a temperature limitation*

By using the formula [2.12], we show that the optimal specific power can be obtained at the compression ratio:

$$\Pi_{WS} = \left(\frac{T_3}{T_1}\right)^{\frac{1}{2\beta}} \quad [2.12]$$

Thus, for a T_1 temperature of 25 °C and a T_3 temperature of 1,100 °C, the optimal compression ratio with respect to the specific power would be 14.

In Table 2.2, the maximum values of the η thermal efficiency and the W_S specific power, as well as the values of the $\Pi_{\eta th}$ compression ratio optimal for efficiency and Π_{Ws} optimal for specific power, which correspond to 1,000 °C, 1,200 °C and 1,400 °C, are presented for a single cycle.

We can see that the $\Pi_{0(\eta th)max}$ compression ratio is higher than $\Pi_{0(Ws)max}$ for the same temperature.

T_3	1000 °C	1200 °C	1400 °C
(η_{th})max	40%	42%	48%
Πηth max	21	26	44
(W_s)max	257 KJ/kg	301 KJ/kg	444 KJ/kg
ΠWs max	9	10	14

Table 2.2. *Characteristic parameters of a simple cycle*

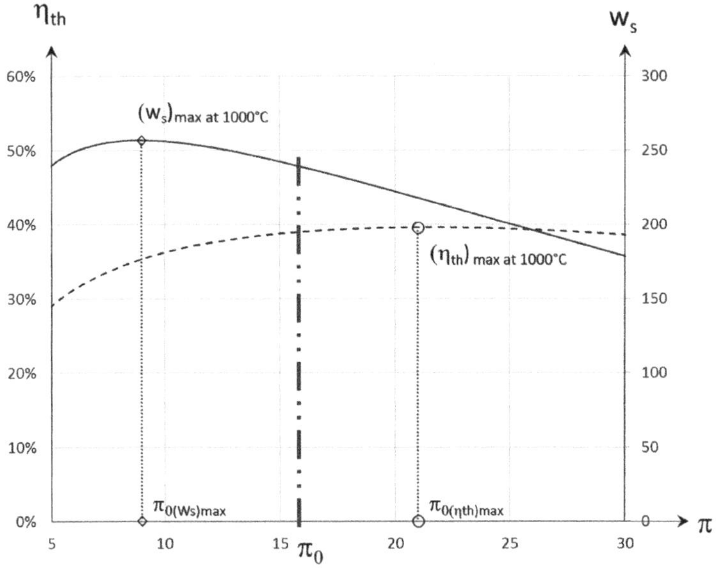

Figure 2.11. *Evolution of efficiency and specific power as a function of the compression ratio. For a color version of this figure, see www.iste.co.uk/krysinski/automotive.zip*

The construction of an engine that promotes density will have a low compression ratio. In this case, with a $\Pi_{0(Ws)max}$ compression ratio, it has a lower efficiency than one that works at $\Pi_{0(\eta th)max}$.

In practice, the Π_o intermediate compression ratio value is chosen between $\Pi_{0(Ws)max}$ and $\Pi_{0(\eta th)max}$. Figure 2.11 shows that the efficiency curves are very flat around the maximum values. The choice of a compression ratio below $\Pi_{\eta th\ max}$ still ensures satisfactory efficiency. The compression ratio of single cycle engines often varies between 8 and 20.

The maximum temperature reached in a cycle at point 3, at the turbine inlet temperature, is limited. Metal turbine blades, currently made of monocrystalline nickel-based superalloys, cannot withstand temperatures above 1,300 K (1,027 °C). Higher temperatures, up to 1,600 K (1,327 °C), can be obtained with oxide ceramic turbine blades. However, they remain fragile, compared to metals and alloys, and a mechanical performance drop occurs rapidly beyond 1,300 °C (Figure 2.12). The use of ceramic matrix composite materials (MCM) could pave the way for new opportunities [AOU 15].

The minimum temperature is defined by the air temperature at the inlet of the engine T_1.

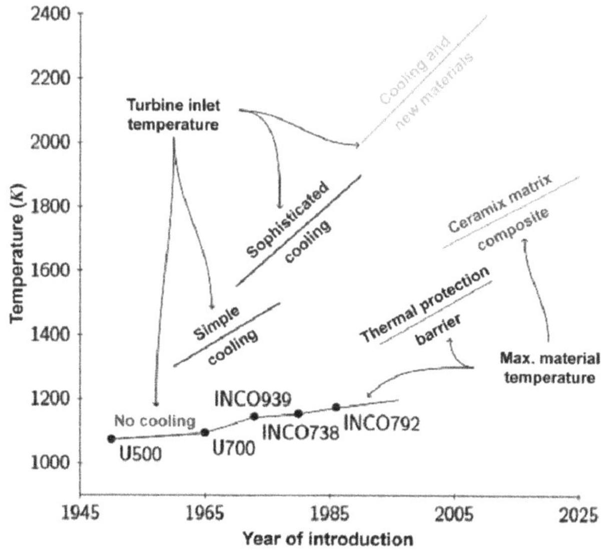

Figure 2.12. *Evolution of temperature resistance capacities of turbine blade materials [AOU 15] . For a color version of this figure, see www.iste.co.uk/krysinski/automotive.zip*

Since the 1960s, blade cooling has increased the permissible turbine temperature T_3 [FAC 05]. Initially, cooling was performed by radial pipes, then through multi-feed systems with film cooling. Current techniques, such as multi-pass channels with turbulence promoters and fine needle type blades cooled by impact jets, allow reaching higher temperatures (Figure 2.13).

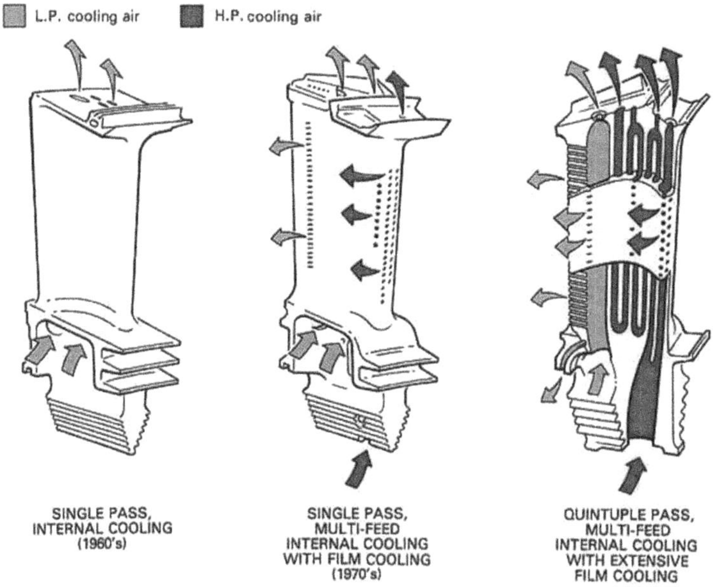

Figure 2.13. *Cooling channels of a turbine blade [ROL 86]. For a color version of this figure, see www.iste.co.uk/krysinski/automotive.zip*

2.2.1.4. *Implementation of gas turbines*

Today, gas turbines are widely used in the aeronautics industry, sometimes in competition with piston engines. They are used for the propulsion of fixed wings, such as in aircraft or missiles, or on rotating wings, as in the case of helicopters.

They are very diversified in non-aeronautical applications. Their applications include turbo-alternators for power plants, turbo-pumps used in pumping and recompression stations for gas and oil pipelines as well as on offshore oil platforms, rail-type land-based traction systems with turbotrains or armored vehicles for military applications, or the so-called total energy industrial installations, where the system can supply three forms of energy simultaneously: electric (alternator), pneumatic (by sampling air from the compressor) and calorific (heat recovery from exhaust gases).

Among gas turbines, we will distinguish turbojet engines, including turbofans, which provide kinetic energy usable for propulsion, turbo propellers supplying energy to a propeller and turboshaft engines supplying mechanical energy available on a rotor, both with a single and a double shaft (Figure 2.14).

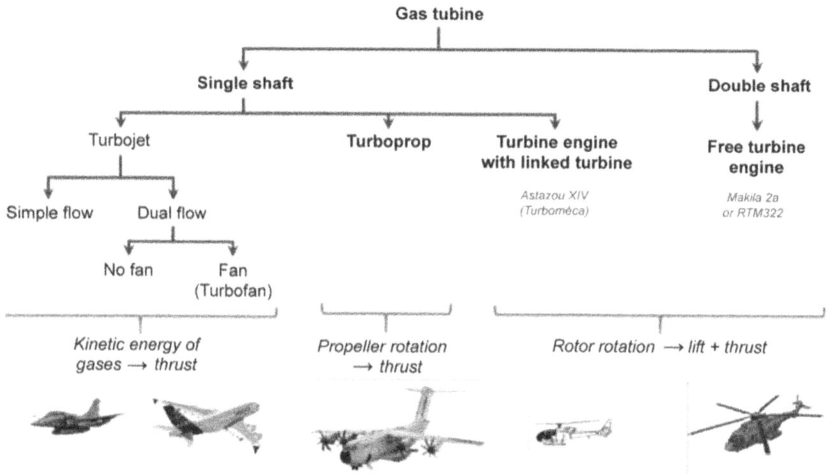

Figure 2.14. *Use of gas turbines in aeronautics. For a color version of this figure, see www.iste.co.uk/krysinski/automotive.zip*

In all these cases, the objective of the gas turbine is to provide thrust to ensure the aircraft's flight, as well as lift to ensure hovering, in the case of a helicopter. The power required for the function of flight or lift is defined as the propulsion or lift power, noted W_p. The overall ratio of the transportation powertrain can then be broken down as the product of the thermal efficiency η_{th} associated with the gas turbine [2.2], divided by the propulsive efficiency η_p, defined as the ratio of the propulsive power W_p, divided by the actual power W_e supplied by the gas turbine:

$$\eta_p = \frac{W_p}{W_e} \qquad [2.13]$$

This propulsive efficiency corresponds to the propeller efficiency and associated mechanical transmission for an aircraft with a turboprop engine and the efficiency of the rotor and associated transmission for a helicopter. The characterization of this efficiency for these cases will be discussed in Chapter 4. Since this part discusses the engine more, turbojet engines will be discussed in the following sections.

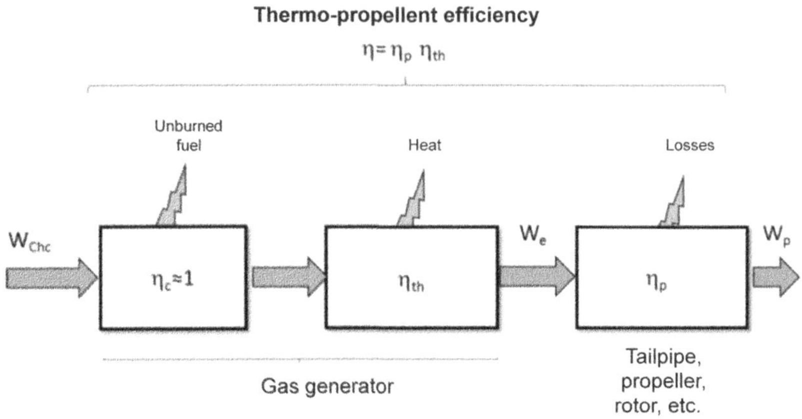

Figure 2.15. *Breakdown of thermo-propulsive efficiency*

2.2.1.4.1. Turbojets

Single in-flight turbojet

The turbojet engine is distinguished by the addition of a nozzle on the side of the exhaust. At the turbine outlet, burned gases are expanded in this nozzle (points 4 and 5, Figure 2.16), which accelerates them and provides a net thrust. The engine fluid itself generates the F_p thrust.

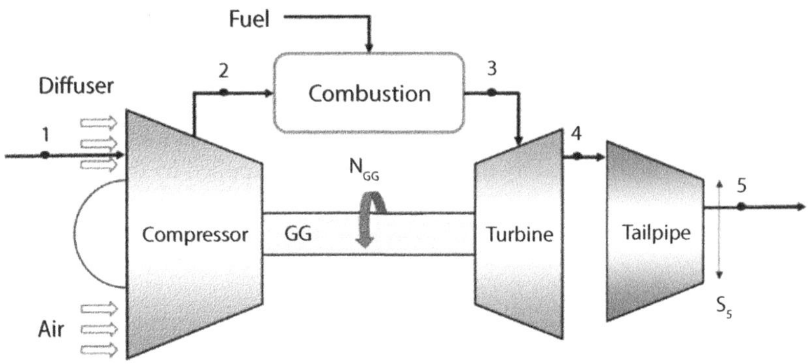

Figure 2.16a. *Turbojet, thermodynamic cycle. For a color version of this figure, see www.iste.co.uk/krysinski/automotive.zip*

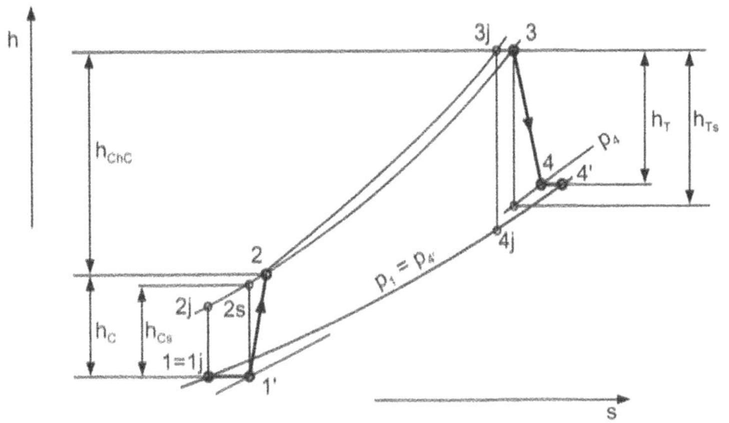

Figure 2.16b. *Turbojet, thermodynamic cycle (continued)*

By definition, the propulsive force is generated as follows:

$$F_p = q_m (V_5 - V_1) + (p_5 - p_1) S \approx q_m (V_5 - V_1) \qquad [2.14]$$

Pressure effects are disregarded (p5 = p1). V_5 is the gas ejection velocity and V_1 is the engine input corresponding to the aircraft speed. The power required for propulsion is thus expressed as:

$$W_p = F_p V_1 = q_m V_1 (V_5 - V_1) \qquad [2.15]$$

In the case of a gas turbine where the energy is transferred to the gas, the provided work can be measured by the energy variation conveyed to the mass of gas. Thus:

$$W_e = \frac{1}{2} q_m \left(V_5^2 - V_1^2\right) \qquad [2.16]$$

The propulsive efficiency, the ratio of the propulsion power to the kinetic power available at the reactor outlet, is then:

$$\eta_p = \frac{W_p}{W_e} = 2\, V_1 \frac{(V_5 - V_1)}{\left(V_5^2 - V_1^2\right)} = \frac{2}{1 + \frac{V_5}{V_1}} = \frac{1}{1 + \frac{V_5 - V_1}{2\, V_1}} \qquad [2.17]$$

It can be seen that at the aircraft's low forward speed, $V_1 << V_5$, propulsive efficiency is low. For example, for an ejection velocity of 1,600 km/h and a forward speed of 560 km/h, the efficiency will be 52%. For the same ejection velocity and a forward speed of 725 km/h, the efficiency will be 62%. It is interesting to note that the difference between V_5 and V_1 is small. Expression [2.17] shows that the higher the propulsive efficiency, the better, as the relative velocity $(V_5 - V_1)$ of the exhaust gas for a given velocity V_1 is lower.

The thermal efficiency η_{th} is defined as the ratio of kinetic power to heat supplied by W_{CCh}. Let:

$$\eta_{th} = \frac{W_e}{W_{CCh}} = \frac{V_5^2 - V_1^2}{2\,W_{CCh}} \qquad W_{CCh} = q_{mc}\ \text{PCI} \tag{2.18}$$

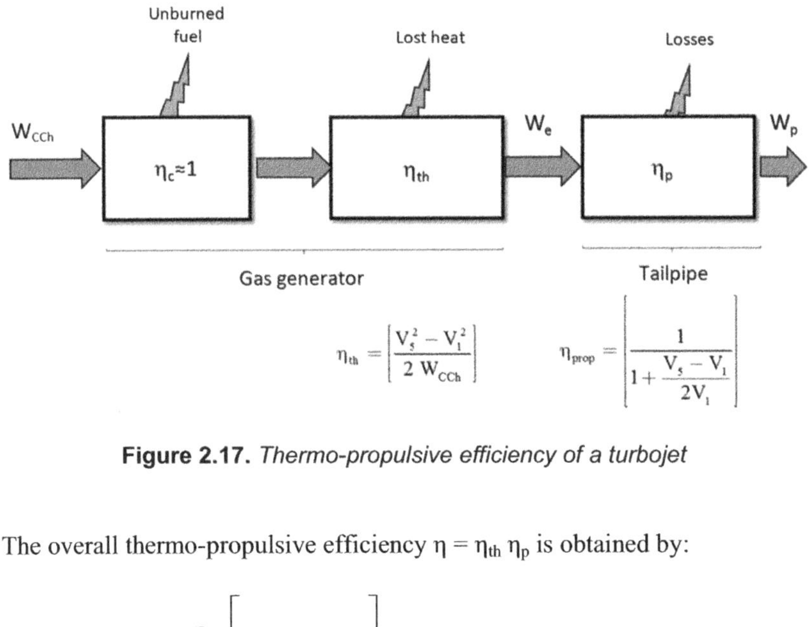

Figure 2.17. *Thermo-propulsive efficiency of a turbojet*

The overall thermo-propulsive efficiency $\eta = \eta_{th}\,\eta_p$ is obtained by:

$$\eta = \left[\frac{V_5^2 - V_1^2}{2\,W_{Chc}}\right]_{th} \left[\frac{1}{1 + \dfrac{V_5 - V_1}{2V_1}}\right]_{prop} \tag{2.19}$$

Using the example of an aircraft with a jet engine, for an ejection velocity of 1,600 km/h (~994 mph) and a forward speed of 725 km/h (~450 mph) at cruising speed, the efficiency will be 52%, the thermal efficiency will be 35% and the thermo-propulsive efficiency will be close to 18%.

On a turbojet, the thermo-propulsive efficiency is relatively low. This is due to the fact that there is a compromise between good thermal efficiency, which implies giving gases a high kinetic energy in the nozzle, a significant difference ($V_5^2 - V_1^2$), and good propulsive efficiency, which requires a low relative velocity of the gases at the nozzle outlet, or a low difference ($V_5 - V_1$) [2.19].

In the case of the single-flow turbojet, the flow rate through the gas generator is the same as that through the nozzle. This problem is generally unresolved.

The dual-flow turbojet

The dual-flow turbojet, or turbofan, makes it possible to reconcile these two requirements by accelerating an air flow rate (known as the secondary flow) that is complementary to that of the gas generator to a speed barely higher than that achieved by the aircraft. This operation is made possible by an additional low-pressure compressor, called a blowing machine, also driven by the turbine. It is the equivalent of a streamlined turboprop engine.

Thus, there are two separate air flows within a turbofan:

– the central flow, sometimes called the "hot flow", is the flow of the engine itself. After combustion, it passes through a turbine, the power of which far exceeds that of the compressor. This excess power is transmitted to the blowing machine;

– the external flow, sometimes called the "cold flow", is slightly compressed by the blowing machine and directly expanded into the nozzle that encompasses the engine's hot body. It is never heated. It is this "cold" air that provides most of the thrust. It is shown that the higher the ratio of cold air to warm air (known as the dilution rate), the more efficient an engine is. The dilution rate of modern engines amounts to approximately 10.

We note the dilution rate reflecting the ratio of air flow rate passing through the fan with the air flow rate passing through the body t_d:

$$\tau_d = \frac{q_{m2'}}{q_{m2}} \quad [2.20]$$

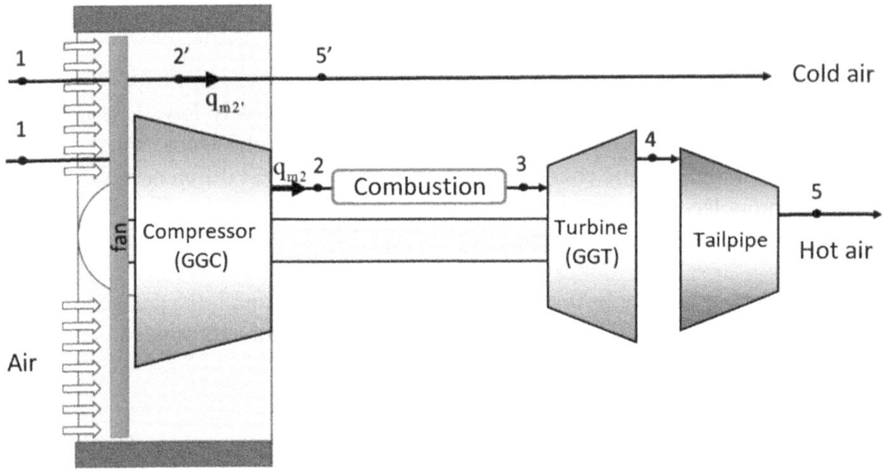

Figure 2.18. *Characteristic flow points of a dual flow turbofan. For a color version of this figure, see www.iste.co.uk/krysinski/automotive.zip*

The overall mass flow rate can thus be expressed by:

$$q_{m1} = q_{m2} + q_{m2'} = (1+\tau_d)\, q_{m2} \tag{2.21}$$

The propulsive force is then given, neglecting the effects of pressure, by:

$$F_p = q_{m2'}\left(V_{5'} - V_1\right) + q_{m2}\left(V_5 - V_1\right) \tag{2.22}$$

The power required for propulsion is then:

$$W_p = F_p\, V_1 = q_{m2'}\left(V_{5'} - V_1\right) V_1 + q_{m2}\left(V_5 - V_1\right) V_1 \tag{2.23}$$

In the case of a turbojet, the work provided can be measured by the variation in kinetic energy given to the air mass. Let:

$$W_e = \frac{1}{2} q_{m2'}\left(V_{5'}^2 - V_1^2\right) + \frac{1}{2} q_{m2}\left(V_5^2 - V_1^2\right) \tag{2.24}$$

The propulsive efficiency can then be given by introducing the dilution rate τ_d (Figure 2.19):

$$\eta_{th} = \left[\frac{V_5^2 - V_1^2}{2W_{CCh}}\right] \qquad \eta_p = \frac{1+\tau_d \dfrac{\left(V_{5'} - V_1\right)}{\left(V_5 - V_1\right)}}{1+\dfrac{\left(V_5 - V_1\right)}{2V_1}+\tau_d \dfrac{\left(V_{5'}^2 - V_1^2\right)}{2V_1\left(V_5 - V_1\right)}} \qquad [2.25]$$

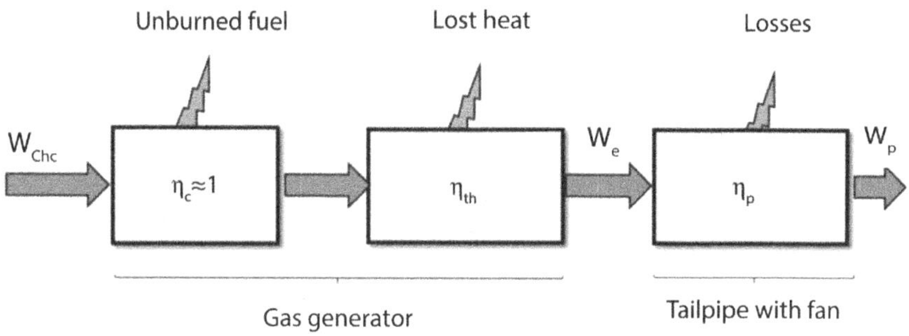

Figure 2.19. *Thermo-propulsive efficiency of a turbofan*

For example, for an ejection velocity V_5 of 1,300 km/h, a velocity $V_{5'}$ of 1,255 km/h, a forward speed of 933 km/h and a dilution rate of 5, the propulsive efficiency will be 85%. With a thermal efficiency of 35%, we obtain a thermo-propulsive efficiency close to 29%.

One possible way to improve this is to install a gearbox between the fan and the low-pressure shaft of the turbomachine. This allows each of these components to rotate at more appropriate speeds, while increasing the dilution rate. Another solution applied for turbojet engines with very high thrust is the twin- or three-shaft gas generator, which corresponds to low, medium and high pressure parts. Each part rotates at the velocity appropriate to its geometry.

We note the compression rates:

$$\begin{cases} \Pi = \dfrac{p_2}{p_1} \\ \Pi' = \dfrac{P_{5'}}{P_1} \end{cases} \qquad [2.26]$$

In practice, we will try to optimize the dilution rate τ_d, the compression ratio Π of the gas turbine compressor and the compression ratio Π' of the fan, to achieve the correct efficiency and thrust while obtaining a minimum specific consumption.

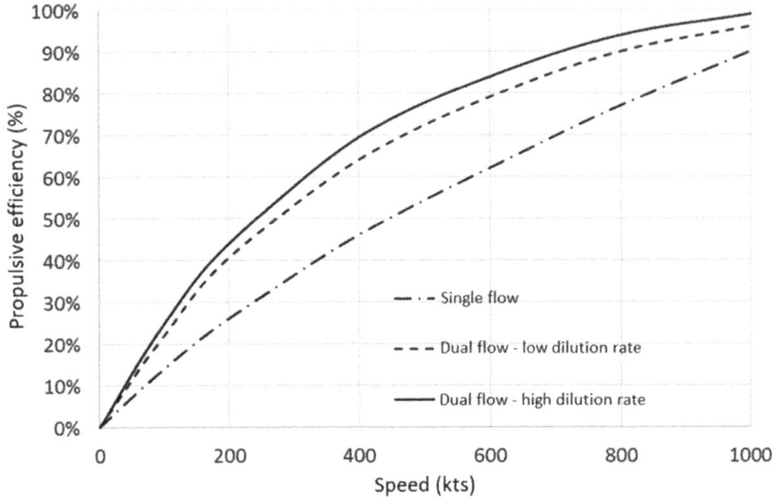

Figure 2.20. *Propulsive efficiency as a function of input velocity*

Another advantage of the dual flow is a lower noise level, due to the cold air flow that envelops the warm air flow.

Among the latest engines being developed, Safran's LEAP (Leading Edge Aviation Propulsion) is a new generation of turbojet engine with a dual-body flow and a dilution rate of around 10, as is the Rolls-Royce Trent XWB, which has a dilution rate of around 9.3.

By comparison, Pratt & Whitney's or Rolls-Royce Spey's JTD8 engines of the 1960s have a dilution rate below 1. Snecma's CMF56 of the 1970s (and later, in its various versions) has a dilution rate of 6. The Pratt & Whitney PW4000 has a rate of 5.3.

These engines are also improved by reducing weight: Snecma developed the ML340, a new alloy for the LEAP turbine shaft, with Aubert & Duval. This new material offers the engine a weight loss thanks to a lower density and a 15% increase in torsional mechanical resistance. Similarly, the blades or protective casings are made of 3D RTM woven composite, which is much lighter than conventional materials. This allows a gain of 450 kg/aircraft. The reduction in the mass of the transportation powertrain has an impact on the sizing of the structure and, by a virtuous effect, on the transportation powertrain itself (Figure 2.22).

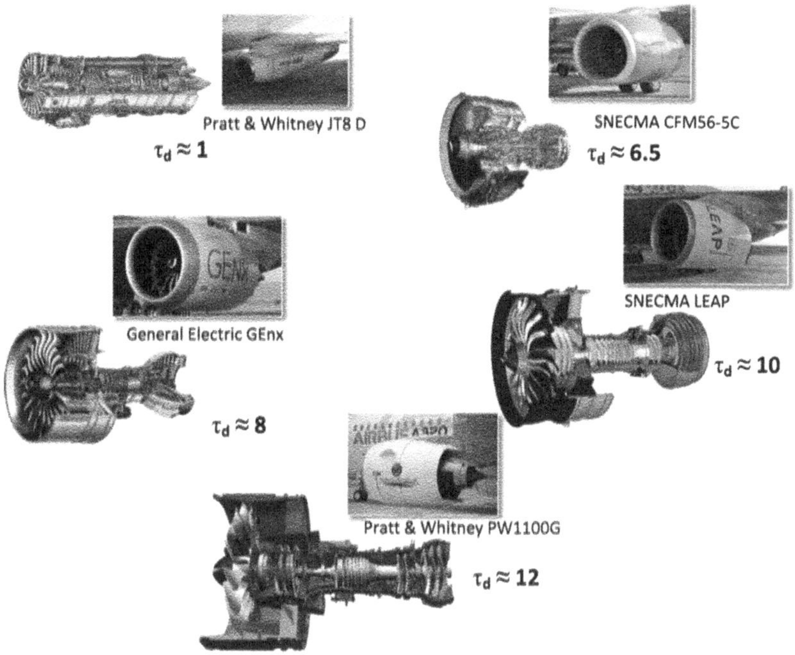

Figure 2.21. *Examples of engines with different dilution rates. For a color version of this figure, see www.iste.co.uk/krysinski/automotive.zip*

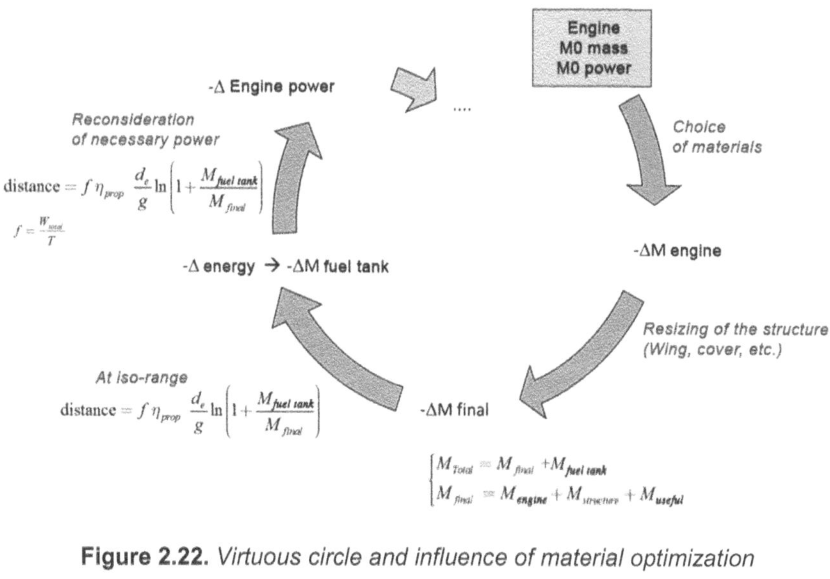

Figure 2.22. *Virtuous circle and influence of material optimization*

2.2.1.4.2. Turbopropeller

Instead of using a nozzle, as in a turbojet engine, it is possible to continue the expansion in the turbine to atmospheric pressure. The power supplied by the turbine is then greater than the power consumed by the compressor. The motor shaft thus provides work which can be used to power a propeller. The term "turboprop" is then used to describe this type of transportation powertrain.

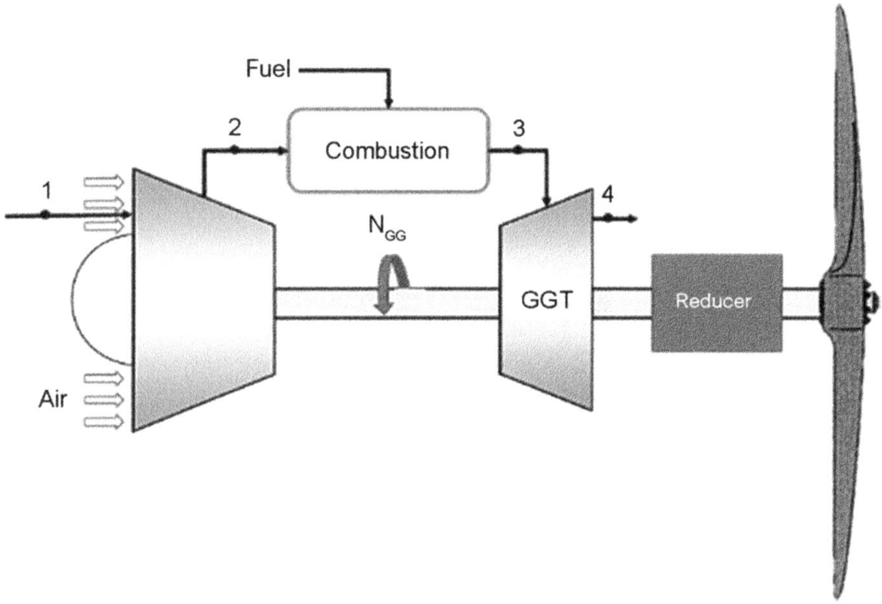

Figure 2.23. *Diagram of a turboprop engine*

Due to the approximate value of the angular velocities linked to the GGT turbines and the angular velocities of the propellers, the addition of a mechanical gearbox is necessary.

The residual exhaust gas thrust is low (less than 10% of propeller thrust), with most of the thrust being produced by the propeller, with much better efficiency than a conventional engine. However, a disadvantage remains in this solution, with it not being able to approach supersonic speeds, due to the risk of exceeding the limit velocity at the end of the propeller blade.

The schematization of energy losses is expressed in Figure 2.24.

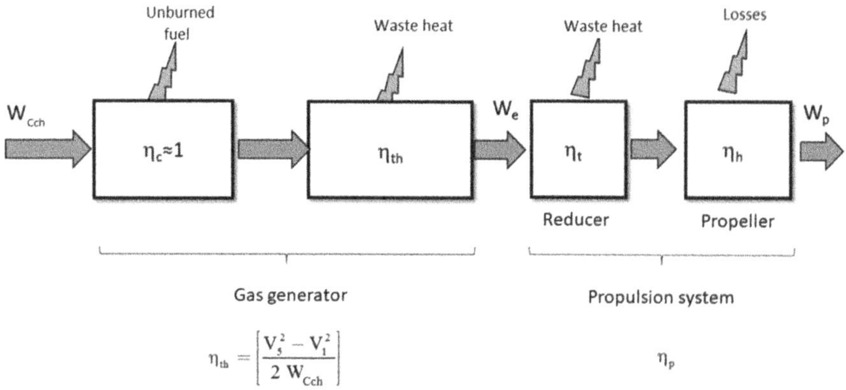

Figure 2.24. *Power chain efficiency of a turboprop engine*

Propulsive efficiency is defined as the product of the gearbox efficiency and the propeller's own efficiency. The efficiency of the reducing factor is generally about 95%:

$$\eta_p=\eta_r\eta_h \quad [2.27]$$

The theoretical efficiency of a propeller can be expressed by:

$$\eta_h=\frac{W_u}{W_d} \quad [2.28]$$

with:

– W_u the useful power to generate the movement at flight speed V with a propulsive effort of T;

– W_d the available power, corresponding to the difference in kinetic power between the propeller's inlet and outlet (increase in kinetic energy of the gases).

We then have:

$$W_u=T\,V$$

$$W_d=\frac{1}{2}q_m\left(V+v\right)^2-\frac{1}{2}q_m V^2=\frac{1}{2}q_m\left(\left(V+v\right)^2-V^2\right) \quad [2.29]$$

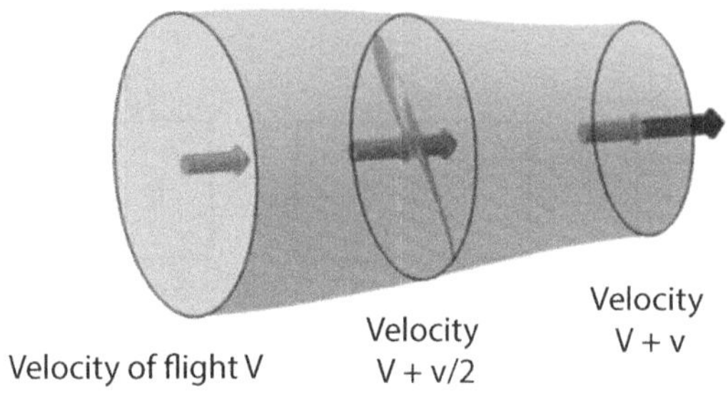

Figure 2.25. *Velocity flow through a propeller's rotor disc*

The efficiency is expressed by:

$$\eta_h = \frac{2TV}{q_m\left((V+v)^2 - V^2\right)} = \frac{2(V+v-V)V}{\left((V+v)^2 - V^2\right)} = \frac{2}{\left(1+\frac{V+v}{V}\right)} \quad [2.30]$$

The thrust force F_p is given by:

$$F_p = q_m(V+v-V) \quad [2.31]$$

Equation [2.28] shows that the propeller's efficiency is better when the propulsion jet's velocity (V+v) is close to the flight velocity V. On the other hand, equation [2.31] shows that a jet velocity too close to the flight velocity leads to a drop in propeller thrust, unless the mass m is significantly increased by increasing the propeller's surface area, that is, having propellers with a large diameter.

In reality, the effects of drag on the blades that, in turn, affect propeller efficiency must be considered in the same way as an aircraft wing. Generally, this efficiency is characterized according to progress, noted J and defined by:

$$J = \frac{V}{nD} \quad [2.32]$$

with:

– V the ground speed;

– n the propeller's angular velocity;

– D the propeller's diameter.

Depending on this parameter, for different blade shapes and angular velocity values, it is shown that the evolution of efficiency is the proposed shape on the curve in Figure 2.26.

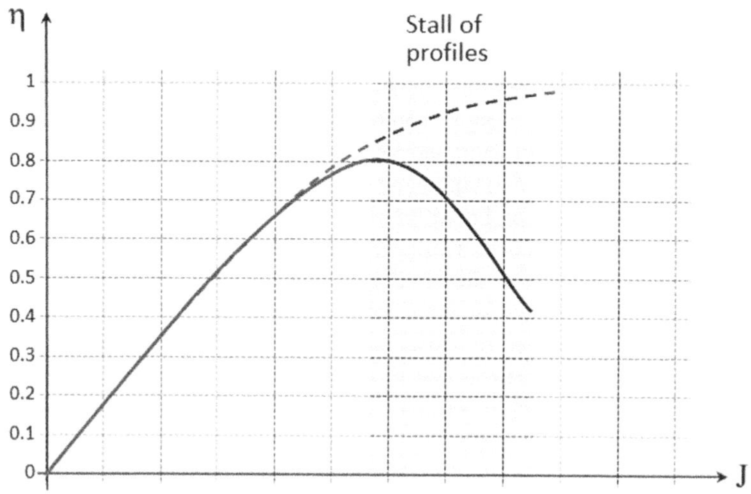

Figure 2.26. *Efficiency of a propeller as a function of the progress parameter. For a color version of this figure, see www.iste.co.uk/krysinski/automotive.zip*

Propeller stalling, the angle formed by the blade chord and the propeller's plane of rotation, has a significant influence on efficiency.

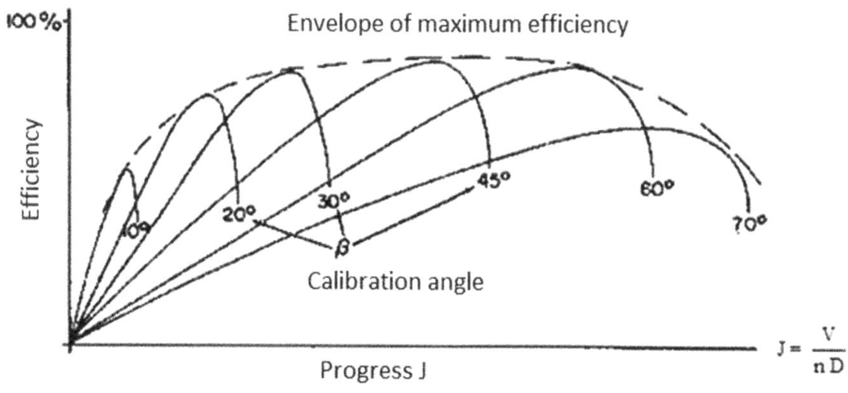

Figure 2.27. *Influence of the calibration angle on efficiency*

Let us take the example of Embraer EBM 312 Tucano, equipped with a Pratt & Whitney PT6A turboshaft engine. For a cruising speed of 400 km/h, the propeller efficiency would be 80%, the mechanical efficiency 0.95% and the thermal efficiency 35%. This results in a thermo-propulsive efficiency close to 26%.

Vertical take-off and landing (VTOL) aircraft models have several rotors or propellers, unlike the conventional helicopter, which operates with a single main rotor and tail rotor. The design combining rotors and propellers allows the separation between the propulsion and lift functions; the main rotor is thus relieved of its propulsive function in favor of its lifting function, an essential solution for achieving high speed. For combined formulas such as the Airbus Helicopters X3, it is necessary to study the behavior of propellers that have the particularity of being able to work under traction or thrust, as well as taking into account the interaction phenomena with the fuselage or main rotor. In terms of numerical modeling, the CFD approach, supplemented by experimental tests, makes it possible to study the propellers' aerodynamic performance in detail (torque, power consumption), their radiated noise and the impact of aerodynamic interaction phenomena.

In hover flight, the tail rotor power is provided by both propellers. In this configuration, the choice was made that one of the two propellers work in traction and the other in reverse (Figure 2.28). In forward flight, due to the decreasing need for power to achieve lift and the fact that the rotor is discharged from its forward function, there is less need for tail power. The two propellers then work under traction. Propeller optimization is thus the result of a compromise between the traction operation linked to the forward flight and the thrust operation required for hover flight.

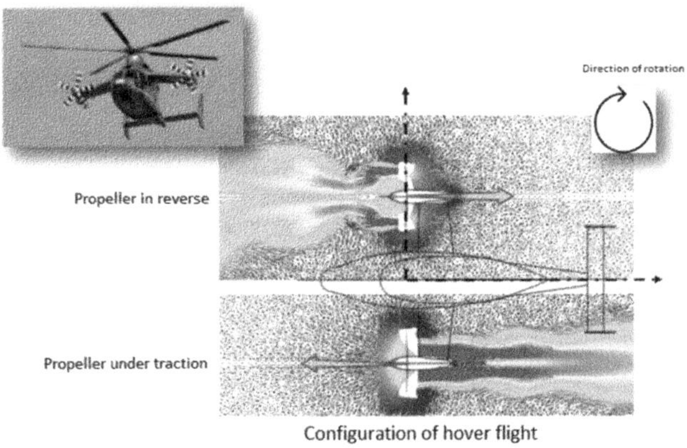

Figure 2.28. *CFD modeling of propeller aerodynamic flow. For a color version of this figure, see www.iste.co.uk/krysinski/automotive.zip*

2.2.1.4.3. Turboshaft engine

Depending on the application, several turbine and compressor arrangements can be used.

With a free turbine (FT) configuration, the mechanical power supplied by the engine is transmitted by a dedicated turbine (linked TGG turbine – see Figure 2.29). This allows each of the two axes to be maintained at different velocities.

The velocity of the turbine/compressor axis is not constrained by the load imposed on the free axis, so it can move at velocities closer to its optimum point and accelerate more easily. This advantage compensates for a greater mechanical complexity of the configuration in applications such as helicopter engines or certain turboprop engines, such as the PW 127G, powering the Airbus C295, where significant power variations are frequently required.

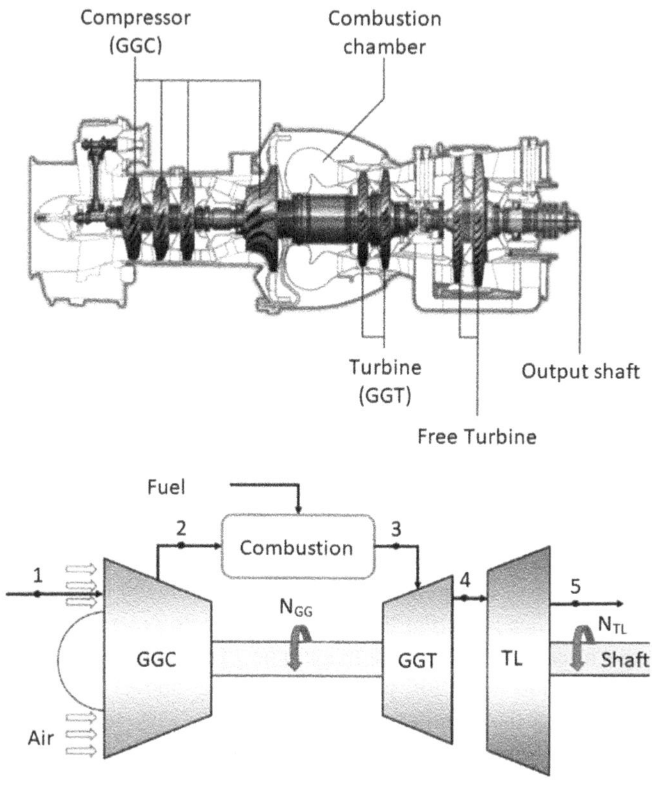

Figure 2.29. *Diagram of a gas turbine with a free turbine. For a color version of this figure, see www.iste.co.uk/krysinski/automotive.zip*

In fact, acceleration is easier, and it has been made possible to eliminate clutches on helicopters. Multi-engine adaptation is also easier to achieve. However, the response time is, in principle, longer than with a linked turbine.

In the case of helicopters, the power chain and its efficiency can be expressed according to the principle illustrated in Figure 2.30.

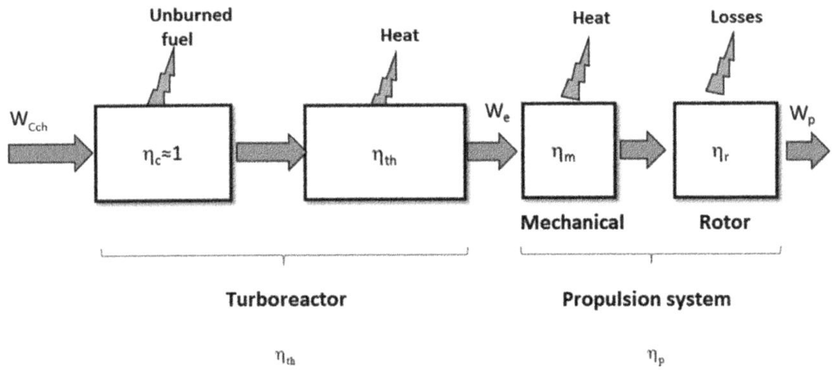

Figure 2.30. *Helicopter power chain*

When measuring the free gas turbine's energy efficiency, the following two effects must be taken into account: the flow of fluid from the linked turbine (GGT) to the free turbine (FT) results in a total pressure loss in the interturbine channel (2% to 3%) and the isentropic expansion efficiency of the free turbine is of the order of 0.85. At the outlet of the free turbine, the gases pass through a nozzle with an isentropic efficiency of about 0.9.

For the Turbomeca TM 333 turboshaft engine, the thermal efficiency is estimated at approximately 0.27.

For rotating wings, rotor angular velocity is limited by the aerodynamic behavior at the blade end. A compromise must be made between the size of the rotor (radius) and the angular velocity, so that the relative blade-air velocity does not approach the speed of sound too closely during the forward phase. In practice, depending on the tonnage of the aircraft, the order of magnitude of the rotor's angular velocity is between 200 and 500 rpm for turbines operating above 30,000 rpm. Unlike turboprop propellers, the gearbox has a very high gear ratio and much more sophisticated technologies. The organic efficiency of this part is close to 0.85.

For rotors used for lift, it is not possible to define efficiency in the same way as in the case of propellers used to provide power for a forward speed V in the presence of drag T. The approach is therefore different and based on the Rankine-Froude theory [ROU 07].

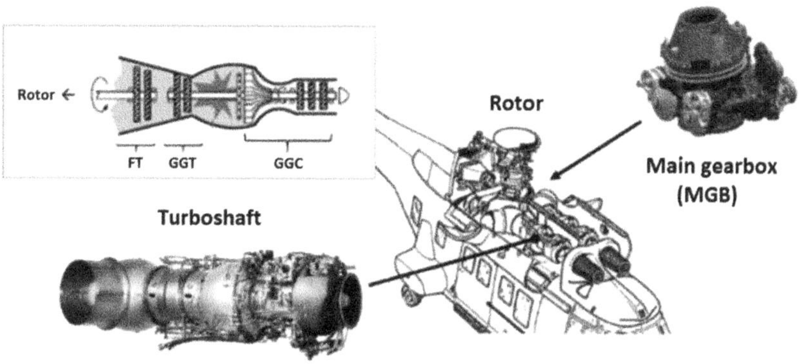

Figure 2.31. *Helicopter power chain (THM Airbus Helicopters)*

The useful power is measured by the induced power required to communicate kinetic energy to the air flow through the rotor. The air flow in return exerts a reaction force on the rotor disc, ensuring the device's lift. Air viscosity has an impact on power consumption, due to the drag effects induced by the blade profile. In addition, with a real fluid, vortices form at the blade end, reducing the efficiency of the blades (lift losses at the blade end). Another phenomenon occurs as well: the velocity induced through the rotor is assumed to be uniform in average terms, when its distribution is actually more complex. Other factors can also influence the interaction between the air blast generated by the rotor and the fuselage: the rotor ejects air at a velocity of about 10 meters per second, which creates a significant offset force at the top of the fuselage.

Rotor efficiency is difficult to define because a large number of parameters are involved. H. Prewitt of Kellett Aircraft chose to introduce the notion of a "figure" of merit (FM) in the 1940s, defined by the ratio between the induced power (Froude power, equation [0.12] Chapter 0) and the total power on the rotor shaft in hover flight [LEI 06]:

$$FM = \frac{\left(P_{induced}\right)_{hover}}{\left(P_{shaft}\right)_{hover}} \qquad [2.33]$$

The "figure" of merit can be predicted using numerical approaches or obtained experimentally. A helicopter rotor's typical "figure" of merit is in the range of 60% to 80%. Since a helicopter spends a lot of time hovering, it may be justified to seek the highest possible figure of merit without degrading the aircraft's performance in forward flight. Hover flight requires strong twisting, which can produce forward flight deteriorations and increases in dynamic forces.

From the measurement of the axial thrust T, it is possible to experimentally define the thrust coefficient:

$$C_T = \frac{T}{\rho\, S\, \Omega^2\, R^2} \tag{2.34}$$

with:

– ρ the density of the air;

– Ω the rotor angular velocity;

– S the rotor surface;

– R the rotor radius.

We note the irregularity coefficient of the induced velocity κ, the profile's drag coefficient c_{d0} and the strength (or plenitude) σ, equal to the quotient of the surface of the blades s and the surface of the rotor disc S. For a rotor with a radius R and of b blades of chord c, the strength is defined by:

$$\sigma = \frac{b\, c}{\pi\, R} \tag{2.35}$$

Then the "figure" of merit can be assessed by:

$$FM = \frac{\dfrac{C_T^{\frac{3}{2}}}{\sqrt{2}}}{\dfrac{\kappa\, C_T^{\frac{3}{2}}}{\sqrt{2}} + \dfrac{\sigma\, c_{d0}}{8}} \tag{2.36}$$

The shape of the blades can influence the "figure" of merit that is usually obtained experimentally. As a result of rotation, the relative velocity of a point on the blade relative to air is not constant along the blade. Consequently, the distribution of lift and drag is not constant along the blade, instead, these forces vary with a square of the radius. To modify this distribution, it is possible to use the twisting of the blade and the change in blade width in order to reduce the forces at the blade end and increase them at the blade base, while maintaining an almost identical overall lift.

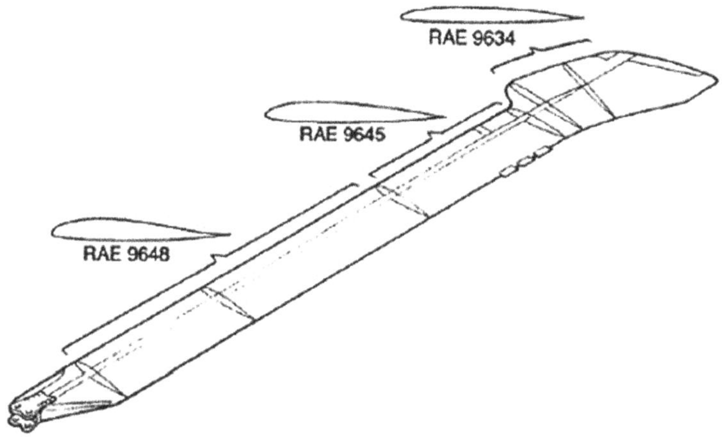

Figure 2.32. *Change of the profile along the blade [BRO 13]*

Twisting is often linear in the order of 8 at 15°. The value is the result of a compromise between the optimization of the "figure" of merit in hover flight at maximum load in general and the degradation brought about in forward flight.

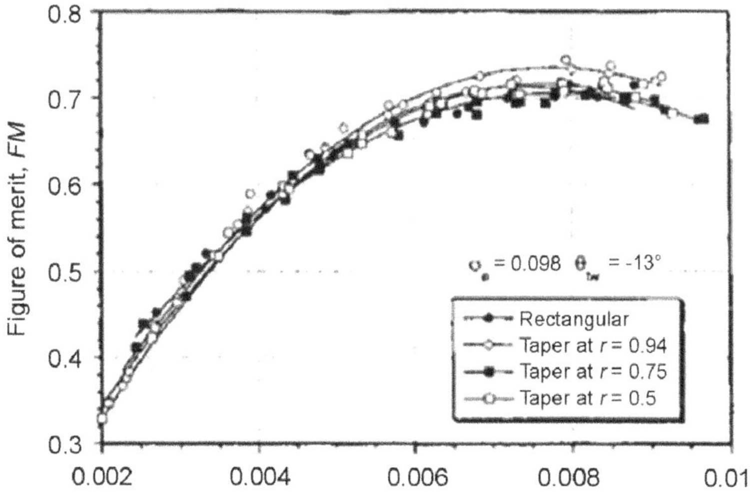

Figure 2.33a. *"Figure" of merit as a function of the thrust coefficient. Influence of blade shape [ALT 90]*

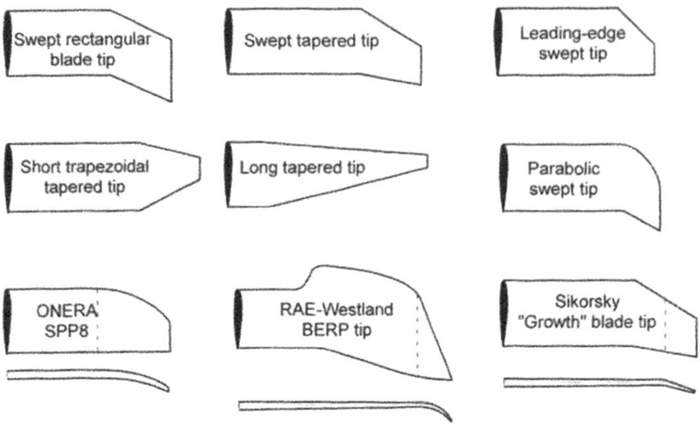

Figure 2.33b. *"Figure" of merit as a function of thrust coefficient. Influence of blade shape [ALT 90] (continued)*

The more or less tapered shapes of the blades have a significant impact on the lift and performance of the aircraft in forward flight. In particular, the angular velocity is limited by the velocity at the blade tip, which can approach the speed of sound and causes the helicopter's characteristic noise. To reduce this phenomenon, it may be worthwhile to modify the shape of the blade at the end. One solution is to use a plane shape similar to a supersonic aircraft wing, such as the blades of the Tiger helicopter. However, due to manufacturing difficulties, the choice may be made to keep the blades straight.

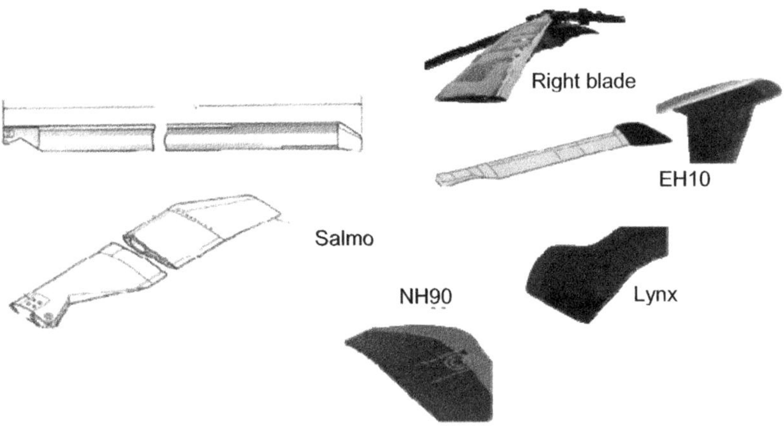

Figure 2.34. *Different blade end shapes*

In order to optimize the blade's acoustic behavior, it is possible to define new blade shapes by optimizing geometric parameters, such as bearing surfaces, twisting and distribution of the torque along the span and line geometry.

As part of the ERATO project, ONERA was able to compare blades with a boomerang-shaped tip with conventional straight blades [VAN 16]. The purpose of this shape is to reduce noise and improve performance, especially when the blade is in the area of turbulence created by the blade at its front (Figure 2.35). Overall, the gain is about 3 dB in terms of sound level. Experimental tests showed this noise gain, particularly in descending flight, and the gain in power consumed by the rotor (Figure 2.36).

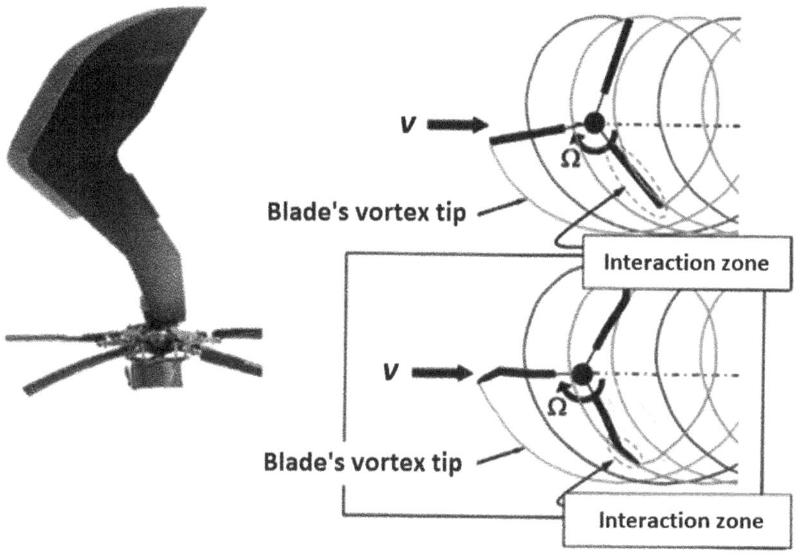

Figure 2.35. *Blade in the shape of a boomerang. ERATO Project [VAN 16]. For a color version of this figure, see www.iste.co.uk/krysinski/automotive.zip*

The operating range limits the performance of gas turbine engines. This has an impact on maximum altitude, the extremes of surrounding temperatures and operating times.

Gas turbine engines, whatever their use is (continuous, take-off, emergency velocities, etc.), have thermal limits (maximum temperature) or mechanical limits (maximum torque or maximum/minimum speed).

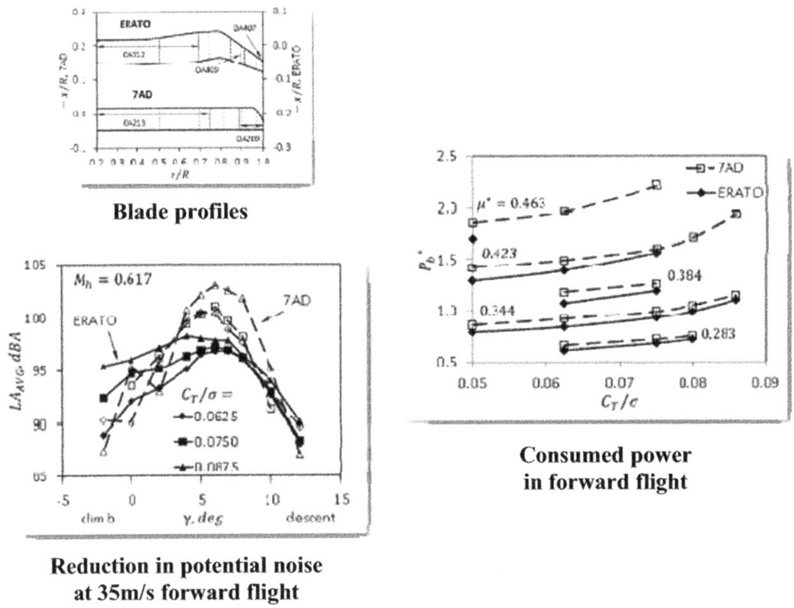

Figure 2.36. *Influence of blade shape. ERATO Project [VAN 16]*

Influence of usage

Until now, the operation of gas turbine engines was described using the study of efficiency and specific power according to manufacturing characteristics (compression ratio, materials, etc.). This sizing was based on the consideration of a continuous operating point.

During their lifetime, engines can operate outside this operating point. In such circumstances, they will be subject to limits. Those limits are determined by the permissible operating temperature inside the turbine.

If the engine must operate at a power higher than the normal continuous limit, in order to prevent permanent damage, a time limit is specified.

For example, for a helicopter, it is considered that emergency situations related to its mission can lead to extremely short periods of use. To illustrate this, typical examples of such power limitations are as follows.

Maximum take-off power rate (MTOP)

This rate is applicable for high power situations, such as operation at a high altitude and/or ambient temperature, particularly for take-off and hover flight. The allowed time amounts to approximately 30 minutes (sometimes 1 hour) before the engine returns to a lower power setting. It is this power that serves as a reference.

Maximum continuous power rate (MCP)

This is the maximum power at which an engine can operate continuously. Therefore, there is no time constraint. The order of magnitude of power is 10% lower than the maximum take-off power.

One Engine Operative emergency rate type, OEI at 2 mins

This power is used in emergency situations, such as in the event of the loss of an engine for a twin-engine helicopter. The authorization period is much shorter and corresponds to a period of 2 to 3 minutes. The order of magnitude of power is 10% higher than the maximum take-off power. It is possible that an engine inspection may be required after use in this configuration.

One Engine Inoperative, extreme emergency rate type, OEI 30 secs

This is a rate that is only used as a last resort, when saving the helicopter is the priority. Damage to the engine is allowed in this situation. The delay is very short, 30 seconds, and the power is approximately 25% above the maximum take-off power.

This extreme emergency situation may occur, for example, for a twin-engine naval helicopter suffering from sea-level engine loss in hover flight with a high weight. If this happens, the pilot may be forced to land immediately. To achieve a take-off with one engine down, this extreme emergency power is necessary. After recovering the helicopter from such a disastrous situation and returning to the base or a boat, the engine will likely require maintenance and refurbishment.

For example [GIF 11], the RTM 322 01/9 for NH Industries' European NH90 helicopter is capable of providing 1,799 kW take-off power (MTOP 60 min) under ISA conditions and 1,664 kW in cruise mode (MCP). Its maximum power in OEI 2 min mode is 1,962 kW and 2,206 kW in OEI 30 sec.

Influence of the environment

The power of a gas turbine engine decreases as the ambient temperature increases. The order of magnitude of this loss is 1% per degree of the T_1 compressor inlet temperature.

In order to be able to calculate the power available for engines used in aeronautics, it is necessary to take the characteristics (pressure, temperature, etc.) of the air in which they operate into account. The International Organization for Standardization (ISO) has arbitrarily defined a standardized atmosphere, called the International Standard Atmosphere (ISA). The International Civil Aviation Organization (ICAO) has also defined its own standard atmosphere, based on the ISA atmosphere model but covering altitudes up to 80 km (262, 467 ft) and does not take humidity into account.

Thus, the reference value – ISA – is a temperature of 15 °C at an altitude of 0 m with a pressure of 1,013.25 hPa per reference for an air density of 1.225 kg/m³. Figure 2.37 shows the engine power curves according to temperature. These curves are used to evaluate the power losses related to altitude climb according to the ISA daily flight conditions.

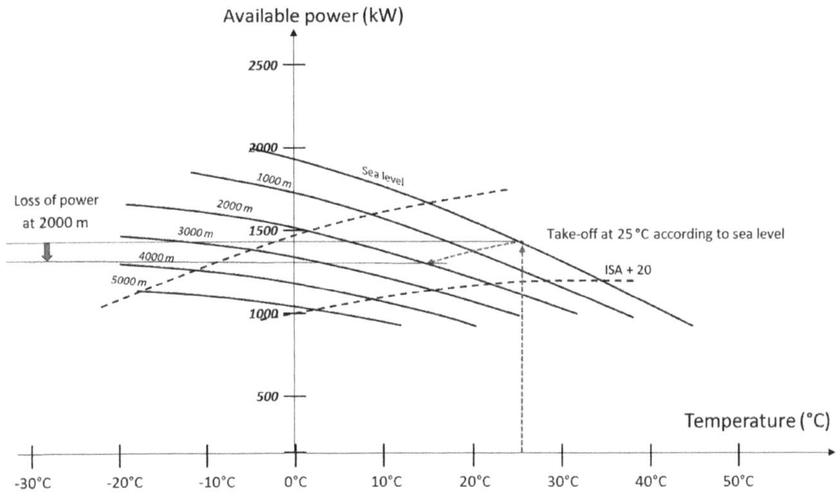

Figure 2.37. *Example of the available power of a turboshaft engine on the shaft as a function of temperature for different altitude and temperature values*

In general, the turbine is characterized by the actual power available on the driveshaft, usually at its maximum continuous power (MCP) operating point. This power is represented according to temperature for different operating altitude values (Figure 2.37).

2.2.1.4.4. New generation engines

Several solutions are possible for an improvement of the performance of aeronautical propulsion systems:

– propfans or Open Rotor: engine with a transonic propeller or two counter-rotating propellers, the second rectifies the air flow discharged by the first. The dilution rate can exceed 35. The blade feet operate like a fan and as a propeller. It has a Mach 0.8 efficiency of 0.85 for the UDF (unducted fan) and FPT (fast propeller turbojet) types;

– superfans or ultrafans: double-flow turbojet engine equipped with a very large fan; the dilution rate can be as high as 70.

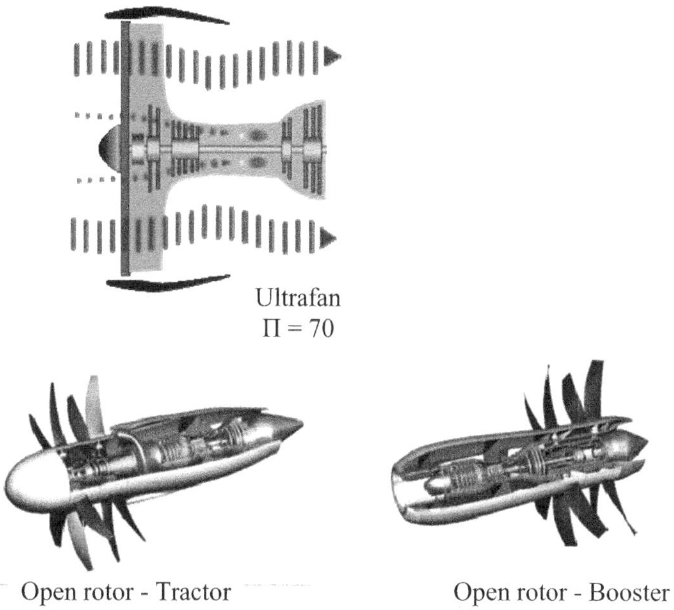

Figure 2.38. *Turbojet engine developments by improving secondary flow. For a color version of this figure, see www.iste.co.uk/krysinski/automotive.zip*

The counter-rotating propeller system, also known as an Open Rotor, improves the secondary circuit of the fluid. The removal of the nacelle and the housings constituting the secondary flow is a radical way of increasing this optimum to a great degree, but at the cost of an increase in noise. The propulsive efficiency can be improved.

It is possible to compare the propulsive efficiency of these different engines according to aircraft ground speed (Figure 2.39).

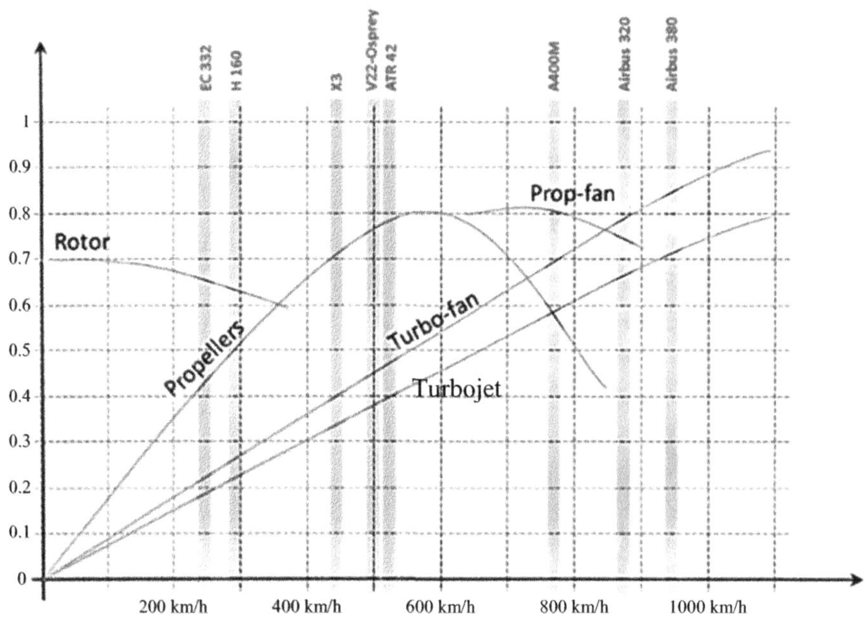

Figure 2.39. *Comparison of propulsive efficiencies for different engine types (source: Airbus Helicopters). For a color version of this figure, see www.iste.co.uk/krysinski/automotive.zip*

Another element by which one can compare engine technologies is the specific fuel consumption (SFC) (C_S). This is the amount of fuel required to provide power or thrust in a given time. It is generally expressed in kg/(kW·h) or kg/(kN·h).

The constant improvement of engine technologies enables an improvement of their specific consumption over time. Figure 2.40 shows the change in specific consumption compared to that of the first turbojet engine.

New technologies, such as heat recovery from exhaust gases or electrical hybridization, make it possible to go even further.

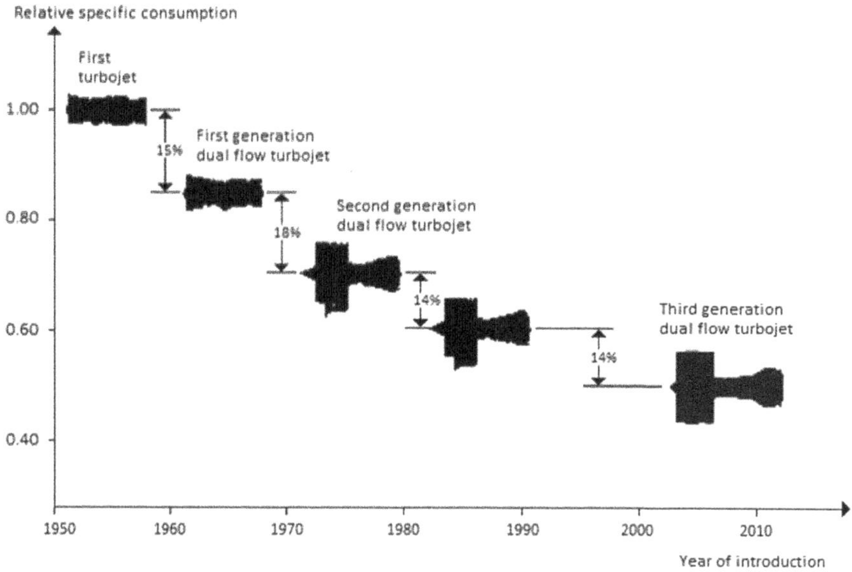

Figure 2.40. *Chronological comparison of gains in terms of specific fuel consumption obtained on propulsion systems [CAL 16]*

For example, the transportation powertrain of a small helicopter of the H120 type requires power of around 370 kW (500 hp), obtained by a Turbomeca Arius 2F turboshaft engine [SAF 11]. It has a specific consumption of 305 g/(kW h). This type of transportation powertrain can be compared with a piston engine with a high compression ratio.

In Figure 2.41, it can be seen that at full load power, the combustion engine has a better specific consumption than the ARIUS turbine. This gain is accentuated at half load, where a piston combustion engine retains a specific consumption while the turbine consumption is significantly increased.

This difference can be explained by comparing respective thermodynamic cycles. For a piston combustion engine cycle with a high compression ratio, the overall efficiency is about 45%, higher than that of the turbine for which it is closer to 30%.

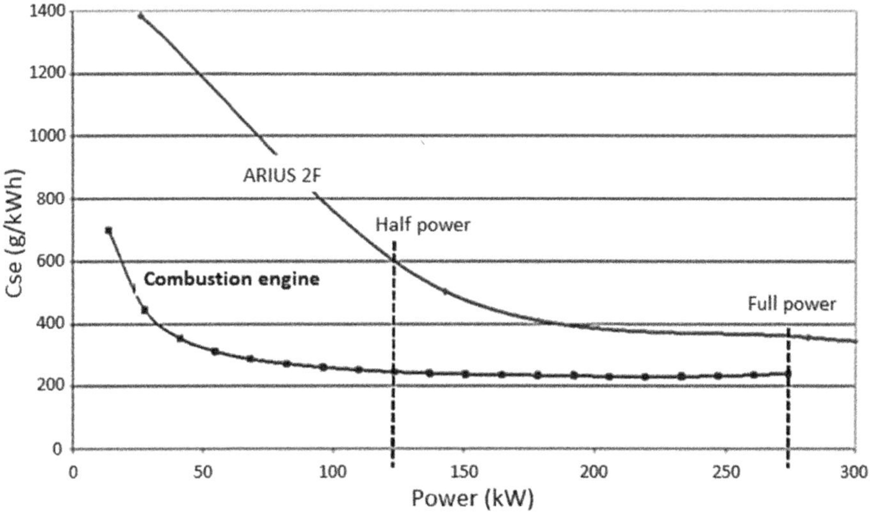

Figure 2.41. *Comparison of the specific consumption of a turbine and a piston combustion engine with a high compression ratio. For a color version of this figure, see www.iste.co.uk/krysinski/automotive.zip*

This raises the question of the weight of the transportation powertrain.

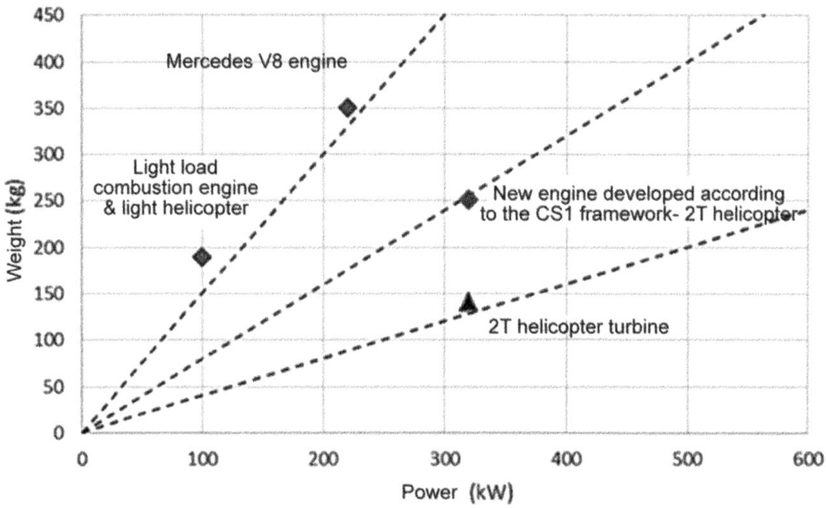

Figure 2.42. *Weight of engines as a function of their power. For a color version of this figure, see www.iste.co.uk/krysinski/automotive.zip*

A light aircraft diesel engine has a mass weight of 1.5 kg/kW. For this type of engine, the order of magnitude of specific consumption is 0.240 kg/(kW·h). These engines are too heavy to be used by a helicopter.

It is thanks to the development of engines with a specific consumption closer to 0.8 kg/kW that this type of transportation powertrain can be considered with an installed power closer to 330 kW (440 hp). The technology used in this case is a 4.6-liter, high compression piston engine with many technologies already implemented in advanced self-ignition engines, using kerosene commonly used in aircraft engines. The V8 engine consists of two cylinder groups inclined at 90 degrees to each other, with a high-pressure (1,800 bar) common rail direct injection system and a turbocharger per cylinder.

2.2.2. *Improvement of gas turbines*

2.2.2.1. *Exhaust gas heat recovery*

The low efficiency of a gas turbine is due to the fact that, as in a piston engine, some of the energy supplied by the fuel is required to drive the compressor and some is lost as heat in the exhaust gases. If we analyze the parameters of the single cycle, we can see that the temperature T_4 of the exhaust gases at the turbine outlet is higher than T_2 of the air at the combustion chamber inlet. This means that internal heat transmission between the exhaust gases and the air leaving the compressor can be achieved and the fluid can be heated before the combustion chamber to increase efficiency.

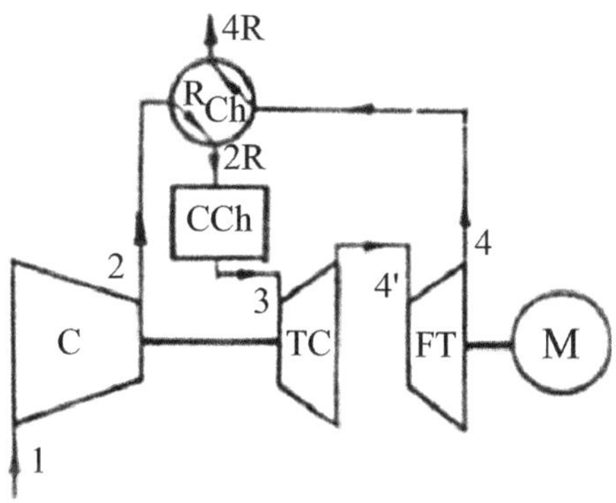

Figure 2.43a. *Diagram of a twin-shaft internal combustion turbine with internal exhaust heat recovery and the associated h-s diagram (source: Jan Krysinski)*

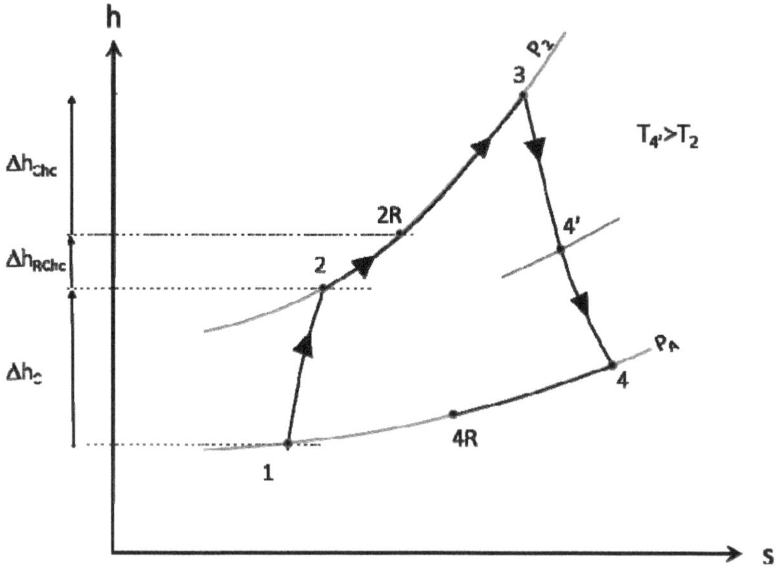

Figure 2.43b. *Diagram of a twin-shaft internal combustion turbine with internal exhaust heat recovery and associated h-s diagram (source: Pr Jan Krysinski) (continued). For a color version of this figure, see www.iste.co.uk/krysinski/automotive.zip*

The air with a T_2 temperature passes through the heat exchanger (collector) of the RCh heat and its temperature increases to T_{2R}. However, the exhaust gases passing through the collector reduce their temperature from T_4 to T_{4R}. Thanks to this exchange, the air temperature at the inlet of the combustion chamber is such that $T^{2R} > T_2$, which leads to a reduction in thermal power of the combustion chamber compared to a simple cycle and a significant increase in the efficiency of the machine.

The heat exchange in the collector theoretically takes place at constant pressure. In reality, there is a slight pressure drop on the air side as well as on the gas side.

The amount characterizing the collector's operation is defined by its efficiency coefficient which is:

$$\varepsilon = \frac{\text{Exchanged heat}}{\text{Theoretical heat possible to exchange}} = \frac{h_{2R} - h_2}{h_4 - h_2} \qquad [2.37]$$

By definition:

$$\varepsilon = \frac{C_{p\,air}\left(T_{2R} - T_2\right)}{C_{p\,fum}\left(T_4 - T_2\right)} \qquad [2.38]$$

If we neglect the difference in heat mass C_p of the air and fumes, we have:

$$\varepsilon = \frac{T_{2R} - T_2}{T_4 - T_2} \qquad [2.39]$$

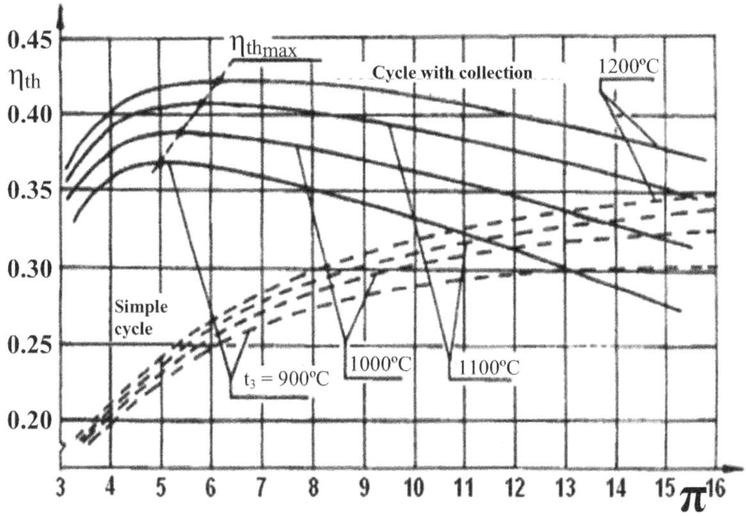

Figure 2.44. *Thermal efficiency of an internal combustion turbine as a function of the compression ratio πc for different temperature values t3. The exchanger's efficiency e = 0.83, the total pressure drop 3.2% (source: Pr Jan Krysinski)*

The cycle with exhaust gas heat collection achieves higher efficiencies than a simple cycle. The maximum efficiency values are achieved for lower compression ratios than in a simple cycle.

It should be noted that there is a limit to heat recovery. If the compression ratio increases, the T_2 temperature at the outlet of the compressor increases and for a certain value of π, it becomes equal to the T_4 temperature at the outlet of the turbine and heat exchange is no longer possible.

2.2.2.2. *Hybridization of a turboshaft engine*

It is assumed that a gas generator rotor, noted GG, rotates at a constant angular velocity, noted N. The power W_e supplied by an electric motor is added to its shaft.

The GG rotor starts to accelerate. To keep the velocity constant, N = Cte, it is necessary to reduce the power of the GG turbine (GGT) by reducing the flow of fuel injected into the combustion chamber.

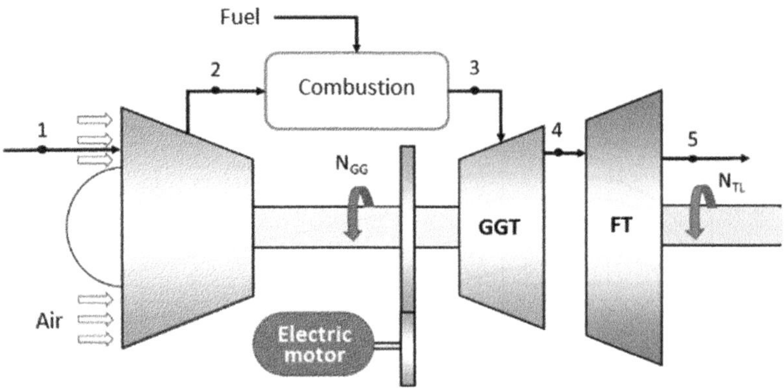

Figure 2.45. *Illustration of microhybridization of a turboshaft engine. For a color version of this figure, see www.iste.co.uk/krysinski/automotive.zip*

The decrease in flow rate causes the temperature T_3 at the inlet of the GGT to drop.

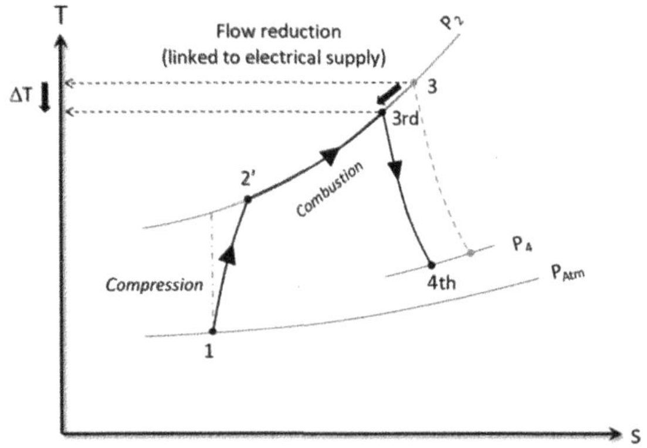

Figure 2.46. *Influence of hybridization on the T-s diagram. For a color version of this figure, see www.iste.co.uk/krysinski/automotive.zip*

2.2.2.2.1. Compressor-turbine operating point

At a steady state with N = constant, through power conservation, the power absorbed by the compressor can be estimated as follows:

$$W_c = W_{GGT} + W_{elect} \quad [2.40]$$

The thermodynamic parameters at the inlet of the GGT are as follows: pressure P_{3e}, temperature T_{3e}. Compressors and turbines are generally calculated for an operating point (mass flow and pressure rate) where the efficiency is at its maximum: this is the nominal operating point. It is defined by the intersection of curves which respectively represent the compressor characteristic (pressure – isovelocity flow) and the turbine characteristic (pressure – operating line flow at constant temperature).

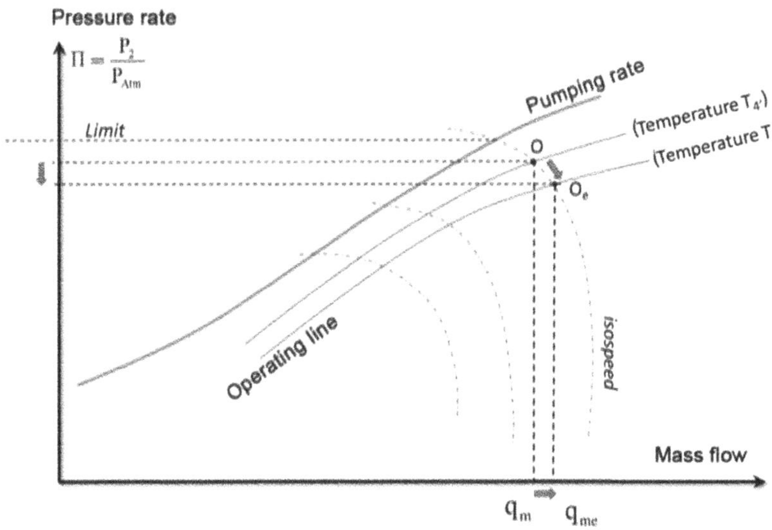

Figure 2.47. *The gas generator's operating point*

The operating point moves from O to O_e. The compression ratio decreases, then P_3 decreases to P_{3e} and the mass flow rate q_m can increase to q_{me}, which essentially depends on the compressor's characteristics (isovelocity curves).

When the compressor's compression ratio becomes too high for a given flow rate, instabilities may appear such as stalls on the compressor blades. The aerodynamic stall no longer allows the gases to be pushed in the right direction, and the "high pressure" part of the compressor (the outlet) empties into its "low pressure" part (the inlet). These large flow fluctuations are referred to as "pumping" because of the nature of this aerodynamic instability phenomenon which gives rise to longitudinal waves. They are obviously to be avoided.

It can be seen that, in the case of microhybridization, the margin against the pumping limit is higher, which is therefore favorable. In reality, the electrical power W_e is low compared to that of the turbine (3–5%) and in fact, the changes in the pressure parameters P_3 and flow rate are not significant.

2.2.2.2.2. Increasing the power of the free turbine FT

The advantage of microhybridization is obtained when the free turbine's power variation is greater than the power supplied by the electric motor (W_e), that is:

$$W_{TLe} - W_{TL} > W_e \qquad [2.41]$$

The analysis can be done on the h-s diagram (enthalpy – entropy).

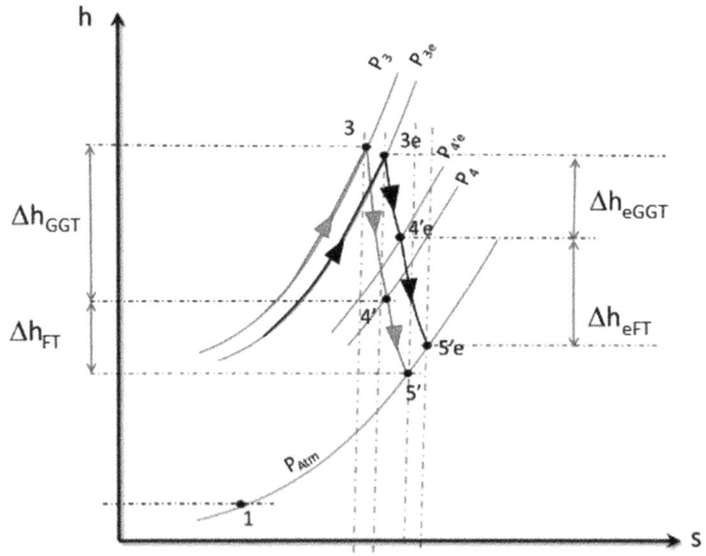

Figure 2.48. *H-s diagram of a turboshaft engine. Influence of microhybridization. For a color version of this figure, see www.iste.co.uk/krysinski/automotive.zip*

The two cases are compared assuming that the capacity of the compressor W_c is the same for both cases. The gas generator's power balance for a constant N is then expressed as follows for both cases:

$$\begin{aligned} W_c &= W_{GGT} \text{ (classical)} \\ W_c &= W_e + W_{eGGT} \quad \text{(microhybridization)} \end{aligned} \qquad [2.42]$$

Thus:

$$W_{GGT} > W_{eGGT} \qquad [2.43]$$

The mechanical power is the product of an enthalpy drop and the mass flow rate:

$$W = \Delta h \text{ x } q_m \qquad [2.44]$$

Thus:

$$\Delta h_{GGT} \text{ x } q_{GGTm} > \Delta h_{eGGT} \text{ x } q_{eGGTm} \qquad [2.45]$$

To simplify, the change in flow rate is neglected, which is a fairly good approximation in the case of the supersonic compressor.

So, we have:

$$\Delta h_{GGT} > \Delta h_{eGGT} \qquad [2.46]$$

This shows that GGT ends its expansion under pressure $p_{4'e} > p_{4'}$ and $T_{4'e} > T_{4'}$. We then have a gain in terms of enthalpy and a power variation of the FT:

$$\begin{cases} \Delta h_{TTLe} > \Delta h_{TTL} \\ W_{TTLe} > W_{TTL} \end{cases} \qquad [2.47]$$

For example, for a 1,000 kW turboshaft engine, a compression ratio Π of 15, standard input characteristics P1 of 101,325 Pa and T_1 of 15 °C, T_3 temperatures are obtained in the order of 1,230 K. The specific consumption SFC amounts to approximately 0.59 kg/kWh.

For a turboshaft engine with a hybridization rate of 6%, that is a contribution in excess of 60 kW, T_{3e} temperatures of the order of 1,217 °K are obtained. The specific SFC consumption amounts to approximately 0.58 kg/kWh.

The change in this specific consumption according to hybridization power is shown in Figure 2.49.

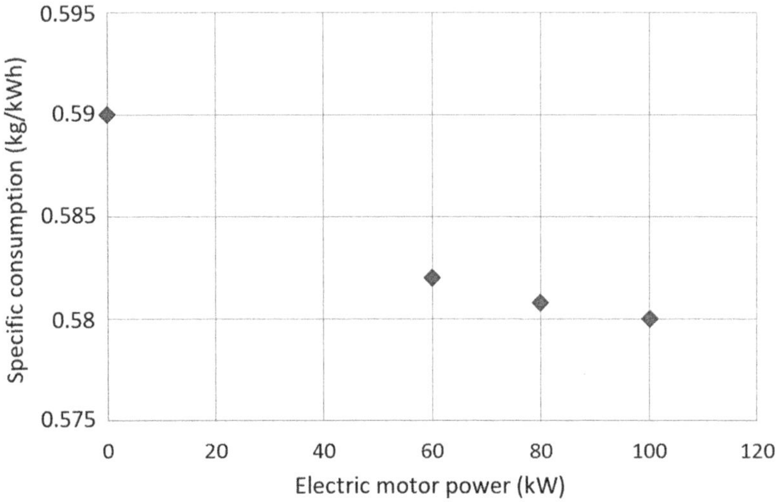

Figure 2.49. *Evolution of specific consumption as a function of hybridization power*

There is a limit to the increase in turboshaft power caused by the addition of power W_e. If W_e increases too much, the available enthalpy drop will be such that $h_{4'e} - h_{5'e} << h_{4'} - h_{5'}$ and consequently $W_{eFT} < W_{FT}$.

In this case, it is preferable to add the power W_e directly to the free turbine shaft.

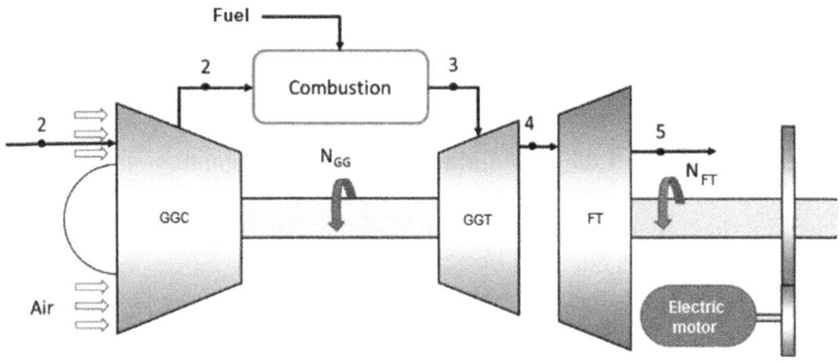

Figure 2.50. *Example of hybridization on a free turbine. For a color version of this figure, see www.iste.co.uk/krysinski/automotive.zip*

The advantage of adding the power W_e to the GG is a faster acceleration of the GG rotor due to the operation that allows the compressor pumping limit to be exceeded.

2.3. Electric motors

There are several ways to use an electric motor for transport. We can distinguish those that are autonomous (usually with an electrochemical storage battery or a fuel cell) and the other prisoners – which we often say are "captive" – that operate on a fixed route and a means of electrical connection to the land network (tram, trolley, etc.). Only the first ones are presented here; they correspond to the possible use on an aircraft or a car.

2.3.1. *Introduction to electric motors*

There are several electric motor technologies available. A distinction can be made between direct current and alternating current motors. These include synchronous and asynchronous motors.

Electric motors are known for their good efficiency, their energy reversibility, but also for the ease of controlling their torque over a very wide range of velocities. This makes it a motor well suited for automotive traction. By using the reversibility of an electric motor and therefore by using it as a generator, we can recover the energy during braking. In traditional systems such as friction brakes, braking energy is lost through the Joule effect. This feature minimizes energy consumption for electric and hybrid cars that combine electric and thermal propulsion.

The first electric motors used on electric vehicles were direct current motors. They had the advantage of being easily adjustable (PSA choice in the 1990s). Nevertheless, their efficiency was limited (90%), they required high angular velocities and their operation induced wear and tear on the control sticks.

Subsequently, asynchronous motors with inverters were developed (Tesla, Mia, Renault Twizy). These engines, very robust, compact and reliable, have the disadvantage of having an efficiency that reaches a maximum of 82% and which quickly drops to 75% over a wider operating range. Due to their low cost and reliability, they are more appropriate for hybrid vehicles.

The third possibility is a synchronous motor. Although they are more difficult to operate and are potentially less robust, they are the most efficient. Their major drawback is that China has a virtual monopoly on the rare earth elements needed for their production which is becoming an obstacle for development in the European automotive industry. There are two families of synchronous motors: a synchronous motor with a wound rotor, which is heavier and larger (ZOE, Fluence and Kangoo Renault cars) or a synchronous motor with permanent magnets (Peugeot Ion, Toyota Prius hybrids).

An electric vehicle's power chain is composed of a battery, an electrical management system consisting of a converter and a velocity variator, an electric motor, a gear box (fixed ratio) and a differential.

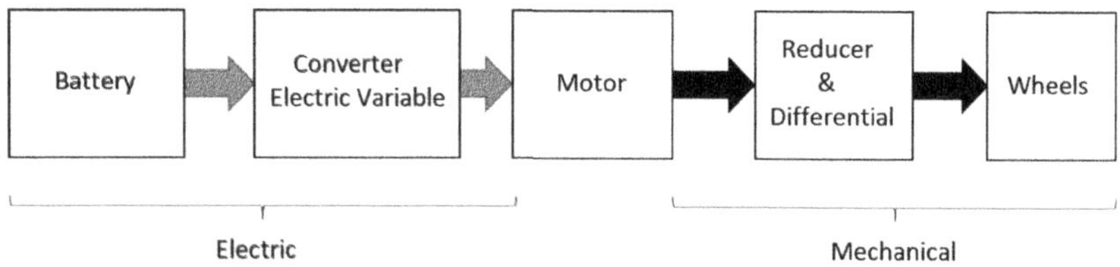

Figure 2.51. *Diagram of an electric car's power chain*

The objective of this section is to present the use and behavior of electric motors. In general, electric motors are electromagnetism-based electromechanical devices that convert electrical energy P_e into mechanical energy P_m (Figure 2.52).

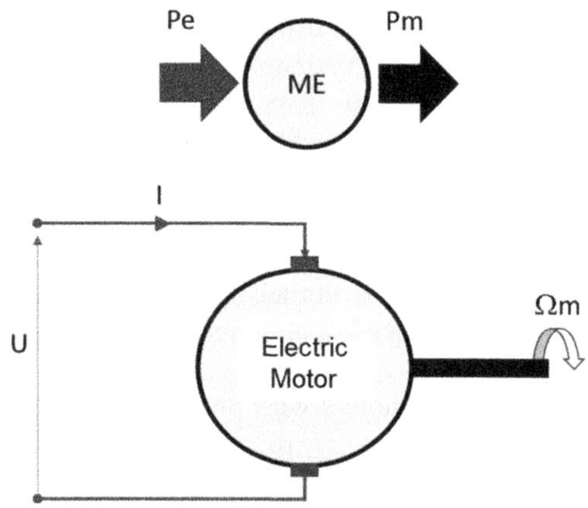

Figure 2.52. *Macroscopic example of an electromechanical energy conversion. For a color version of this figure, see www.iste.co.uk/krysinski/automotive.zip*

As with a combustion engine, the curve showing the torque-velocity characteristic of an electric machine describes its operation.

Figure 2.53 shows a general case of an electric motor powered by a battery with a voltage limited to U_{max}.

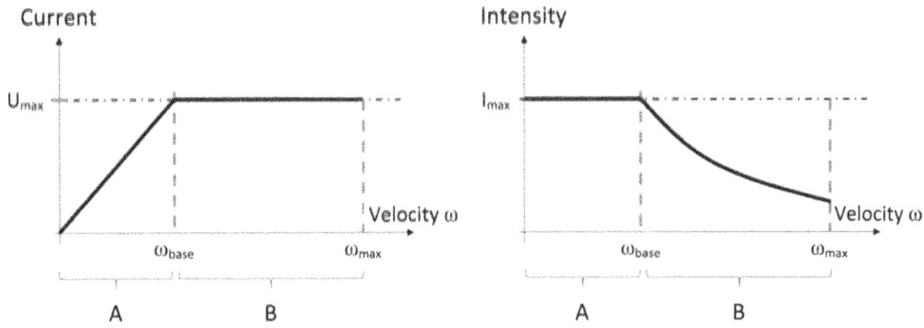

Figure 2.53. *Characteristic curves in voltage and intensity of electric motors*

There are two distinct areas. Zone (1) corresponds to an operation where the velocity is controlled by the voltage. In this area, the counter-electromotive force will increase linearly with velocity. The motor supply voltage will increase until it reaches the maximum voltage allowed by the U_{max} battery. The velocity reached, noted ω_{base}, is dependent on the characteristics of the motor and its power supply. In zone (1), the intensity will depend on the resistive torque applied to the motor. At maximum torque, the intensity will be maximal, defined by the following relationship:

$$I(t) = I_{max} \quad [2.48]$$

From an electrical equivalent point of view, we have the diagram in Figure 2.54.

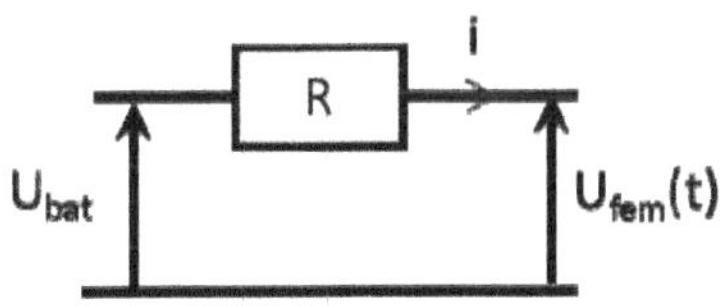

Figure 2.54. *Electrical diagram of the motor supply*

We then have:

$$U_{bat}(t) - U_{fem}(t) = Ri(t) \quad [2.49]$$

The battery control is done such that:

$$U_{bat}(t) = U_{bat}(0) + \beta\,\omega(t) \quad [2.50]$$

So:

$$\begin{aligned} &U_{fem}(t) = \alpha\,\omega(t) \\ &U_{bat}(0) + \beta\,\omega(t) - U_{fem} = R\,i_{max} \end{aligned} \quad [2.51]$$

Then we show that:

$$\alpha = \beta \quad [2.52]$$

and:

$$U_{max\,bat} - \alpha\,\omega(t) = R\,i_{max} \quad [2.53]$$

This defines the basic velocity:

$$\omega(t) = \frac{\left(U_{max\,bat} - R\,i_{max}\right)}{\alpha} \quad [2.54]$$

In the operating zone (2), above the basic velocity, the motor runs at constant power. The torque drops in a hyperbolic way. This zone is obtained by maintaining the voltage delivered by the battery at its maximum value, by reducing the magnetic flux in the machine (defluxing).

The engine, due to its technology, is limited to maximum velocity (ω_{max}). It thus defines the notion of flexibility, noted e, of the motor by the following relationship:

$$e = \frac{\omega_{max}}{\omega_{base}} \quad [2.55]$$

Consequently, the mechanical characteristics provided by the electric motor are given in Figure 2.55. They include a constant torque (1), a constant power (2) and a limit of motor velocity (3) operation.

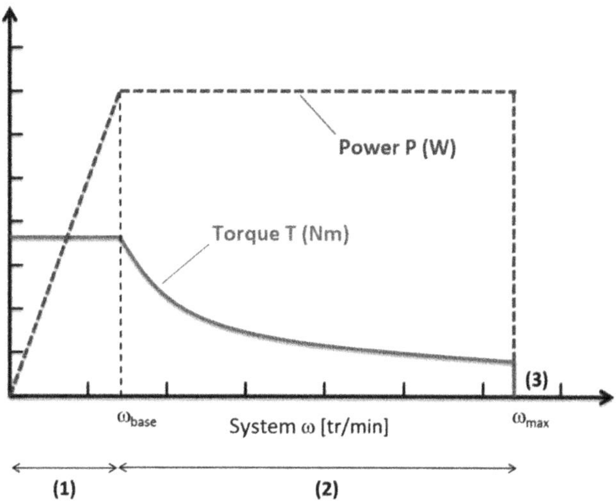

Figure 2.55. *Operating characteristic of an electric machine. For a color version of this figure, see www.iste.co.uk/krysinski/automotive.zip*

As in the case of internal combustion engines, it is important to consider efficiency. The efficiency depends on the velocity and torque that the motor delivers. There are three main classes of dissipative phenomena occurring in a motor system: Joule effect losses in conductive materials, losses in magnetic materials, commonly referred to as iron losses (by hysteresis or eddy current) and losses by mechanical friction. The electric motor's efficiency can be assessed by analyzing energy losses.

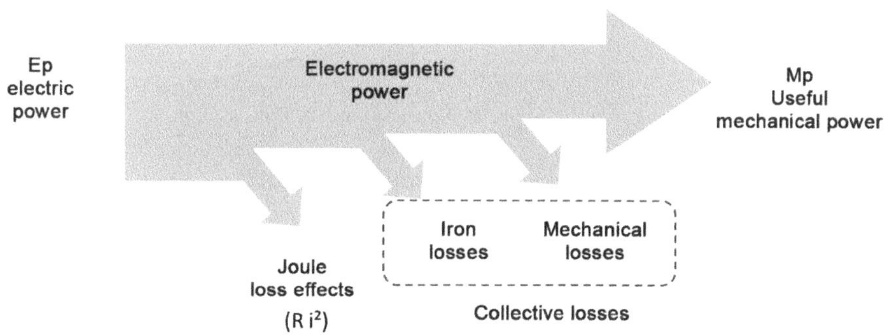

Figure 2.56. *Energy losses in an electric motor*

The efficiency can be evaluated for motors with no control sticks as follows:

$$\eta = \frac{P_m}{P_e} = \frac{C_m \Omega_m}{C_m \Omega_m + R i^2 + P_{iron} + P_{meca}} \quad [2.56]$$

The expression can then be in the form:

$$\eta = \frac{P_m}{P_e} = \frac{1}{1 + \frac{\left(R i^2 + P_{iron} + P_{meca}\right)}{C_m \Omega_m}} \quad [2.57]$$

For a given torque C_m, there is always a maximum efficiency related to velocity. It can be seen that at a low speed, in zone 1, at constant torque, the intensity is constant. In this case, the iron power is a fixed lost power regardless of velocity. At very low velocities, efficiency is poor and improves with increasing velocity. From a theoretical point of view, the efficiency is given by Figure 2.57 according to the operating point (ω_m, C_m).

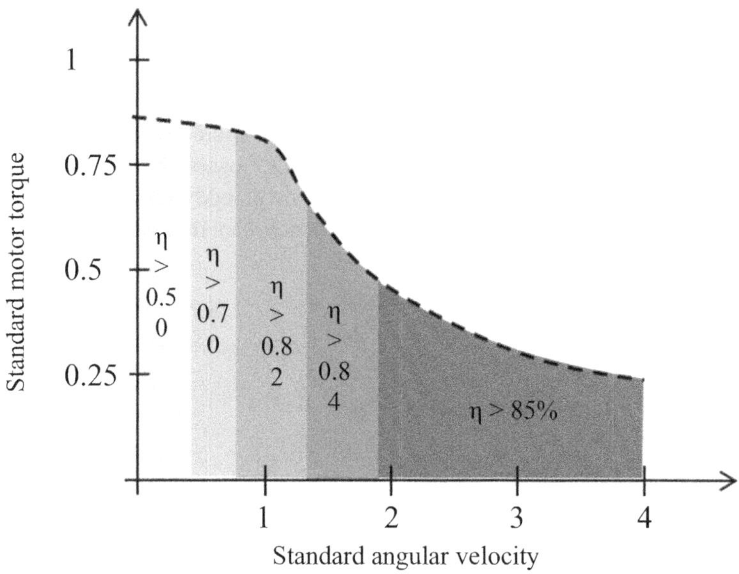

Figure 2.57. *Efficiency of a wound rotor electric motor according to the operating point (ωm, Cm) at a constant voltage*

In practice, other phenomena appear, and the mapping of engine efficiency is more complex. It is generally obtained experimentally (Figure 2.58).

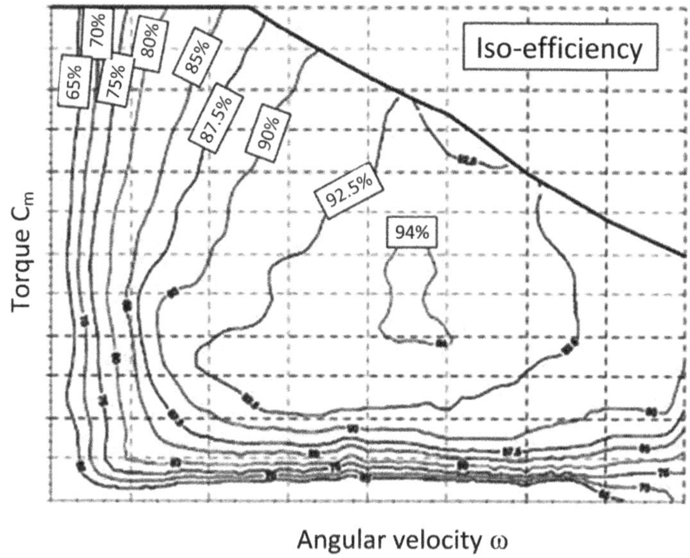

Figure 2.58. *Experimental mapping of the efficiency of a synchronous electric motor mission profile with no control sticks. For a color version of this figure, see www.iste.co.uk/krysinski/automotive.zip*

2.3.2. *Use of electric motors and mission profile*

2.3.2.1. *Cars*

It is interesting to make a link between the operating characteristics of electric motors and the practical use of a car. For this purpose, a driving cycle is used, generally representing a set of vehicle velocity points according to time.

Homologation cycles can include many variations in velocity, with linear cycles as proposed by the European NEDC (New European Driving Cycle) for the Euro 6 standard or more complex variations such as the new WLTP (worldwide harmonized light vehicle test procedure) for the Euro 7 standard. The Artemis cycle is based on a statistical study carried out in Europe as part of a research program (Common Artemis Driving Cycle). It is composed of three different configurations, plus an additional variant: urban cycle (Figure 2.59), rural, 130 km/h highway and 150 km/h highway. This type of cycle is better suited to understand real driving conditions and to evaluate the real performance of vehicles.

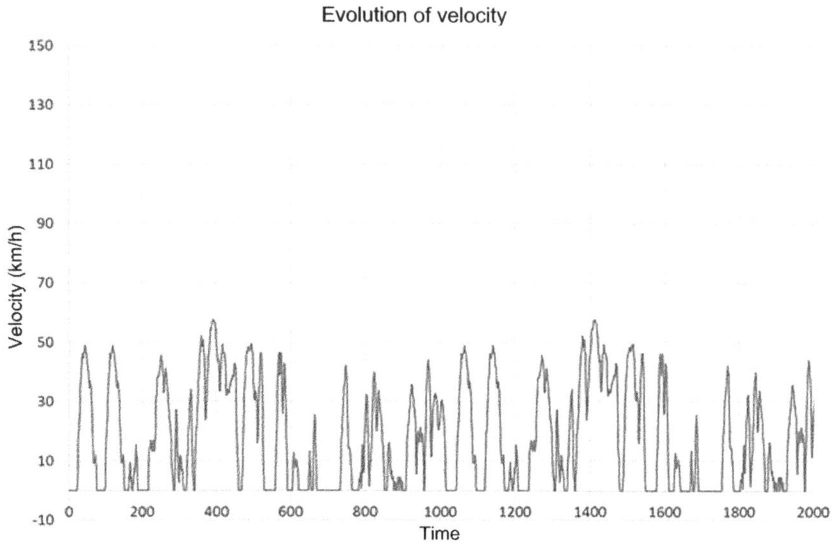

Figure 2.59. *Statistical cycle of vehicle use in an urban cycle (Artemis project)*

Figure 2.60 shows the principle of simulating the use of an electric motor vehicle during an Artemis cycle.

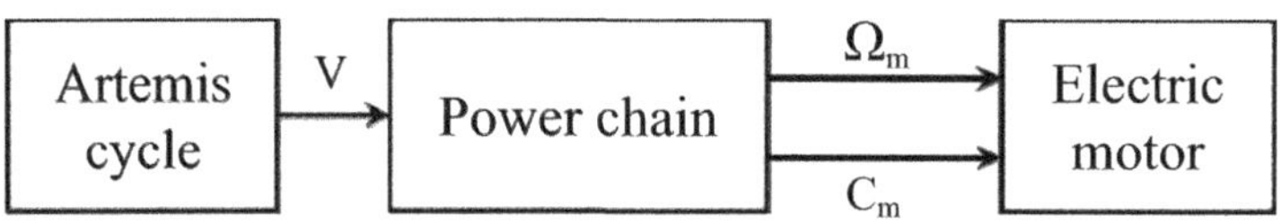

Figure 2.60. *Simulation of the use of the electric motor for an Artemis cycle*

Figure 2.61 shows the use of this type of motor on land vehicles in an urban cycle. We note that the cycle very often requires the engine to run at maximum torque when starting at a low speed, which requires use in this area over short periods of time. It can be seen that the efficiency in this area is poor.

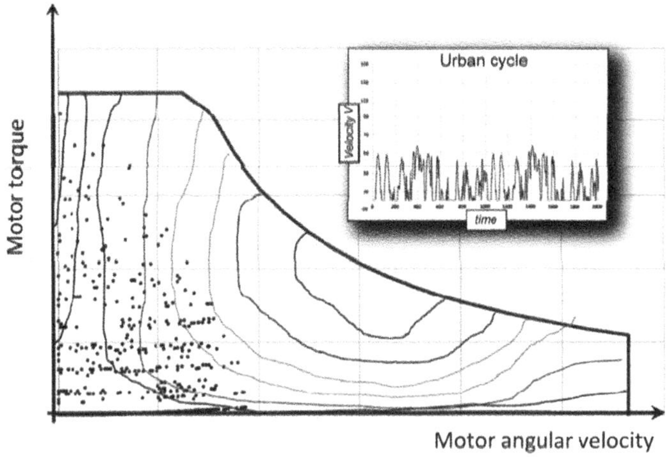

Figure 2.61. *Use of an electric motor on an urban Artemis cycle. For a color version of this figure, see www.iste.co.uk/krysinski/automotive.zip*

Figure 2.62 shows the use of this type of motor for a rural cycle. It can be seen that since the torque required is lower at high speed, the high velocity use of the motor makes it possible to meet the power requirement. In this high velocity operating zone, the efficiency is better.

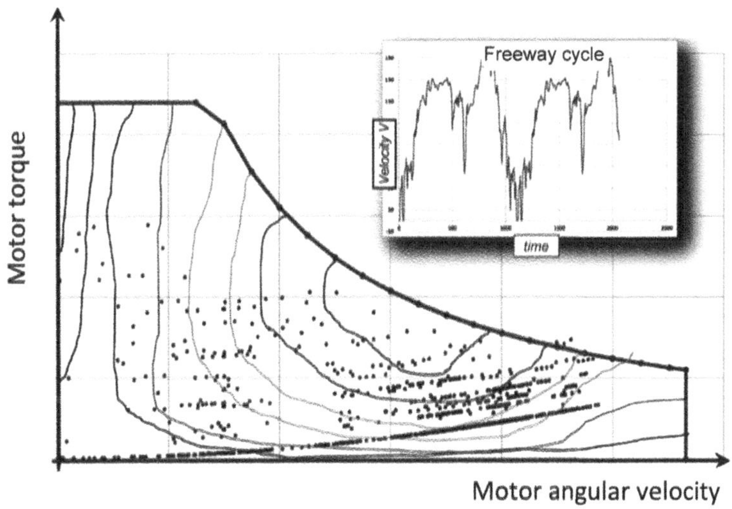

Figure 2.62. *Use of an electric motor in an Artemis urban cycle. For a color version of this figure, see www.iste.co.uk/krysinski/automotive.zip*

All these results show that the use of an electric motor is adapted to the motor vehicle according to the torque requirement or power requirement. This is because the electric motor:

– has a high torque at low speed, as well as a good efficiency adapted to city traffic;

– can meet the power and torque requirements at high speed without changing the reduction ratio (fixed reduction ratio).

2.3.2.2. *Helicopters*

For a helicopter, the problem is different. The use profile depends on the type of device and the type of mission for which it is intended. Figure 2.63 shows the desired mission profile for a medium helicopter (medium range) for a rescue mission (SAR) and an offshore transport mission (OPS). In both cases, it is observed that the mission is essentially composed of periods of hover flight, vertical climb or descent and level flight.

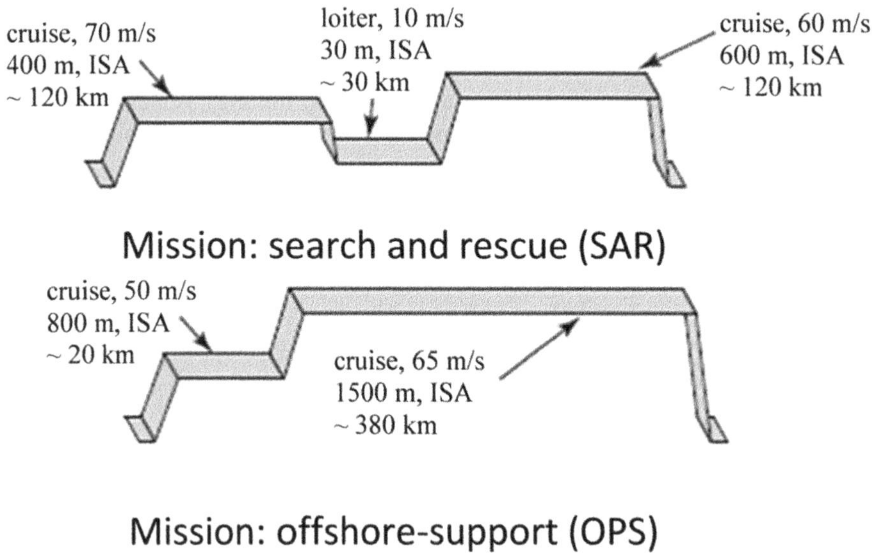

Figure 2.63. *Typical missions of a medium helicopter (source: Airbus Helicopters)*

Due to the operation of the helicopter rotor (see Chapter 1), the power requirement depends on the aircraft's velocity (Figure 2.64). We can observe that the power requirement is high in hover flight and lower in cruise flight.

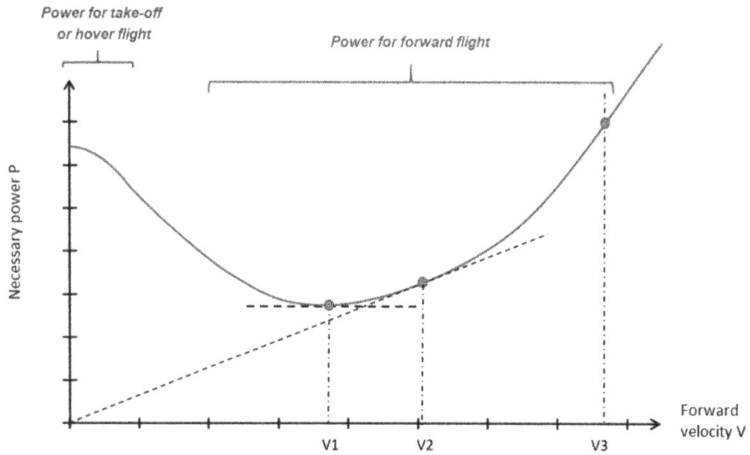

Figure 2.64. A helicopter's *power requirement as a function of the forward speed V. For a color version of this figure, see www.iste.co.uk/krysinski/automotive.zip*

Another feature of helicopter operation is associated with the rotor operation mode. Its angular velocity is regulated around a reference velocity (NR). The helicopter is flown by means of step control to change the lift, which consequently changes the resistive torque seen by the motor. By considering transportation powertrains on a helicopter, the area of use of the motor during a mission can be traced in the torque/velocity characteristics plane (Figure 2.65).

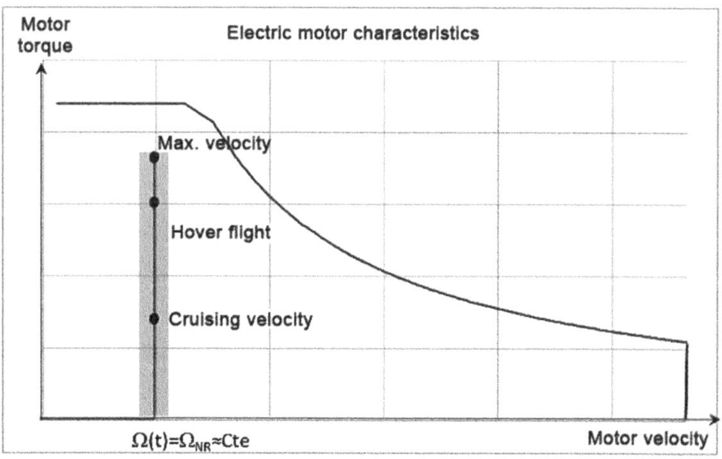

Figure 2.65. *Use of an electric motor for a helicopter mission. For a color version of this figure, see www.iste.co.uk/krysinski/automotive.zip*

Due to a peripheral velocity limitation, angular rotor velocities are generally limited (< 600 tr/min). The heavier the unit, the larger the rotor diameter and, in fact, the lower the angular velocity to maintain the same peripheral velocity at the end. Let us take a small power helicopter whose rotor rotates at 530 rpm for a maximum resistance torque of 1,950 Nm as a reference. Two solutions can then be compared (Figure 2.66).

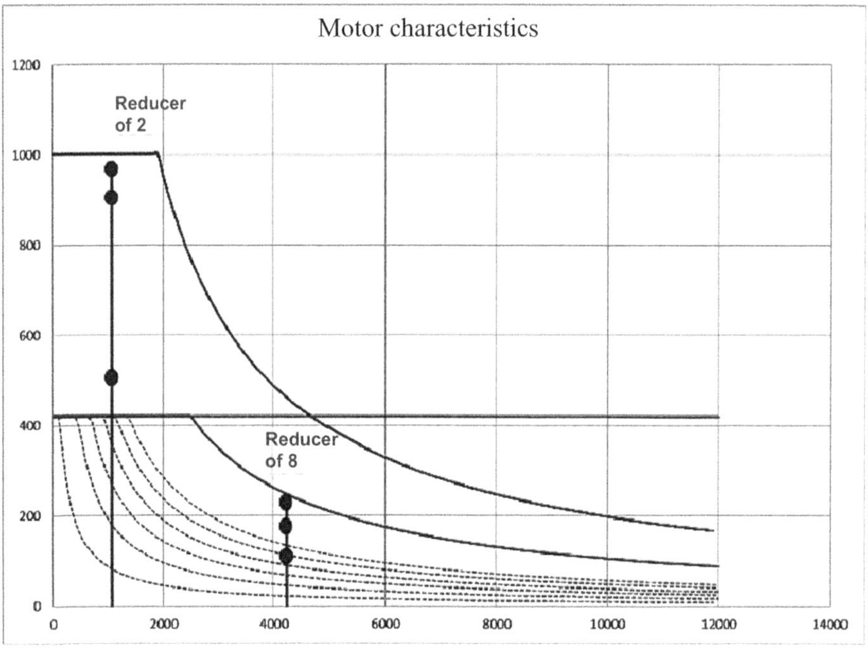

Figure 2.66. *Comparison of two electric transportation powertrain solutions for a typical helicopter mission. For a color version of this figure, see www.iste.co.uk/krysinski/automotive.zip*

A first solution with a reducer ratio of 2 requires a high maximum torque in the range of 975 Nm to 1,066 rpm. For a basic velocity electric motor of about 2,000 rpm, the installed power will be about 210 kW.

A second solution with a reducer ratio of 8 requires a high maximum torque of 240 Nm to 4,240 rpm. For a basic velocity electric motor of about 2,000 rpm, the installed power will be about 110 kW.

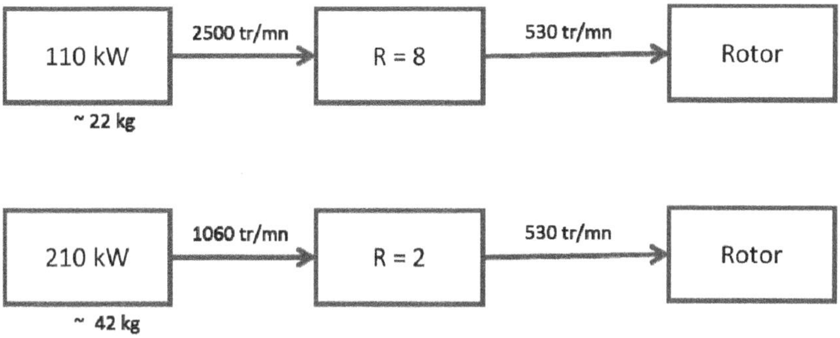

Figure 2.67. *Comparison of the two power chains*

Using the reference data of a 5 kW/kg automotive battery, the mass of the two engines can be compared: 22 kg for a solution with a high velocity reduction (x8) and 42 kg for a solution with a low velocity reduction (x2). The first solution also has the advantage of running the electric motor in a rotation range where the motor has a better efficiency.

The difference in the reduction of the two solutions does not bring about a great variation in the reducers' mass and does not call into question the advantage of running the motor at high speed.

In conclusion, the advantage for an electric motor to have a high torque at low speed is interesting on a car following an urban route but because of its operation, this advantage disappears for a helicopter.

In both cases, it will be necessary to jointly study the problem of energy storage to justify and compare an electrical solution with another mode of propulsion.

2.3.2.3. *Aircraft*

As with a helicopter, a typical aircraft mission can be characterized by take-off, cruise flight and landing phases.

For propeller aircraft, the problem is similar to that of a helicopter. The velocity at the blade end of the propeller must be limited. Since propellers have smaller dimensions than a helicopter rotor, propeller angular velocities are generally less than 3,000 rpm. It is thus possible to have an electric motor in direct drive or with a low ratio gearbox.

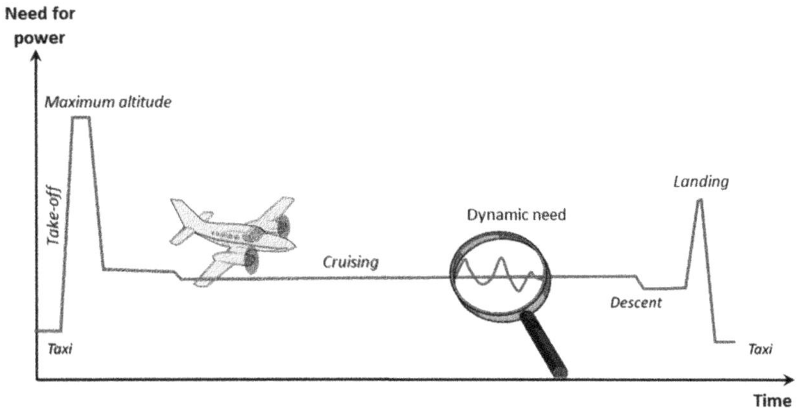

Figure 2.68. *An aircraft's typical mission*

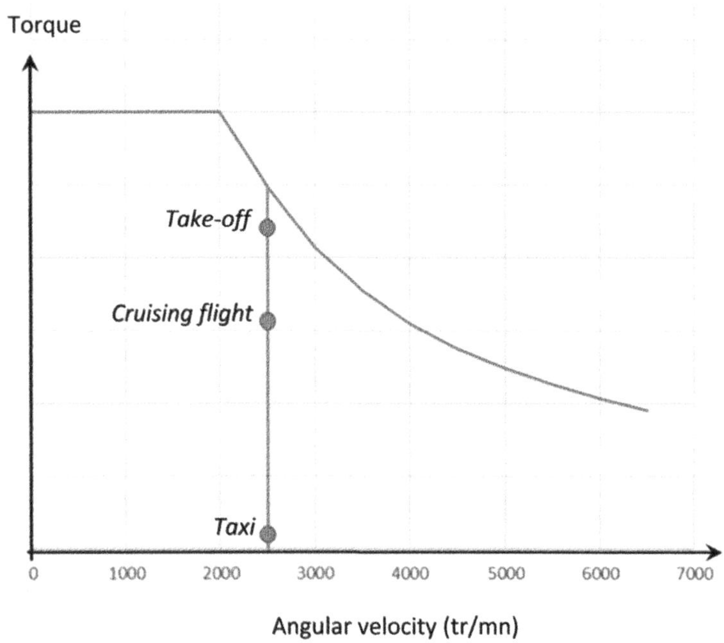

Figure 2.69. *Example of electric powertrains for an aircraft mission. For a color version of this figure, see www.iste.co.uk/krysinski/automotive.zip*

This solution makes it possible to operate the electric motor at parameters at which the efficiency is the best, avoiding an overpowered motor as well as a gearbox that would bring about extra weight.

2.3.3. *Electric motor technologies for propulsion*

As mentioned in section 2.3.1, for motor vehicles and rotary wings, manufacturers mainly focus on asynchronous motors or synchronous motors, with a wound rotor or a permanent magnet.

It is important to be able to describe the principle of internal functioning of each of these technologies and their main characteristics in broad terms.

2.3.3.1. *Synchronous motors*

Synchronous motors are the most commonly used type of delivering transportation powertrain in vehicles today and have returned to the forefront in recent years thanks to the development of variable frequency drives from which they are inseparable.

There are different technologies and design architectures that lead to significant performance disparities (efficiency, weight, etc.). Manufacturing costs and reliability criteria are also to be taken into consideration by car, helicopter or aircraft manufacturers.

A synchronous motor consists of a stator and a rotor separated by an air gap. The rotor is composed of protruding poles made of permanent magnets (permanent-magnet synchronous generators, PMSG) or direct current supplied electromagnets (MSRB) (Figure 2.70).

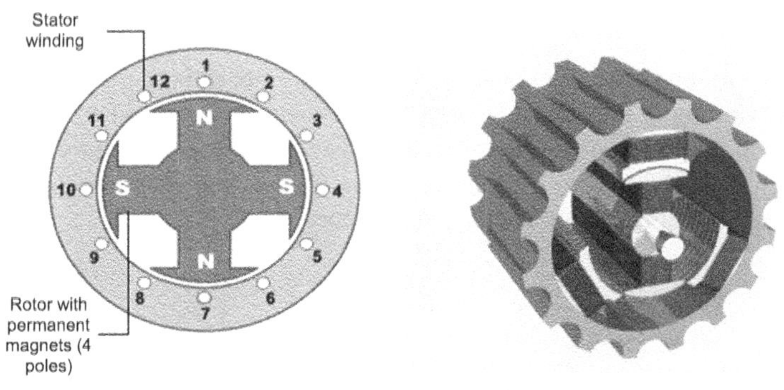

Figure 2.70. *Synchronous motors. Permanent magnet and wound rotor technologies (source: Mirsad Todorovac)*

After starting, the motor runs in synchronicity with a rotating field. When empty, the axes of the poles of the rotating field and the rotor come together. Under load, the axes are slightly offset.

A motor's stator is supplied with three-phase power (the most common in medium and high power). It mainly consists of the body, bearings, bearing flanges, motor cooling fan and fan protection cover. When the stator windings are traversed by a three-phase current, they produce a magnetic field rotating at the synchronization velocity. The synchronization velocity is a function of the power supply frequency and the number of pole pairs.

The rotor consists of magnets or coils powered by direct current.

2.3.3.1.1. Permanent magnet rotors

These are motors that can accept high overload currents to start quickly. By working on high-powered permanent magnets, such as neodymium-iron-boron alloys, and new geometries, it is possible to design more compact and lightweight motors.

Thanks to the use of axial flow, efficiencies can be improved (Phoenix engines for Renault vehicles).

2.3.3.1.2. Wound rotors

For medium and high powers, synchronous motors with wound rotor, combined with a variable frequency drive, are high-performance machines. The rotor consists of a stack of ferromagnetic discs. Like in the motor stator, windings are housed in slots on the rotor and electrically connected to the shaft end bands. The direct current is supplied via the band ring assembly.

2.3.3.2. *Asynchronous motors*

An asynchronous machine consists of two main elements: a stator, a part that includes a winding connected to a variable frequency drive, and a rotor, which includes a winding made of short-circuited conductors traversed by currents induced by the magnetic field created by stator currents. This is the main difference in the case of a synchronous machine with a wound rotor.

Figure 2.71. *Squirrel cage asynchronous motor (source: Mirsad Todorovac)*

The behavioral characteristics of such a motor are quite different from those of a synchronous motor. In general, the starting current is 5 to 7 times the nominal current. Thanks to current power electronics (inverters), it is possible to control the motor to reduce the current and ensure that a high torque is available during start-up. For some motors, it is possible by means of shunting to limit the starting current and at the same time obtain a high torque at low speed. For vehicles, start-up occurs by keeping the frequency voltage ratio (U/f) constant, which allows the magnetic flux values to remain the same for a sinusoidal supply voltage. There is therefore no risk of magnetic saturation. After starting and increasing the motor's velocity, the nominal operation is reached. If we want to increase the angular velocity further, we will increase the frequency, but at constant voltage. The magnetic flux then decreases. The angular velocity will increase, but the available torque will decrease: this is then a variation at constant power.

2.3.3.3. *Variable reluctance motors*

The rotor of this type of motor does not contain any magnets or excitation windings. The torque is created only by the variation effect of the reluctance.

Reluctance characterizes the "resistance" of the magnetic circuit to the passage of the flux. It is a notion similar to the notion of resistance in Ohm's law. The inverse of reluctance corresponds to magnetic permeance.

The variation in reluctance is due to the variation in the air gap caused by the presence of teeth on the rotating part (rotor).

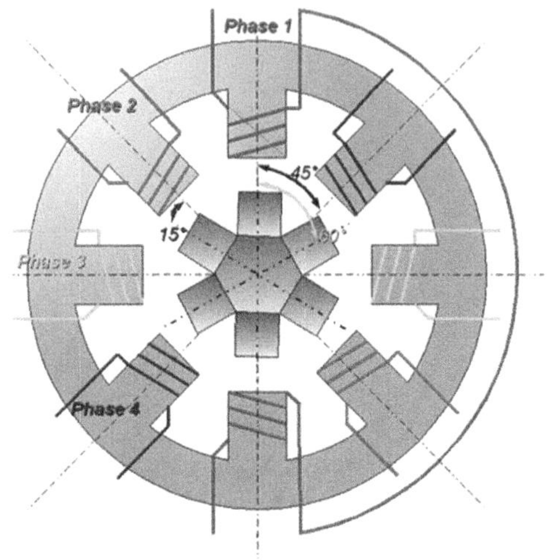

Figure 2.72. *Variable reluctance motor (source: MDP). For a color version of this figure, see www.iste.co.uk/krysinski/automotive.zip*

The rotor is designed in such a way that the ratio between the inductance in the direct axis and the quadrature axis is as high as possible. The operating range at constant power velocity is directly related to this ratio.

The disadvantages for this type of machine are the fragility of the manufacturing process because of the need for a high salience ratio, the low power factor and the complexity of the control electronics requiring a position sensor. This type of motor has the advantage of being able to run very fast, but has the disadvantage of having an engine torque with many acyclisms and a very high acoustic noise.

2.3.3.4. *Converters*

We have seen that electric motors on transport systems require an electronic velocity variator, the function of which is to adjust the velocity by varying the frequency and voltage, and the torque supplied by varying the current. This box, also known as the power electronics box, includes three major components: an inverter, a charger and a voltage converter.

2.3.3.4.1. Inverters

The role of the inverter is to provide the electric motor with a current the characteristics of which vary according to the torque required by the conductor. Electric motors and inverters are inseparable, similar to how a gasoline engine is controlled by ignition and injection systems. As the electric motor operates in a three-phase mode, the inverter converts the direct current of the battery into an alternating current that varies in frequency and intensity.

2.3.3.4.2. Chargers

The charger is a system that allows the battery to be recharged, a function similar to filling fuel tanks in conventional combustion vehicles. For electric cars, it performs the reverse work of the inverter: the AC voltage of the mains is converted into a DC voltage of a value compatible with that of the battery.

The charger performs this function by taking into account the safety constraints associated with charging the batteries, by constantly communicating with the battery management computer.

2.3.3.4.3. Converters

The converter transforms the high voltage of the traction battery into 14 volts to power a car's various comfort and driving systems. It is called a DC/DC converter because it only converts DC voltages. It has a function similar to the alternator of conventional combustion cars, since it is the one that maintains the charge of the 12 V accessories battery.

2.3.4. *Examples of specific propulsion systems and applications*

2.3.4.1. *Cars*

Electric vehicles have an accumulator as an energy source that is currently of the electrochemical type (battery). Battery technologies (see Chapter 4) are those of conventional (lead-acid and cadmium-nickel) or more advanced (nickel-metal-hydride and lithium) batteries.

Regenerative braking, possible on a vehicle, significantly increases range, especially in urban cycles (by about 20%). It also provides an engine brake. For these reasons, it is necessary to have a reversible conversion system.

The use of electric motors makes it possible to consider new architectures compared to piston combustion engines.

2.3.4.1.1. Single-motor solutions

Manufacturers may want a solution to use an existing platform and electrify a vehicle that is usually thermal. This is the case with Citroën's C15 and C25 electric vehicles. This results in a power transmission consisting of a sequential gearbox, a differential and a clutch (Figure 2.73). For a vehicle whose architecture has been specifically designed, the opening of the engine (minimum velocity-maximum velocity) makes it possible to avoid a gearbox and the associated clutch. A fixed gear reducer is sufficient to cover the vehicle's operating velocities (Figure 2.73).

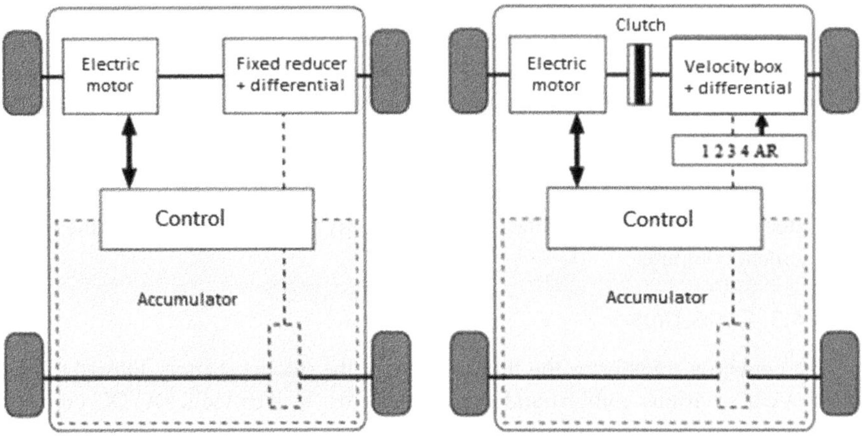

Figure 2.73. *Single-engine architecture*

The Renault ZOE vehicle transmission consists of a synchronous motor (R 240) with a 65 kW wound rotor and a maximum velocity of 11,300 rpm, and provides a constant torque of 250 Nm up to 2500 rpm. The motor is associated with a fixed gear reducer of 9.32.

The Nissan LEAF has an 80 kW synchronous permanent magnet electric motor (EM 57) for a maximum velocity of 10,500 rpm and a constant torque of 254 Nm up to 3,000 rpm, with a fixed gearbox of about 8.20.

2.3.4.1.2. Multi-motor solutions

Some manufacturers choose to use several motors, each associated with either a train (Tesla architecture) or a wheel (Michelin engine-wheel system) (Figure 2.74).

Tesla S vehicles (85D, 60D and 40D) have two four-pole asynchronous motors with a copper rotor, one per axletree, for electric all-wheel drive. The system controls the torque transmitted to the front and rear axletrees independently, via the on-board system, improving traction.

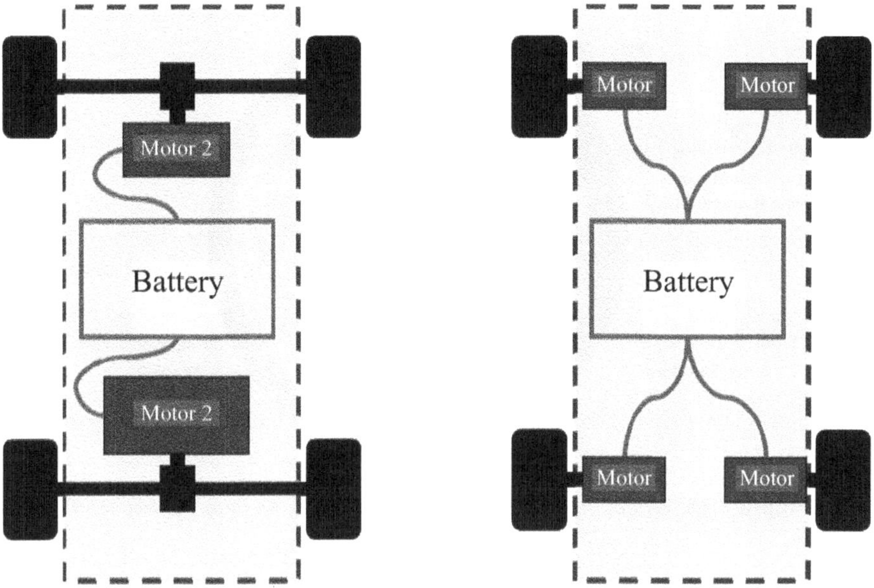

Figure 2.74. *Multi-motor architecture*

For example, on the 85, the rear motor has 285 kW of power and the front axletree 386 kW, a total of more than 650 kW, for torques of 441 Nm and 658 Nm, respectively. The reduction ratio is fixed and in the order of 9.73.

The possibility of imagining a solution where each wheel would have an independent electric motor is attractive. The motor-wheel solution then consists of an assembly that includes a motor incorporated in a wheel. The main advantage of such a system is its small size and the fact that it would not require a transmission component.

Michelin's Active Wheel project (2004–2014) is an example of this type of possibility. In addition, the system includes a set of functions such as an electric active suspension and a forced ventilation braking system. The proposed system consists of a traction motor developing up to 30 kW and a second motor for controlling the active suspension.

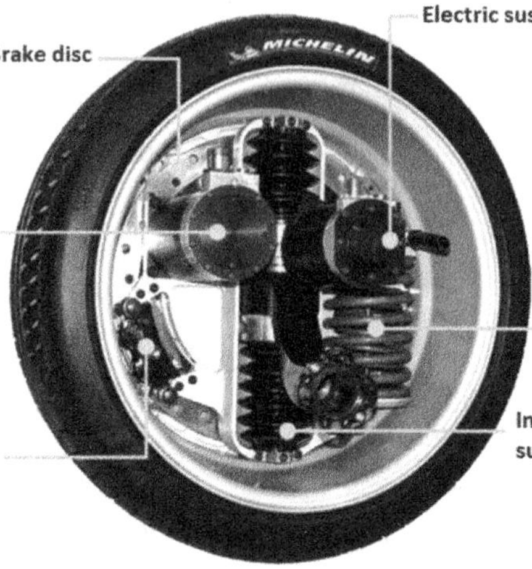

Figure 2.75. *Wheel motor of the Active Wheel project (source: Michelin)*

This type of solution can have several disadvantages. The volume is limited by wheel size. If there is no gearbox, the motor must be of high polarity and low velocity (large and difficult to cool). In the presence of a gearbox – which must be integrated – electrical machines can be fast with low polarity (less bulky motor). Permanent magnet synchronous motors are the most suitable for this type of system.

It should be noted that this type of motor-wheel solution is found on other types of systems: the asynchronous motor-wheel (Trolleybus Civis – Alstom transportation energy train) of 120 kW for a maximum torque of 500 Nm, the TM4 small vehicle wheel motors with permanent magnet synchronous motor on the surface of 80 kW for a torque of 180 Nm, the VAL metro wheel motor, permanent magnet synchronous motor and 65 kW external rotor (Fiat and Matra then Alstom).

2.3.4.2. *Examples of applications in aeronautics*

As far as aircraft are concerned, the use of an electric motor may be considered. These can be powered by batteries or solar panels.

At present, the maturity of technologies does not allow this energy to be used for air transport, whether for airliners or helicopters. The mass energy of the electrical storage is currently too low.

Nevertheless, in some areas where the payload is lower, such as leisure and pilot training (short missions), research or industrial experimentation or the specific use of drones, the use of electrical energy for propulsion has emerged.

As with cars, it can be observed that the systems offer several architectures: single-propeller single-motor systems or multi-rotor systems with one motor associated with each rotor.

2.3.4.2.1. Example of the Extra 330LE: single motor

The Extra 330 LE is a propeller aircraft equipped with a 260 kW Siemens engine (SP260D) with a mass of 50 kg. The aircraft is basically an aerobatic aircraft with a take-off weight of approximately 1,000 kg. The storage system consists of two 18.6 kWh Li-ion batteries with a weight of approximately 300 kg. The maximum power can be delivered for five minutes at most. When operating in endurance mode, the autonomy is estimated at a maximum of 20 minutes.

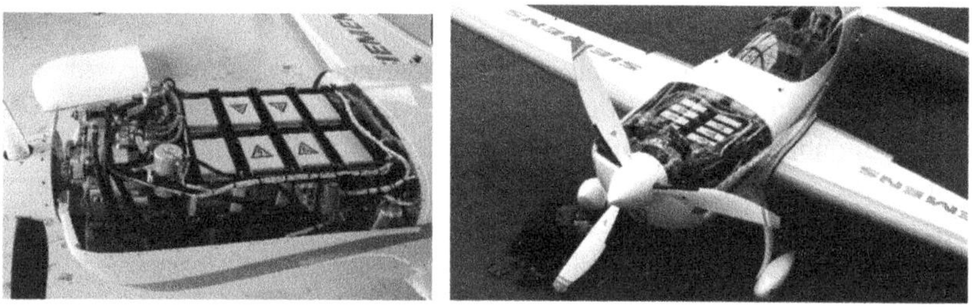

Figure 2.76. *Extra 330LE prototype (source: Siemens). For a color version of this figure, see www.iste.co.uk/krysinski/automotive.zip*

2.3.4.2.2. Example of the E-fan (Airbus): multi-motor

The E-Fan is an electrically powered aircraft developed by Airbus Group Innovations. It is intended for short missions (30 minutes to 1 hour) such as pilot training or glider towing. It is equipped with two 30 kW electric motors supplying two enclosed propellers. The system develops for each 1.5 kN of thrust. The maximum take-off weight is 550 kg.

Figure 2.77. *E-Fan project (source: Airbus). For a color version of this figure, see www.iste.co.uk/krysinski/automotive.zip*

2.3.4.2.3. Example of the NASA X-57 Maxwell project: multi-motor

The X-57 Maxwell is a prototype electric aircraft developed by NASA. The concept uses two 9 kW motors at the tip of the wing dedicated to drag, and twelve motors and propellers for a cumulative power of 60 kW, distributed along the wing, and dedicated to the take-off and landing phases.

2.3.4.3. *Examples of electric rotating wings*

As with fixed-wing aircraft, projects are emerging for rotary wing systems. Several possibilities of single- or multi-motor architecture can be identified again.

2.3.4.3.1. Example of the Volta project: single motor

The Volta project, led by Aquinea in cooperation with ENAC, consists of using the classic architecture of the helicopter, main rotor and rear rotor, with a 100% electric transportation energy train. The target missions are of short duration, the displayed autonomy is 30 minutes in cruise flight and 20 minutes in hover flight. The prototype is equipped with a dual 22 kW permanent magnet synchronous electric motor. The second version is equipped with a 90 kW transportation powertrain for a maximum take-off weight of 640 kg.

2.3.4.3.2. Example of the EHang 184 project: multi-motor

The Low Altitude Flying Transport (LAFT) project is led by EHang. Its objective is to introduce a new transport system in high-density urban areas, such as Hong Kong. The project explores possible new architectures, in particular by proposing a multi-rotor electric drone (8 rotors for a payload of 100 kg or 16 rotors for 280 kg).

The system is designed to transport people. The drone is fully electric, charges in 2 or 4 hours, depending on the mode chosen, and can travel at an average speed of 100 km/h. The announced autonomy is 25 minutes. Inside, there is a new generation seat connected to a 4G tablet, heating and even a trunk for luggage.

2.3.4.3.3. Example of the Volocopter project: multi-motor

The Volocopter proposed a two-seater prototype called VC200. It is distinguished by its 18 propellers and associated engines. It is intended for short missions with a maximum take-off weight of 450 kg, for a maximum of two passengers. All rotors provide 45 kW.

2.4. Internal combustion engine pistons

Internal combustion engine pistons are currently the most widely used on vehicles worldwide. In aeronautics, they can drive a propeller or rotor. They can be found on almost all light aircraft, ultralights or small aircraft and on some light helicopters.

For these engines, combustion takes place within the engine itself. They are thus defined as alternative internal combustion engines. It is the same fluid that pushes the piston back and undergoes combustion.

The design of internal combustion engines dates back to the second part of the 19th Century. The first industrial combustion engine was the gas engine, built by Lenoir in 1859. Its only became efficient when Otto applied the compression, designed by Beau de Rochas in 1877. These engines are two- or four-stroke.

Combustion is caused:

– by a spark at a given moment (gasoline engine);

– by spraying fuel into hot air under pressure (diesel engine).

2.4.1. *Theoretical thermodynamic cycles*

2.4.1.1. *Reminder: thermodynamic cycle/closed cycle engine*

A closed system or closed cycle, as opposed to an open system, is a system that does not exchange matter with its environment, but only heat and energy. The equivalence law for a "closed" transformation, or thermodynamic cycle, states that the sum of the energies received, work W and heat Q, is zero (Figure 2.78).

Regardless of the technology, the internal combustion engine is characterized by:

– a displacement which is the volume swept by the piston between the top dead center TDC and the bottom dead center BDC;

– a low compression ratio for gasoline engines (8 to 13), higher for diesel engines (16 to 23).

The following definitions can be recalled. In the particular case of an internal combustion engine, the displacement is the product of the volume swept by a piston between the top dead center (TDC) and the bottom dead center (BDC), and the number of pistons in the engine. In addition, the compression ratio of a piston engine – also known as the volumetric ratio – is a theoretical comparison ratio between the volume of the combustion chamber when the piston is in the top dead center, and the volume of the cylinder when the piston is in the bottom dead center.

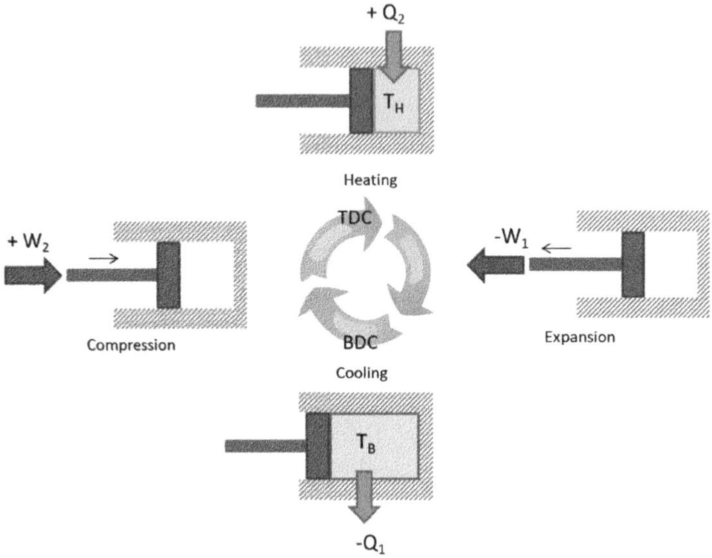

Figure 2.78. *Thermodynamic principle of a closed cycle. For a color version of this figure, see www.iste.co.uk/krysinski/automotive.zip*

The thermodynamic cycle for both engine types is characterized by the four phases illustrated in Figure 2.79.

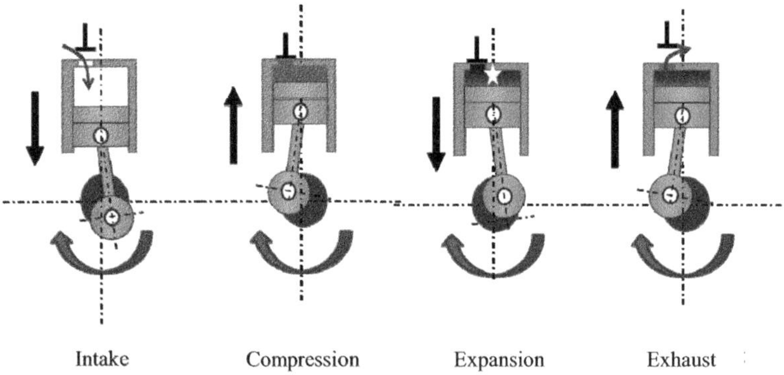

Figure 2.79. *Diagram of the four-stroke engine cycle. For a color version of this figure, see www.iste.co.uk/krysinski/automotive.zip*

These four phases are as follows:

– *Intake phase*: the cylinder descends and sucks in air through the intake valves;

– *Compression phase*: the exhaust valves are closed. The piston rises and compresses the fresh gases (air and fuel in the case of a gasoline engine, only air for a diesel engine);

– *Combustion/expansion phase*: this is the engine phase. For a gasoline engine, the compressed fresh gases are ignited by the spark from the spark plug and a flame spreads. For a diesel engine, the fuel is injected and, given the thermodynamic conditions in the combustion chamber (pressure and temperature), it ignites directly. In both cases, this combustion causes an increase in pressure and temperature and pushes the piston down;

– *Exhaust phase*: as it rises, the piston pushes the burned gases through the exhaust valves.

2.4.1.1.1. Gasoline engine

The Beau de Rochas or Otto cycle, schematizes the change in gases in internal combustion engines (Figure 2.80). It consists of two constant volume transformations of CD and EB, combined by the adiabatic BC and DE. The isobaric transformations AB and BA correspond respectively to exhaust and intake.

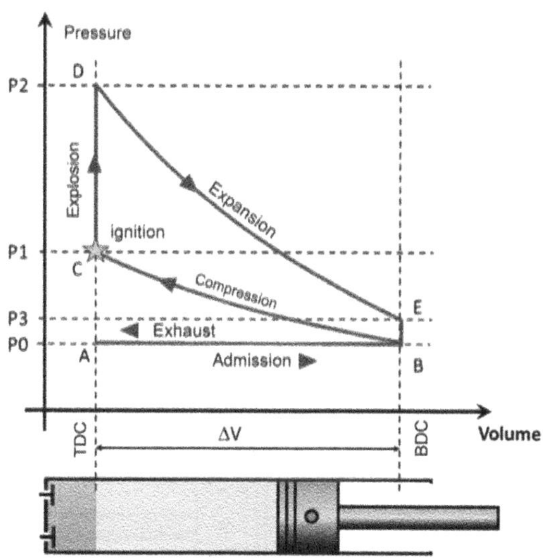

Figure 2.80. *Beau de Rochas cycle. For a color version of this figure, see www.iste.co.uk/krysinski/automotive.zip*

It can be shown that the efficiency of such a cycle depends on the rate of compression τ [MOR 12]:

$$\eta = 1 - \frac{1}{\tau^{\gamma - 1}} \quad [2.58]$$

The polytropic coefficient γ reflects the character of thermodynamic transformation. In the adiabatic case – purely theoretical for an internal combustion engine – it is equal to the ratio of the specific calorific capacities, C_p and C_v, or mass heat, respectively, at constant pressure or volume. It also depends on the operating conditions of the engine and, according to [GIB 97], it is in the order of 1.22.

$$\gamma = \frac{C_p}{C_v} \quad [2.59]$$

2.4.1.1.2. Diesel engines

The diesel cycle is different from the Otto cycle in that it is considered that heat is added not at isovolume but at isopressure (from C to D) (Figure 2.81).

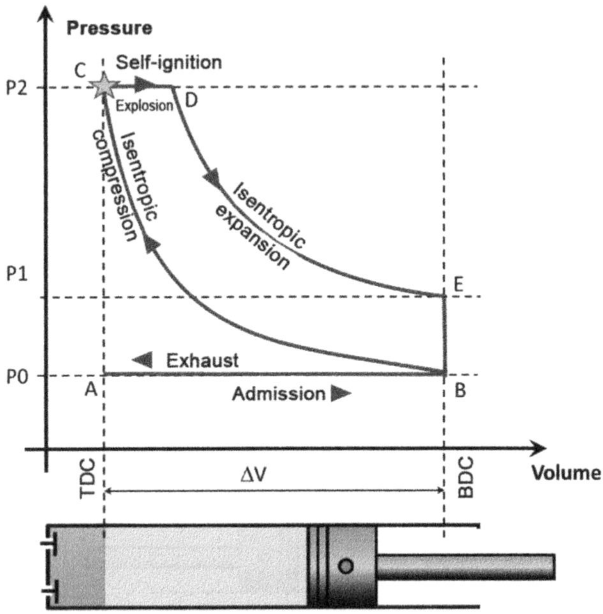

Figure 2.81. *Diesel engine cycle. For a color version of this figure, see www.iste.co.uk/krysinski/automotive.zip*

If during this heat addition phase the volume is multiplied by a factor k (k > 1), it can be shown that the efficiency of the diesel cycle is:

$$\eta = 1 - \frac{1}{\gamma\,\tau^{\gamma-1}}\frac{k^{\gamma}-1}{k-1} \qquad [2.60]$$

with τ the compression ratio and k the volume ratio VD/VC.

As k > 1, the efficiency of a diesel cycle is lower than that of an Otto cycle for the same compression ratio τ.

A real mechanical cycle of a four-stroke engine differs considerably from the elementary models just presented:

– the processes are not completely reversible;

– the mixture does not behave like a perfect gas of constant specific heat;

– the air quantity is not fixed, since there are intake and exhaust phases;

– the heat supply does not come from the outside, since combustion is "internal".

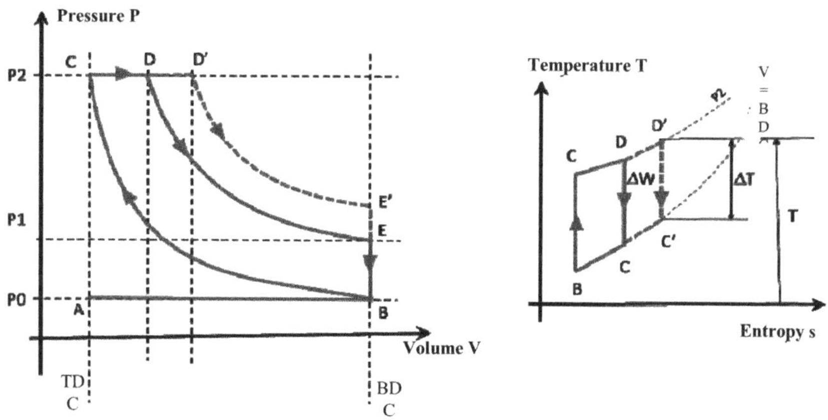

Figure 2.82. *Influence of ratio k on the diesel engine cycle. For a color version of this figure, see www.iste.co.uk/krysinski/automotive.zip*

It is important to analyze the influence of the k ratio (V_D/V_C) on cycle efficiency. The increase in k increases the volume V_D, which for the same volume V_C, reduces the decompression rate. The pressure ratio $p_{D'}/p_{E'}$ being lower than the ratio p_D/p_E, decreases the ratio ΔT/T by causing the cycle efficiency to be reduced. This shows that a transition from C to D is required as soon as possible.

2.4.1.2. *Gasoline/diesel comparison*

The overall efficiency of a diesel engine is higher than that of a positive-ignition engine (Figure 2.83).

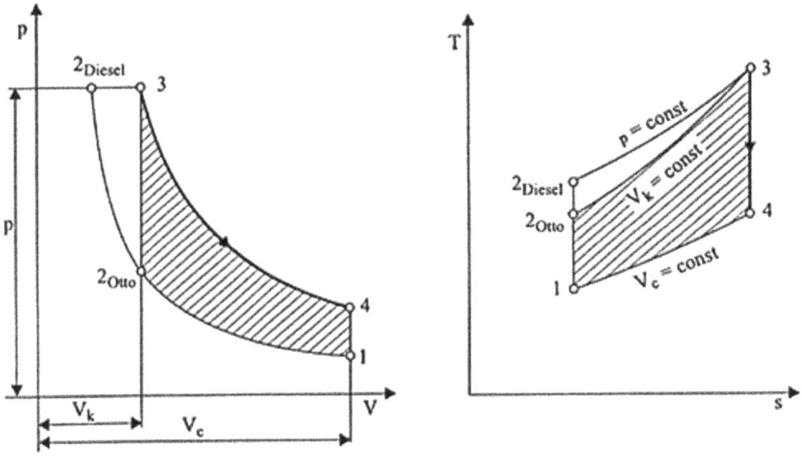

Figure 2.83. *Comparison of the diesel cycle and the Otto cycle*

Figure 2.83 demonstrates the comparison of the Otto cycle and the max isobar diesel engine in the cylinder. It can be seen that the efficiency of diesel is higher.

There are several reasons for this: the different volumetric ratio, the knocking phenomenon, the shape of the combustion chamber, the presence of a butterfly valve, poor adjustment, the volumetric energy of the fuel or supercharging.

2.4.1.2.1. Comparison of volumetric ratios

The volumetric ratio of a diesel engine is higher than that of a gasoline engine. Indeed, we have seen previously that the main specificity of a diesel engine lies in the fact that the ignition of the fuel depends on the thermodynamic conditions prevailing in the engine (pressure, temperature). The higher the volumetric ratio, the higher the efficiency.

For gasoline engines, the volumetric ratio is limited by the thermodynamic efficiency and the rattling phenomenon. The efficiency of the gasoline thermodynamic cycle is expressed as follows:

$$\eta = 1 - \frac{1}{\tau^{\gamma - 1}} \qquad [2.61]$$

The efficiency η of the engine does not depend on the pressure value but depends on the compression ratio τ (engine design parameters) and γ. These two parameters are linked because the increase in the compression ratio makes the thermodynamic transformation less adiabatic and γ decreases and approaches the isotherm (limit on τ design parameter).

From the relationship [2.61], we can obtain:

$$\frac{d\eta}{d\gamma} = \tau^{1-\gamma}\ \ln(\tau) \qquad [2.62]$$

This relationship shows the sensitivity to a variation of γ, that is a variation in the nature of heat exchange. For perfect diatomic gases (oxygen or nitrogen), the parameter γ is for an adiabatic transformation of 7/5 (1.4). The decrease in the parameter γ, which is similar to an isothermal transformation, has a significant impact on engine efficiency, more particularly for higher compression ratios.

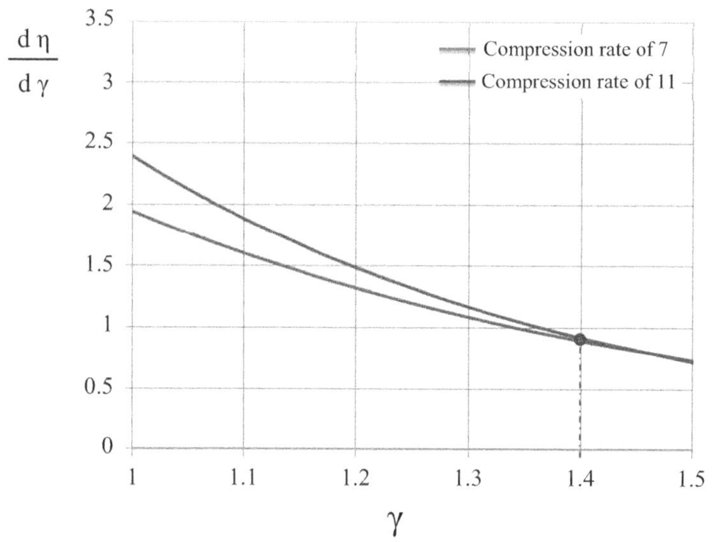

Figure 2.84. *Evolution of efficiency as a function of the parameter γ. For a color version of this figure, see www.iste.co.uk/krysinski/automotive.zip*

2.4.1.2.2. Knocking

The knocking phenomenon corresponds to abnormal combustion resulting from the self-ignition of a still unburned portion of the air and fuel charge under the effect of the temperature and pressure in the combustion chamber [BAD 13]. This results in chaotic overpressure excitation on the walls of the combustion chamber and piston. These resonances can be detected at low velocities by an expert ear. This is a metallic noise well known to engine manufacturers.

There is no knocking in an engine running on the diesel cycle, heat being the trigger for the generalized ignition.

To limit the risk of knocking on gasoline engines, the engineer must limit the volumetric ratio and thus affect, by construction, the thermodynamic efficiency.

2.4.1.2.3. Combustion chambers

The combustion chamber also has a role to play. The two pistons, diesel or gasoline, have completely different shapes (Figures 2.85 and 2.86). The lower piston (diesel engine) has a special bowl shape in which combustion will take place. The piston of the gasoline engine is flatter and combustion takes place more when in contact with the interior lining.

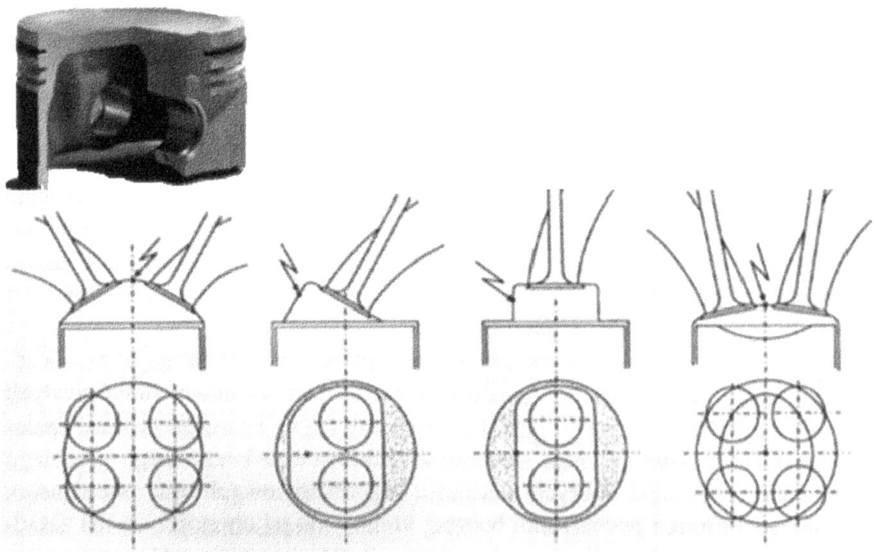

Figure 2.85. *Forms of a gasoline engine piston*

In an engine, only one third of the energy contained in the fuel will be useful, i.e., it will be used to turn the crankshaft. Another third will be dissipated as heat in the coolant. The final third is found in the exhaust gases.

When combustion takes place in the bowl (for the diesel engine), heat exchange with the cylinder, and therefore with the coolant, is reduced. So we save useful energy!

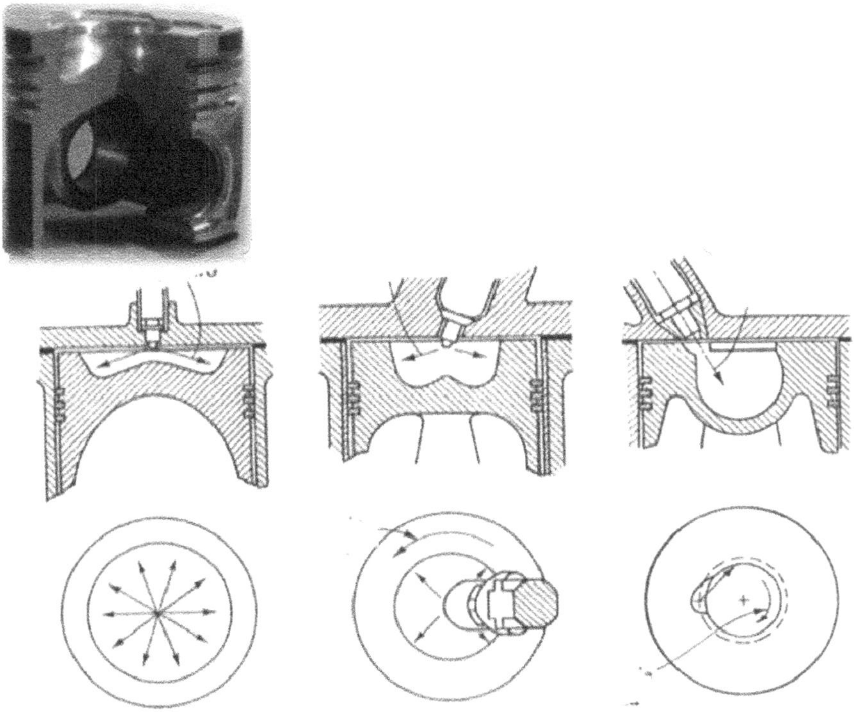

Figure 2.86. *Forms of a diesel engine piston*

2.4.1.2.4. Influence of the butterfly valve

For gasoline engines, the amount of energy introduced into the cylinder in each cycle is regulated by the amount of fuel mixture prepared outside the cylinder. This type of gasoline engine always works with an air-fuel ratio close to 1 for operation, to ensure the proper functioning of its emission control system.

This operation, which previously, was mainly carried out by the so-called "carburetor" mechanical system, is now carried out by electronic injection. For the carburetor system, the introduction of the mixing quantity can only be controlled by a valve called a butterfly valve, to regulate the flow rate.

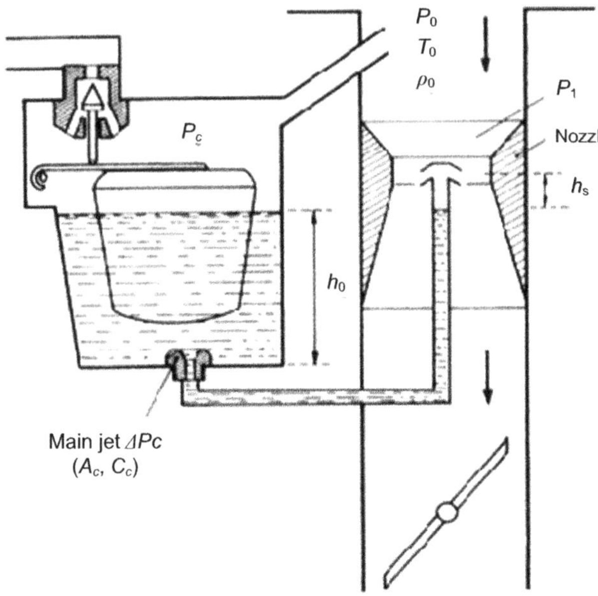

Figure 2.87. *Principle of the old carburetor*

The valve can open and close and thus allows the pressure to drop and the air mass, necessary for the combustion of the entire introduced fuel mass (stoichiometric mixture, richness = 1), to be introduced. This butterfly valve introduces what is called an inlet pressure drop (the inlet is somehow blocked and the piston that descends sucks in the air). The engine therefore has "difficulty" filling up and has a direct effect on efficiency.

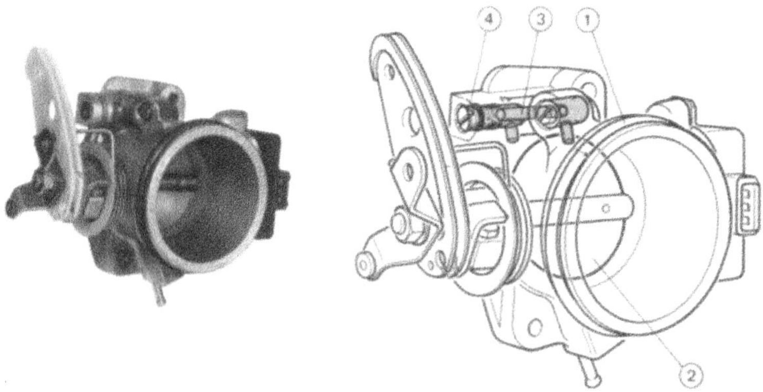

Figure 2.88. *Example of a butterfly valve*

In the diesel engine, the amount of energy is regulated by the amount of fuel, the amount of air is always the same, hence the absence of a butterfly valve.

2.4.1.2.5. Lean burn mode

For gasoline, the ratio that ensures total combustion (stoichiometric ratio) is about 14.5, i.e., it takes approximately 14.5 kg of air to burn 1 kg of fuel completely. The air coefficient l is the ratio of the air mass to the fuel mass, divided by 14.5. Its inverse defines richness.

A poor mixture contains more air, a rich mixture contains less air.

The engine efficiency is affected by the richness or poverty of the adjustment. In particular, it is shown that the indicated efficiency is maximum for a poor setting (Figure 2.89). This technique, developed on some Japanese engines (Honda, Toyota, etc.), allows efficiency gains of 10–15%; however, there is a loss of maximum power at full load.

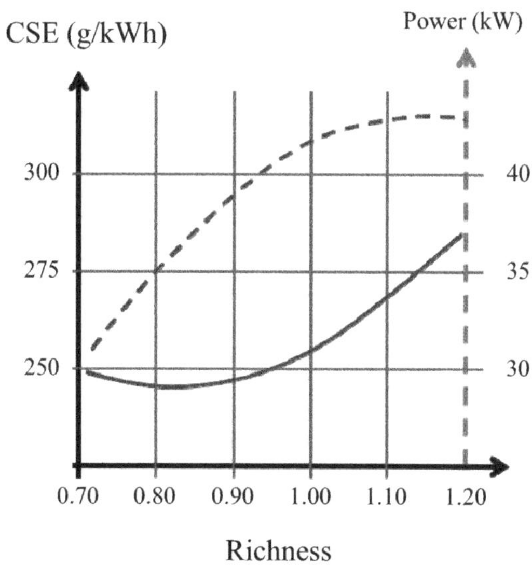

Figure 2.89. *Consumption of a combustion engine as a function of richness*

The diesel engine, on the other hand, always operates in a poor mixture (richness < 1, i.e., the cylinder contains more air than is necessary to burn a given quantity of fuel). There is therefore no need to regulate the air mass in the cylinder, hence the absence of a butterfly valve and the resulting pressure drop.

Current gasoline engines, on the other hand, operate almost everything in a stoichiometric mixture (although this is changing). Indeed they are equipped with an exhaust gas pollution control system – the famous catalyst – which only processes pollutants when the engine is running in a stoichiometric mixture. As consumption is optimal for a richness slightly less than 1, the diesel engine is still at an advantage.

2.4.1.2.6. Energy density

The energy released by a kilogram of fuel during its reaction with oxygen is characterized by the lower calorific value (LCV). As an indication, the LCV for gasoline is 47.3 MJ/kg and for diesel is 44.8 MJ/kg. We observe that the combustion of 1 kg of gasoline releases more energy than 1 kg of diesel.

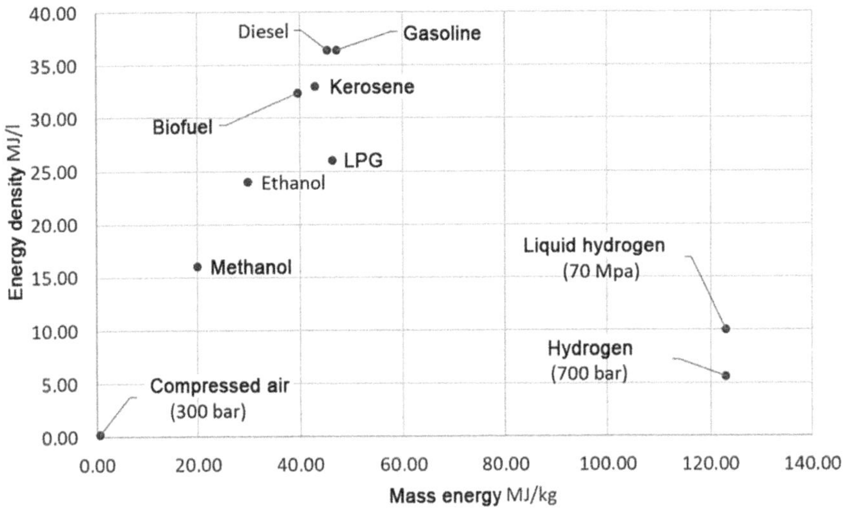

Figure 2.90. *Comparison of the energy densities of different fuels*

The trade and delivery of fuel is carried out in liters. Knowing that the density of diesel is about 850 kg/m^3 and that of gasoline about 750 kg/m^3, we can deduce that with 1 liter of diesel in a car, we have 38 MJ of energy against only 35.5 MJ in the case of gasoline. It can be concluded that for the same power to be supplied, the volume consumption is lower in the case of a diesel engine than in the case of a positive ignition engine.

2.4.1.2.7. Supercharging

Supercharging is a process that aims to increase the efficiency of an internal combustion engine, without increasing its angular velocity. This is done by using a turbocharger. These are used in almost all diesel engines and a small (and growing) number of gasoline engines.

To properly fill the engine with air, and thus achieve correct performance, diesel engines have been equipped with supercharging systems that allow a larger mass of air to enter the cylinder. For a given angular velocity, a higher amount of fuel can therefore be burned with a supercharged engine, than with an atmospheric engine.

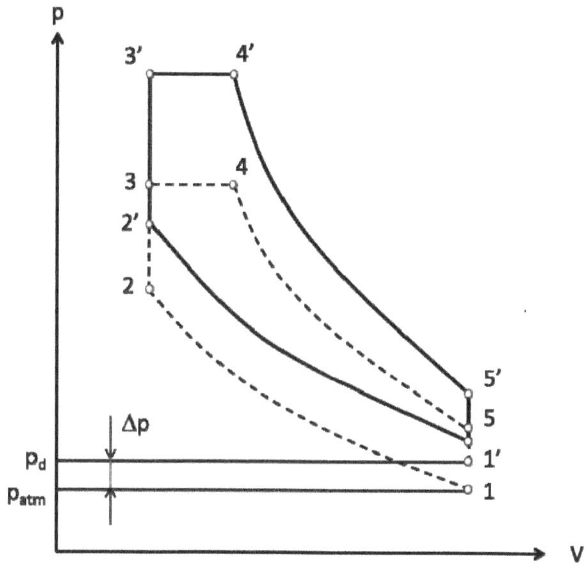

Figure 2.91. *Influence of supercharging for a combustion engine*

The supercharging principle consists of generating the superpressure 1/1′ before the compression starts. The pressure difference Δp allows a higher pressure and temperature value to be obtained in the 3′ point, and therefore a higher power for the same displacement. This supercharging can also lead to an improvement in engine efficiency, given that the energy required for compression comes from the energy contained in the exhaust gases.

One of the applications of supercharged engines is downsizing. The engines used are of a smaller displacement but equipped with a supercharging system (turbocharger, compressor, etc.). So, we end up with "small" engines in "big cars". The engine is subjected to higher loads (load = engine torque/maximum torque available at this velocity), allowing the engine to operate in its lowest specific consumption area (fuel consumption/delivered power).

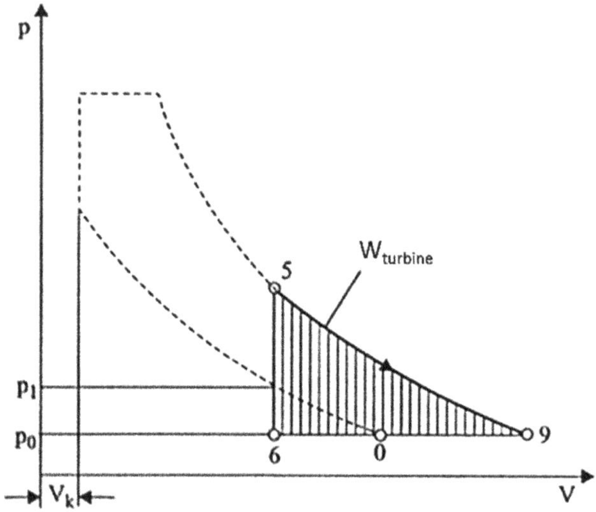

Figure 2.92. *Energy saving through the use of a turbocharger*

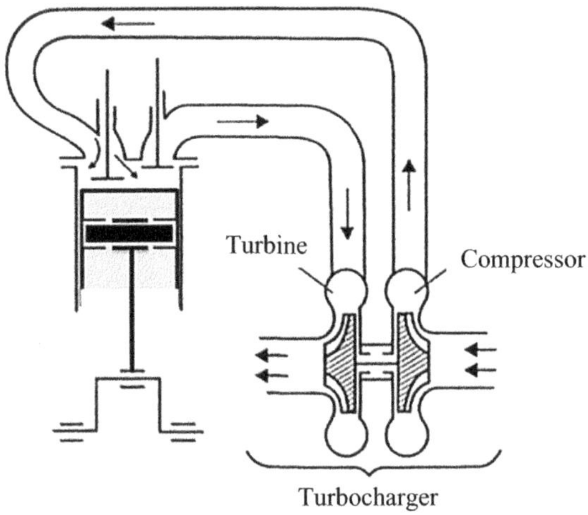

Figure 2.93. *Example of the implementation of a turbocharger*

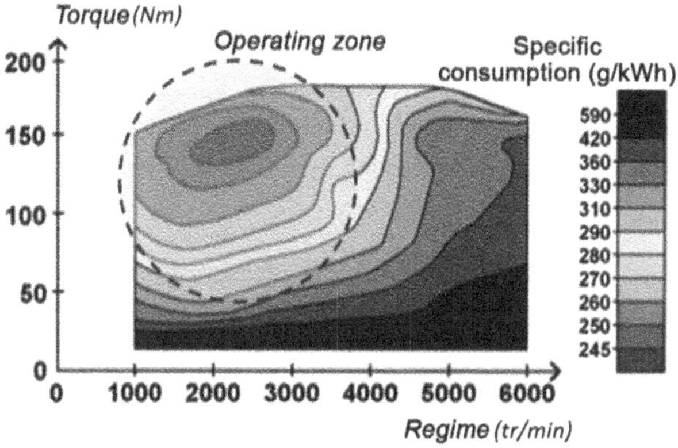

Figure 2.94. *Specific consumption of a combustion engine*

In short, the diesel engine consumes less because:

– its volumetric ratio is higher;

– combustion is more "confined";

– it doesn't have a butterfly valve;

– it works in a poor mixture;

– the fuel for the diesel engine has a higher energy density than gasoline;

– it is always equipped with a supercharging system.

2.4.1.2.8. CAMLESS system and cylinder deactivation

In order to reduce engine consumption, various technical solutions have been developed [BAD 13]. In general, the opening and closing of piston engine valves is controlled by a camshaft driven in rotation by a timing belt, the valvetrain, or increasingly, by a chain in order to essentially increase reliability.

A number of innovations have been developed to reduce engine consumption and pollution by controlling this valve opening and closing phase [SRI 18]. After introducing variable timing systems such as Variable Valve Timing (VVT) on some Renault engines, to adapt to optimal fuel consumption depending on the operating phase (idle, low load or maximum power), the idea came to eliminate the camshaft and introduce control by individual actuators at each valve [NAM 12]. Each one is electronically controlled [LEV 04]. The system reduces the number of parts and optimizes the valve opening and closing moments to adapt to each engine velocity and load [SHE 96].

This optimized control system of the opening and closing angles of the valves at full load allows an optimization of the acoustics over the entire engine rpm range, and a control of the waste gases (resistance to knocking). At partial load, it is possible to reduce pumping losses, and thus influence consumption, as well as control increased combustion by managing internal aerodynamics and waste gas recirculation.

Beyond lower fuel consumption, driving pleasure is increased, with the engine offering more flexibility at low speeds and more torque at high speeds, as needed.

However, this type of technology has the disadvantage of being active, and not passive like camshafts, because of the actuators and consequently of consuming electrical energy. The power consumption is estimated at 300 W when idle and 2,000 W at maximum power. On the other hand, it is difficult to use this system to the best of its ability by eliminating the intake vacuum to avoid pumping, for example. Indeed, this vacuum is useful for the engine brake phases and is not compatible with certain pollution control functions, developed elsewhere.

Another innovation is the ability to temporarily neutralize valves to switch to the same engine of a four-cylinder active system with three active cylinders. The first configuration would provide full power for road or highway routes, while having an engine more adapted to an urban route in the second configuration [LEO 01].

For the deactivated cylinder, no combustion occurs inside but the pistons continue to move and the oil circulates. The fuel savings achieved outweigh the amount of energy required to maintain piston movement and oil circulation.

To make this disruption, car manufacturers offer a camshaft with two cam profiles (Figure 2.95). A first classic profile, therefore ovoid, controls the opening and closing movement of the valve and thus allows the cylinder to operate. The second profile, round, leaves the valves closed, and therefore the concerned cylinders at rest. It is important that the valves are closed to avoid pumping losses. The change from one state to another is carried out by a lateral sliding of the cams, controlled by electromagnetic actuators, steered by the engine control unit. The latter then cuts the injection into these cylinders.

On an eight-cylinder engine (V8), the system allows switching to an active four-cylinder engine configuration. This is not without consequences. In particular, engines optimized for eight cylinders have a noise and vibration level equivalent to a four cylinder configuration. It is then necessary to accompany these motors with special vibration and acoustic treatments such as active motor supports, capable of adapting to both operations [MAT 04].

The advantage of this process is the ability to operate the system at a higher efficiency depending on the load.

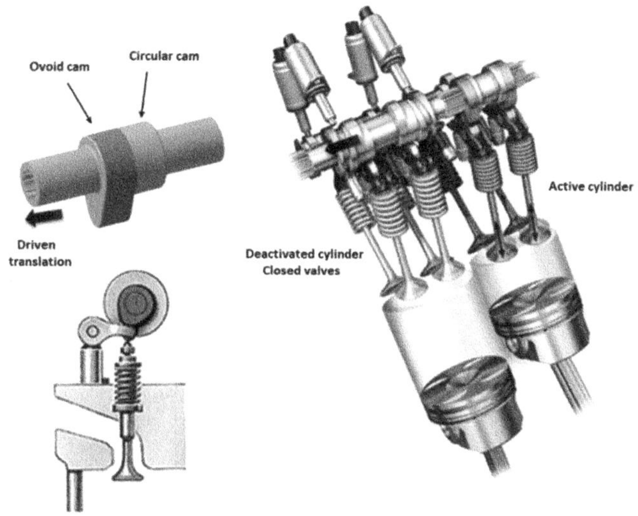

Figure 2.95. *Principle of cylinder deactivation [MAT 04]. For a color version of this figure, see www.iste.co.uk/krysinski/automotive.zip*

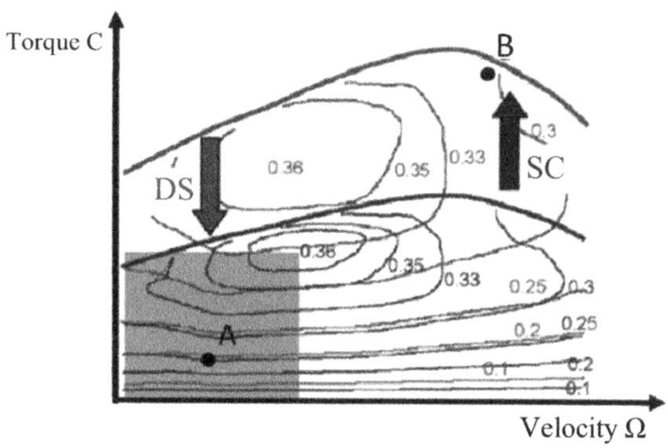

Figure 2.96. *Illustration of the use of the deactivated cylinder engine [GUZ 09]. For a color version of this figure, see www.iste.co.uk/krysinski/automotive.zip*

For the proposed approach, we can observe that the use of the n-1 cylinder engine (downsizing – DS) at low load and low velocity (point A) allows the engine to work on efficiencies of 0.25 instead of 0.2 (which would lead to a gain of 25%), but that the passage to n cylinders (supersizing – SC) enables finding high torque and power at high velocity (point B).

2.4.2. *Real cycles*

A real four-stroke engine cycle is approximately similar to the diagram in Figure 2.97.

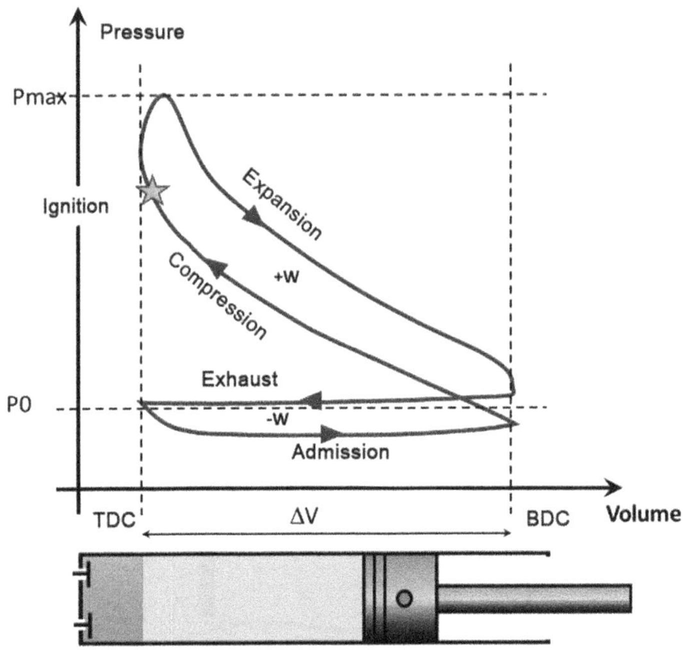

Figure 2.97. *Real cycle of a four-stroke engine. For a color version of this figure, see www.iste.co.uk/krysinski/automotive.zip*

The gap between the intake and exhaust phases is clearly exaggerated for the sake of clarity. At partial load, the pressure can drop to 70 kPa on intake, while it is in the order of 120 kPa for the exhaust. The difference is therefore about 50 kPa. At full load, the exhaust pressure increases slightly, but not to the same extent as the intake pressure. The difference is then reduced to about 40 kPa.

If we reduce this to the maximum 5 MPa of pressure obtained at full load at the end of compression, this difference represents only a very small percentage.

When idle, however, the difference can reach 80 kPa, while the maximum combustion pressure can drop to 235 kPa, or 135 kPa above atmospheric pressure. The ratio thus reaches 60%, which means that these losses are then very significant.

It can be noted that engine efficiency can be arbitrarily but practically defined by:

$$\eta = \frac{-W}{\Delta H} \quad [2.63]$$

where W is the work provided and ΔH the enthalpy variation during combustion.

For the reaction $C + O_2 \rightarrow CO_2$, we have for example $\Delta H = 32.77$ MJ per kilo of C.

2.4.2.1. *Comparison of engines*

The mean effective pressure or MEP represents the constant effective pressure which, when applied to the piston, leads to the same work as that provided by the engine. The most frequently used formula is as follows:

$$PME = \frac{1200\ P_e}{V\ N} \quad [2.64]$$

with MEP in bar, P_e power in kW, V in dm^3, N in rpm.

The MEP is, like the fill rate, a variable that makes it possible to express the load of an engine. It will, for example, be from 0 to 1 bar at very low load. At full admission, the MEP provides information on the compactness of the engine and its intrinsic power performance. It is in the order of 8 to 10 bar on large series engines, 12 to 14 on supercharged engines and 30 to 40 bar in Formula 1.

As indicated above, the constant power Pmax makes it possible to guarantee very good performance at low velocities and to guarantee a sufficient maximum velocity without oversizing the motor.

2.4.2.2. *Power and torque curves*

Torque and power are two distinct characteristics of engine performance. Thermal engines have characteristics far removed from what we would ideally like to have. Classically, for a combustion engine, we generally speak of torque at low rpm and power at high rpm, as shown in Figure 2.98 in a very general form. It is often associated with a specific consumption characteristic curve, Cse.

The specific consumption characteristic, noted as Cse, is the fuel mass (in grams) that the engine would use to deliver 1 kW of power for one hour (i.e. 3,600 kJ of work).

The specific consumption is calculated by dividing the hourly consumption by the actual power:

$$Cse[g\ kW^{-1}\ h^{-1}] = \frac{m_c\,[g]}{P_e\,[kW] \times t[h]} \quad [2.65]$$

For example, if an engine delivers 100 kW of power and consumes 0.5 kg of fuel in 1 minute, its Cse is:

$$Cse = \frac{500}{100 \times \left(\frac{1}{60}\right)} = 300\ g/kW/h \quad [2.66]$$

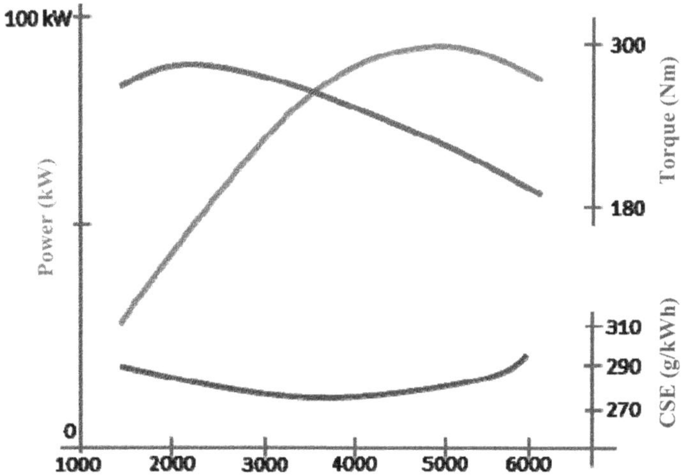

Figure 2.98. *Typical performance characteristics of a combustion engine. For a color version of this figure, see www.iste.co.uk/krysinski/automotive.zip*

The performance curves allow the engines to be compared with each other. As an indication, Figure 2.99 shows the difference between a gasoline engine and a diesel engine. For engines of the same power range, the maximum torque of a diesel engine is obtained for lower angular velocities and these engines climb less in rpm. Since diesel fuel burns less quickly than gasoline, engines cannot run very fast.

For the same power, diesel is considered to have a lower consumption for two reasons:

– the efficiency of diesel's thermodynamic cycle is slightly better and the volumetric ratio is higher;

– consumption is counted in volume (liters per 100) whereas physically, it is the Cse, therefore the mass that counts (g/kW/h); however, since diesel fuel is heavier than gasoline, there are fewer liters of fuel than gasoline at the same mass.

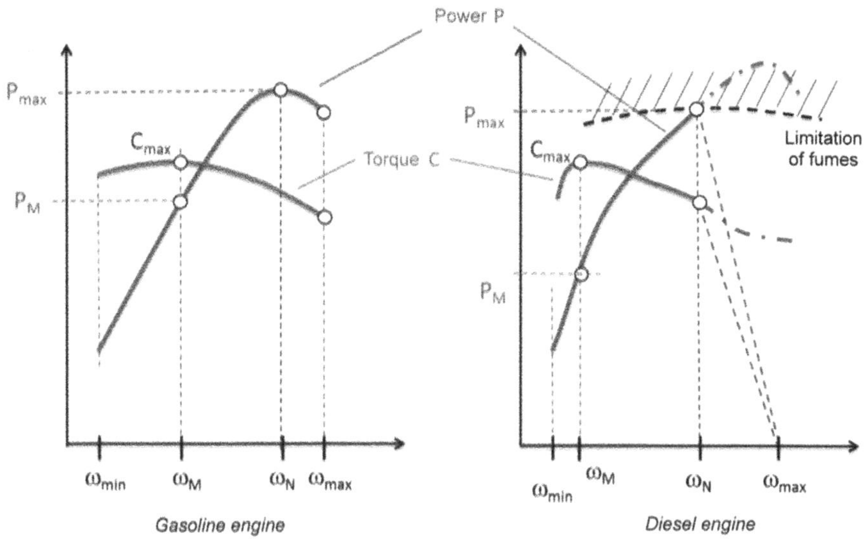

Figure 2.99. *Comparison of diesel and gasoline engines. For a color version of this figure, see www.iste.co.uk/krysinski/automotive.zip*

Another element of engine comparison is the notion of elasticity. This can be defined by e such that:

$$e = e_M \, e_N \tag{2.67}$$

with:

$$e_M = \frac{P_{max}}{P_M} \text{ and } e_N = \frac{\omega_N}{\omega_M} \tag{2.68}$$

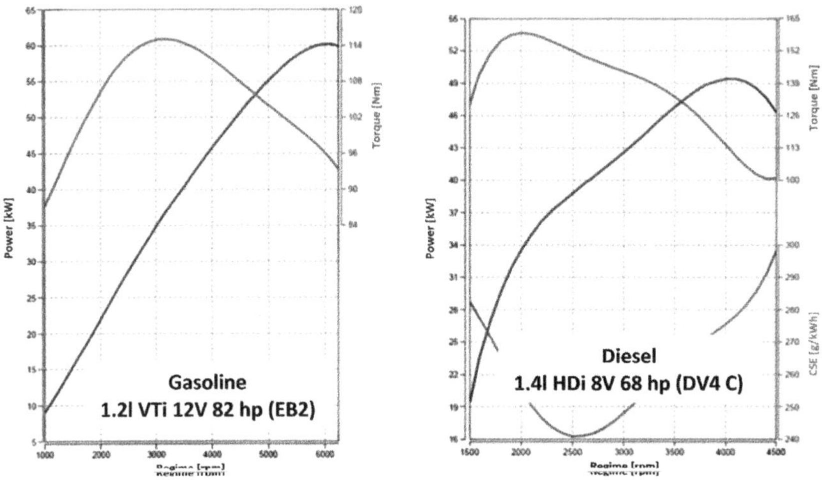

Figure 2.100. *Comparison of gasoline and diesel engines, notion of elasticity (source: PSA). For a color version of this figure, see www.iste.co.uk/krysinski/automotive.zip*

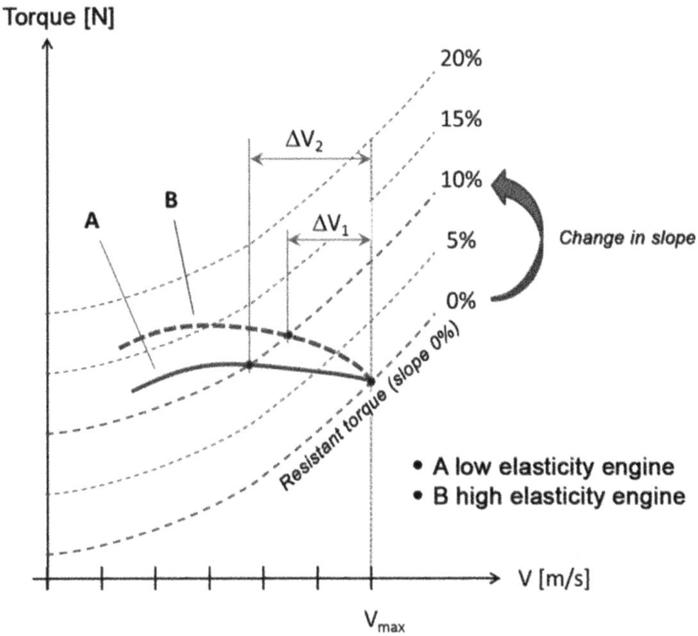

Figure 2.101. *Influence of elasticity on vehicle performance. For a color version of this figure, see www.iste.co.uk/krysinski/automotive.zip*

For comparison, Figure 2.100 quantifies the case of PSA gasoline engines 1.2l VTi 12V 82 hp (EB2) and Diesel 1.4l HDi 8V 68 hp (DV4 C). This gives:

– gasoline: e = 1.62x1.88 = 3.04;

– diesel: e = 1.47x2.00 = 2.94.

The advantage of this feature is shown in Figure 2.101. It can be seen that during a change of slope (example of a change from 0% to 10%) and everything being identical, the engine with lower elasticity will impose a greater velocity variation, and therefore a deceleration, than the engine with higher elasticity.

2.4.2.3. *Limitations of internal combustion engines*

The limitations of internal combustion engines come from the following points:

– the moving parts are subjected to deceleration acceleration cycles which, because of their mass, generate significant mechanical stress on the entire engine. Balancing by adding counterweights is necessary and often difficult to achieve. The use of counterbalancing shafts that rotate in the opposite direction is one solution, but it consumes power and slows down the system. These kinematics associated with the effects of combustion, result in high levels of vibration and acoustics despite balancing precautions;

– the displacement of an engine is not easily variable. It would be better to adapt to the velocity and torque demand according to the use. As mentioned above, this can be achieved by deactivating cylinders. But it remains a cumbersome and costly solution;

– the optimization of valve openings and closures is dependent on operation (idle, full load, engine brake, etc.). The definition and timing of the conventional camshaft is defined by a compromise for these different modes. The solution of a variable valve timing is possible, provided that the new Camless technology mentioned above is used;

– the volumetric ratio, affecting the engine efficiency, is not easily variable. One solution is to introduce engine supercharging to improve the low volumetric ratio;

– due to the mobile crew and the combustion operating cycle, the instantaneous torque on the engine output shaft is highly fluctuating, causing engine acyclisms that can affect the power chain elements and reach the wheels, resulting in driving inconveniences;

– depending on the chosen architecture (in line, V-shaped or flat) the size, or compactness, remains relatively large compared to electric motors for example.

The mathematical definition of the compactness of an engine is:

$$C = \frac{V_{int\,ake}}{V_{engine}} \qquad [2.69]$$

with:

– V_{engine}: the volume of the engine including:

- the entire engine block,

- the volumes to be allocated to allow the movement of all moving parts: piston, connecting rod, crankshaft, valves, rotating shafts, etc.,

- we exclude specific volumes such as those intended for the installation of an alternator, various pumps, etc.;

– V_{intake}: the volume generated by the combustion chambers only during intake, on two crankshaft revolutions, and more generally, two revolutions of the *first* rotating shaft rotated by the moving parts of the engine (excluding the reduction gear device):

- this two-turn reference is necessary to directly use the displacement of four-stroke cylinder engines,

- this volume corresponds to the volume of fresh air drawn in over two revolutions by a non-supercharged four-stroke engine (under external temperature and pressure conditions), assuming that the filling of the chambers goes smoothly.

Whether they are Wankel engines or traditional cylinder engines, their design makes this compactness a number well below 1 (< 0.1). However, *a priori*, nothing prevents a much higher compactness on two turns! By calling P the maximum power of such engines, they would then have a very high-power density, defined by P/Vmot, which would make them ideal candidates for a hybrid architecture.

More generally, any gain in power at the same volume is beneficial in all motorized applications (vehicles, motorcycles, lawnmowers, chainsaws, aircraft, trucks, etc.).

Figure 2.102 shows the engine torque available for different engine types according to a vehicle's velocity for a single gear available in the transmission chain. Assuming that all engines provide identical torque at maximum velocity (arbitrarily chosen), we observe that the combustion engine is penalized at low velocity compared to other technologies.

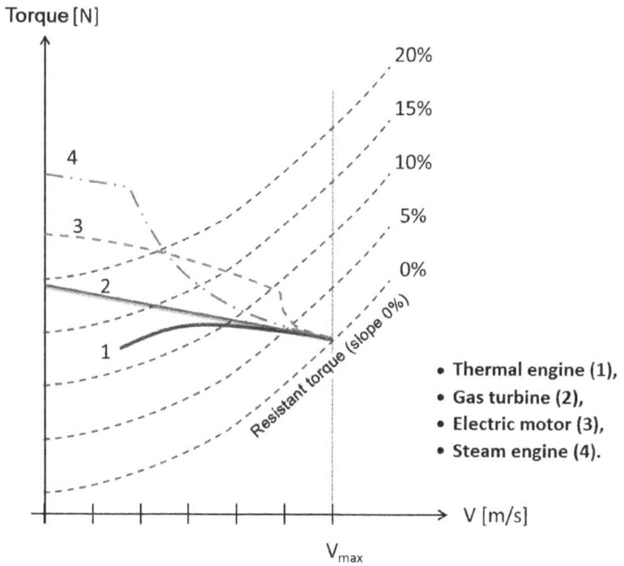

Figure 2.102. *Comparison of torques of different engine types. For a color version of this figure, see www.iste.co.uk/krysinski/automotive.zip*

2.4.2.4. *Downsizing of engines*

Downsizing is a technique used by car manufacturers to reduce CO_2 emissions. It involves reducing the engine size without degrading its final power and the qualities felt by the driver such as flexibility and available torque (engine approval).

Thus, instead of offering a four-cylinder engine on select city cars, manufacturers offer a smaller three-cylinder engine. On some four-cylinder propulsion systems, an engine with a smaller total displacement is available (Figure 2.103). For a of 2 liter sized engine, the power and torque performance of a 2.5 liter can be achieved. In more powerful vehicles, instead of offering an eight-cylinder engine, manufacturers offer a four-cylinder engine with an electric motor.

Downsizing is necessarily accompanied by additional improvements such as the use of one or more turbochargers or the use of multi-valve techniques. By increasing the flow of air and fuel into the combustion chamber, the efficiency of engines equipped with it is greatly improved. This is accompanied by various technological devices to optimize full-load combustion to reduce knock sensitivity.

The main defect of this type of engine is that, due to their high angular velocity, they are less reliable.

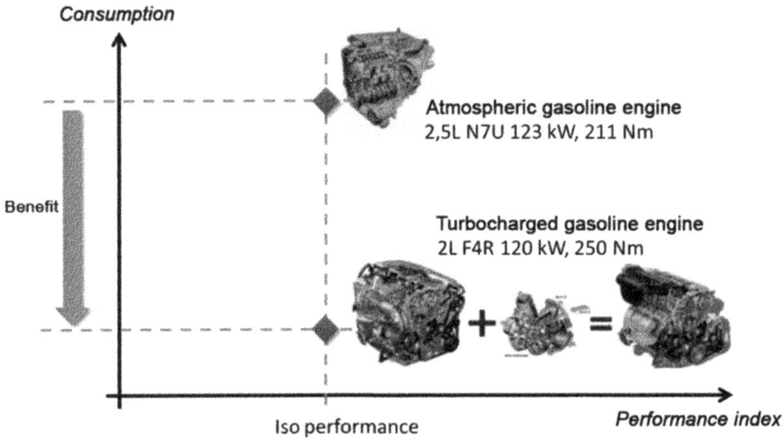

Figure 2.103. *Illustration of downsizing of an engine. For a color version of this figure, see www.iste.co.uk/krysinski/automotive.zip*

Downsizing also has indirect effects. As the mass and volume of the engine decreases, the mass of the structure can be reduced. The reduction of this mass has virtuous effects on the vehicle's performance, on the stresses in the structure and/or on the sizing of certain components such as brakes [KRY 12]. Since the structure and its components can also be downsized, the weight gain obtained allows the engine to be downsized in power while maintaining its performance qualities (Figure 2.104).

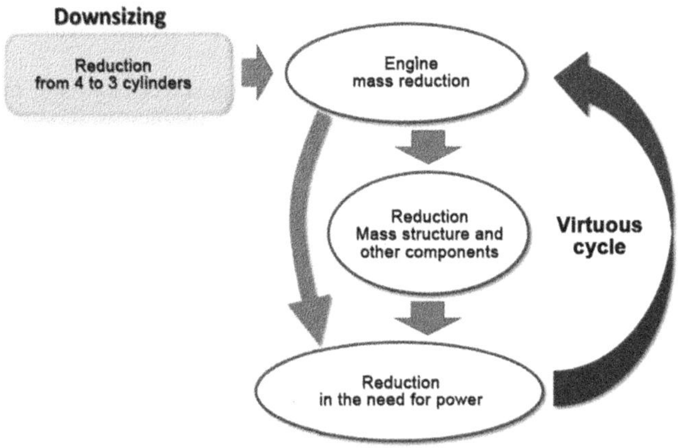

Figure 2.104. *Virtuous effects of downsizing. For a color version of this figure, see www.iste.co.uk/krysinski/automotive.zip*

In addition to downsizing, a choice can be made to reduce the engine's angular velocities to reach operating points where efficiency is better (downspeeding) via so-called long gearbox tiering.

2.4.2.5. *Pollution control of internal combustion engines*

Combustion of internal combustion engines is an exothermic oxidation-reduction reaction. During complete combustion, the reagent reacts with the oxidizer until it forms products that can no longer be oxidized, that is these products can no longer react with the oxidizer: the products have reached a degree of stability that a combustion reaction cannot modify.

In the case of a hydrocarbon reacting with oxygen, the combustion products are carbon dioxide and water. Complete combustion makes it possible to obtain the maximum amount of energy available via a substance and this energy is defined as the calorific value. In this case, the complete combustion of an alkane is governed by the following stoichiometric equation:

$$C_nH_{2n+2} + \frac{3n+1}{2} O_2 \Rightarrow n\,CO_2 + (n+1)H_2O + heat$$

The combustion equation shows that the complete combustion of a hydrocarbon of the formula C_nH_{2n+2} with dry air (which is composed of oxygen and nitrogen) produces exclusively carbon dioxide and water. The nitrogen in the air emerges unmodified from this reaction.

In terms of this equation, we also note that for every mole of fuel burned, n mole(s) of carbon dioxide (CO_2) is produced. Thus, for the different fuels we can deduce that:

– for every 1 kg of diesel consumed, 3.12 kg of carbon dioxide is produced;

– for every 1 kg of C_7H_{16} gasoline consumed, 3.08 kg of carbon dioxide is produced;

– for every 1 kg of $C_{10}H_{22}$ kerosene consumed, 3.10 kg of carbon dioxide is produced.

In reality, combustion is never complete and pollutants appear in the exhaust gases. A pollutant is any compound that can degrade the environment, essentially carbon monoxide (CO), unburned hydrocarbons (HC) and nitrogen oxides (NO_x). The emission of pollutants is not exclusively due to the combustion of fuel with dry air. Indeed, there can be losses due to evaporation, but also solid emissions due to the wear and tear of automotive mechanical parts. Figure 2.105 shows the general mechanisms of air pollution generated by vehicles.

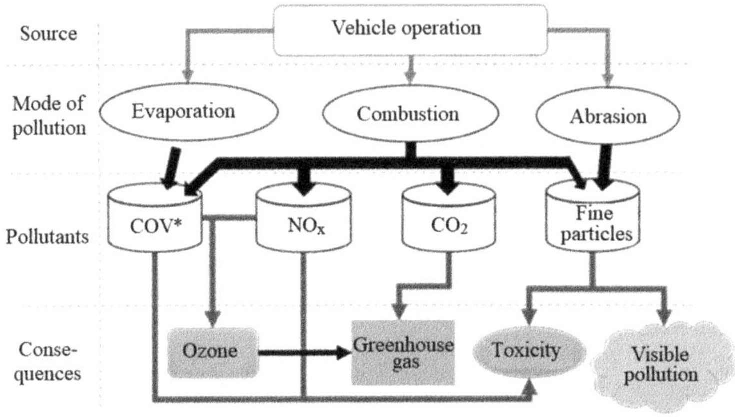

Figure 2.105. *Emissions of pollutants from vehicle operation*

Automobile and engine manufacturers offer pollution control systems that significantly reduce nitrogen oxide (NO_x) emissions. To limit these quantities of pollutants, two approaches are available to manufacturers:

– upstream treatment: acting on the fuel (e.g. additives), air intake, engine geometry, injection;

– downstream treatment: action on the gases at the engine's outlet before escaping.

We propose three possibilities: the influence of richness, the establishment of a solution by using a catalytic converter and urea supply, and finally, the solution of reinjecting exhaust gases.

2.4.2.5.1. Influence of richness

Richness is the ratio between the amount of fuel theoretically needed and the amount of fuel actually used. Thus, a richness greater than 1 gives a rich mixture, a richness equal to 1 gives a stoichiometric mixture and a richness less than 1 gives a poor mixture.

We also use the excess air coefficient, called lambda (**λ**), which is the opposite of richness. It is therefore the ratio between the amount of air actually used and the amount of air theoretically required. Thus, for $\lambda > 1$, the mixture is said to be poor. Operation in a lean mixture has several advantages: less polluting, lower consumption. The problem comes from the rise in temperature generating NO_x. The spark plug must be in a central position, the energy and arc time must be increased. It is also necessary to manage the stability limit (λ close to 1.7): if it is too poor the flame goes out and the engine will misfire, generating a lot of unburned hydrocarbons (HC). To counter this, it is necessary to increase the turbulence in the cylinder, especially the Y tumble (in order to facilitate the initiation of combustion).

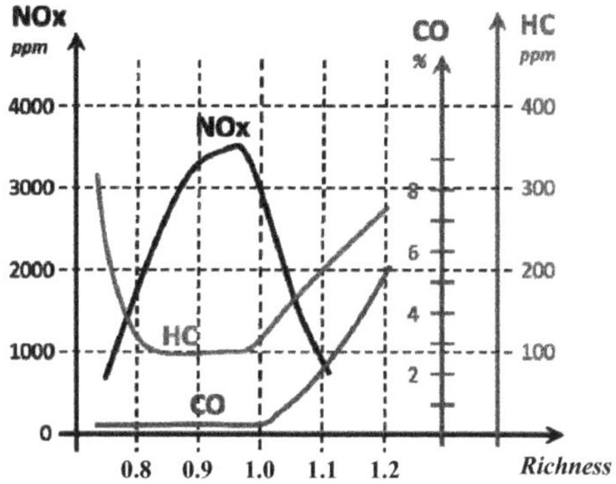

Figure 2.106. *Influence of richness on NO_x, HC and CO emissions. For a color version of this figure, see www.iste.co.uk/krysinski/automotive.zip*

When the air-fuel ratios are optimal, carbon monoxide (CO) and hydrocarbon emissions are relatively low, but the very high temperature in the cylinders leads to high nitrogen oxide (NO_x) emissions.

When the mixture is rich in fuel, combustion is incomplete due to lack of air and CO and hydrocarbon emissions are high, but the lower temperature in the cylinders results in less NO_x formation.

When the mixture is poor, CO and hydrocarbon emissions are low, but NO_x emissions are higher; this is the case with current engines.

As we can see, the best richness is 0.9, that is a slightly poor mixture. It is also the most efficient area, the one where the car will consume the least. Manufacturers, relying on the catalytic converter which allows complete combustion in the exhaust pipe, currently have a richness of 1 which has the additional advantage of being easily detected at a lower cost by the lambda sensors required by the catalytic converter.

2.4.2.5.2. Treatment by catalytic converter and introduction of urea

These devices (catalyst, filter, etc.) are of the post-treatment type. They are located within the exhaust line.

Catalytic converters

One of the techniques consists of using a selective catalytic reduction system, the principle of which involves injecting a urea solution into the exhaust line. This is then transformed into ammonia, which combines with NO_x in the exhaust to form water and harmless nitrogen. Another technique consists of using a NO_x trap (adsorption on barium and then elimination by running the engine in rich mixture). This technique is more energy intensive and efficient over a smaller temperature range.

A NO_x trap loses its effectiveness at high temperatures (less than 40% elimination rate around 350°C), which corresponds to driving on a highway, while for a selective catalyst, from 175°C onwards, 70% of NO_x is removed, a proportion that stabilizes at 90% between 250°C and 350°C.

A widespread and effective means of depollution at present is the catalytic treatment of exhaust gases: oxidation of CO and unburned gases, reduction of nitrogen oxides. Multifunctional catalysis (or "3-way" catalytic converter) carries out all three operations simultaneously, with the constraint of being able to operate only with a stoichiometric fuel mixture. The efficiency of the catalysis is highlighted by the curves presented in Figure 2.107, which show, as a function of the air factor λ (richness), the reductions achieved in the emissions of the various pollutants.

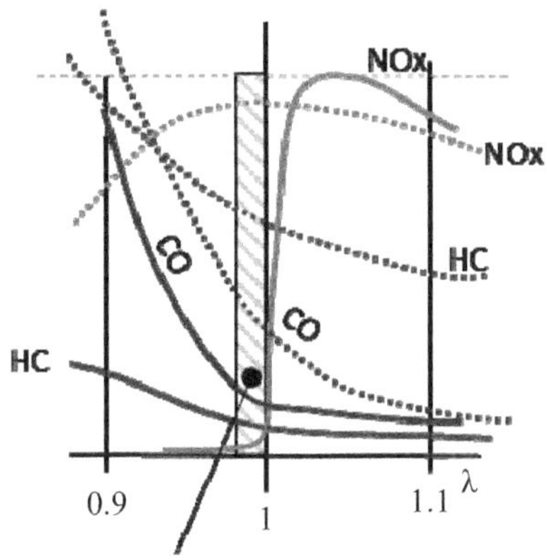

Figure 2.107. *Influence of the catalytic converter on richness. For a color version of this figure, see www.iste.co.uk/krysinski/automotive.zip*

BlueHDI System (PSA)

The BlueHDI system was developed by PSA to reduce polluting emissions including nitrogen oxides and particulate matter. This *system* integrates into the traditional exhaust system.

Firstly, it is the oxidation catalyst that receives the combustion gases directly into the engine. Its role is to remove unburned hydrocarbons (HC) and carbon monoxide (CO) by converting them by chemical reaction into water (H_2O) and carbon dioxide (CO_2).

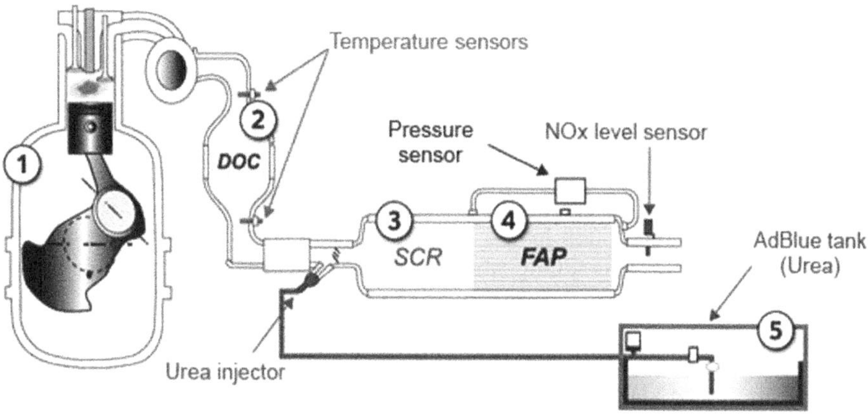

Figure 2.108. *Mixed BlueHDI technique (source: PSA). For a color version of this figure, see www.iste.co.uk/krysinski/automotive.zip*

Then, it is the turn of the selective catalytic reduction (SCR) catalyst, located between the oxidation catalyst and the particulate filter (PF), to carry out the depollution. By adding Adblue® (a mixture of urea and water) to the exhaust system at the SCR, nitrogen oxides (NO_x) are transformed into water (H_20) and nitrogen (N), which are the main components of air.

Finally, the PF will remove 99.9% of the remaining particles in the exhaust gases. As a reminder, it was the PSA group that first launched the PF in 2000 on these diesel engines. The PF accumulates the particles emitted by the engine over a period of 500 to 1,000 km. The PF regenerates automatically to clean itself. One specificity that we wish to clarify is that the PF is effective whatever the vehicle's conditions of use: cold, hot, in the city or on the highway.

2.4.2.5.3. Exhaust gas recirculation (EGR)

Exhaust gas recirculation, also known as EGR, appeared in the 1980s and is one of the most advantageous ways for cars to meet the severity of Euro 5 and 6 standards without imposing any additional penalty costs. This technique consists of redirecting part of the combustion engine exhaust gases into the intake manifold to reduce nitrogen oxide (NO_x) emissions.

The reduction of pollutant emissions can be achieved in several ways: upstream by controlling the richness of the fuel mixture, in the cylinder by intervening in the formation processes and in the exhaust by transforming the pollutants into harmless effluents.

Controlling the richness of the fuel mixture can be carried out in such a way that it is located in the lean mixing zone where there is no CO, no NO_x, and still little unburned fuel. The difficulty lies in obtaining correct combustions with such low richness (0.6 to 0.8), which requires in-depth studies on combustion and its relationship with the processes inside the cylinder. The limitation of pollutant formation during combustion has so far been applied only to nitrogen oxides. The most common process is EGR, which consists of sampling exhaust gases and sending them to the inlet downstream of the butterfly valve. The action on NO_x formation is very effective (up to 50% overall reduction); it is manifested by a decrease in flue gas temperatures due to dilution and a consequent slowing of the pollutant formation kinetics. The quantity of recycled gas can reach 20% in terms of mass of the admitted flow rate.

2.5. Conclusion

To summarize this chapter, the following elements can be retained.

For high powers, as is the case for aeronautics, turbines are essential. The improved method nowadays consists of increasing the dilution rate, which leads to the development of large engines that appear on commercial aircraft. The development of electrical hybridization on turbines to improve efficiency by playing around with pumping limits and more particularly via transient regimes is another route for improvement.

For helicopters, the stakes are different. For low powers, less than 300 kW, internal combustion engines are interesting because of their mass power. They still have a future if we approve their efficiency and pollutant emissions. Standards are currently less stringent for aeronautics than for the automotive industry. For higher powers, turbines are still the solutions of the future. Their improvement requires the introduction of recovered thermodynamic cycles, in particular the use of exchangers and the use of electrical hybridization to treat pumping limits and transients.

As far as electric motor propulsion is concerned, development is currently very important for automotive transport where energy recovery is an asset for the justification of this type of energy. For aeronautics, there is no such possibility. An electric transportation powertrain can then find its place in small aviation where the power and range required is lower or for hybrid use in transitional phases such as take-off or safety phases such as autorotation.

3

Power Transmission Elements

It has been seen that regardless of engine type (electric, piston or gas turbine), the engines' velocity range did not directly correspond to the car wheels' velocity range and that of aircraft propellers or helicopter rotors. As the velocities of these receivers are generally lower than those of the engines, it is necessary to introduce a reduction to adapt the proper operator of the engine to the receiver's needs.

For multi-rotors, as the power per receiver is lower, it is possible to use smaller propellers that can rotate faster. In this case, the use of gearboxes is not necessary.

The modes of use concerning receivers, rotors, propellers or wheels are not identical in the aeronautics and automotive industries, particularly because of their mode of operation and regulation. The purpose of this chapter is to describe and justify the components used to adapt the kinematics of the engine-receiver for applications in aeronautics and automotive fields.

3.1. Transmission system for rotating wings

3.1.1. *Conventional helicopters*

As discussed in previous chapters, helicopters of a certain tonnage (about 1T) or more mainly use free-turbine gas turbines. To ensure the proper operation of these turbines in terms of efficiency and effectiveness, the engine is regulated in terms of velocity. The regulation is done via the Q fuel flow using the measurements of the velocity of the primary and secondary shafts (Figure 3.1).

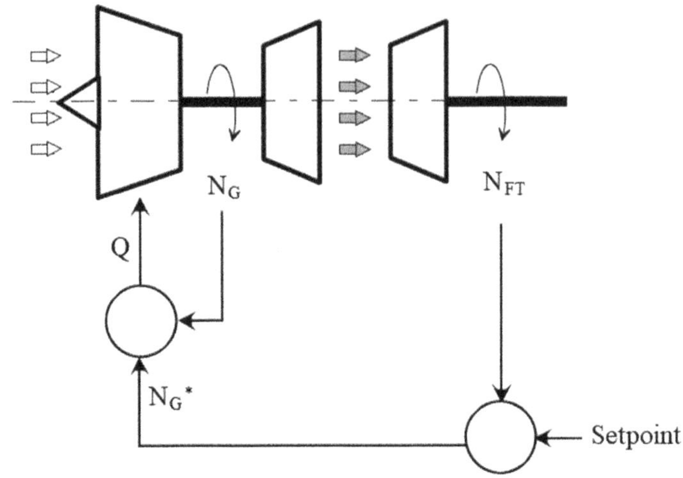

Figure 3.1. *Speed control of a free turbine on a helicopter*

As the power demand evolves during its mission (Figure 3.2), the power management is done by the pilot using collective pitch control through the modification of the resistive torque.

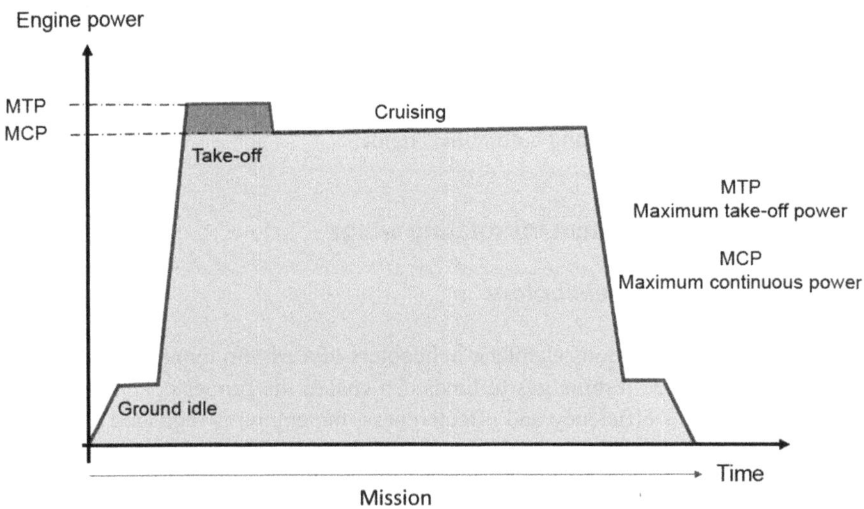

Figure 3.2. *Graph describing the evolution of maximum engine power of a conventional helicopter during a mission*

The flight controls affect the pitch angle of the main and tail rotor blades (Figure 3.3).

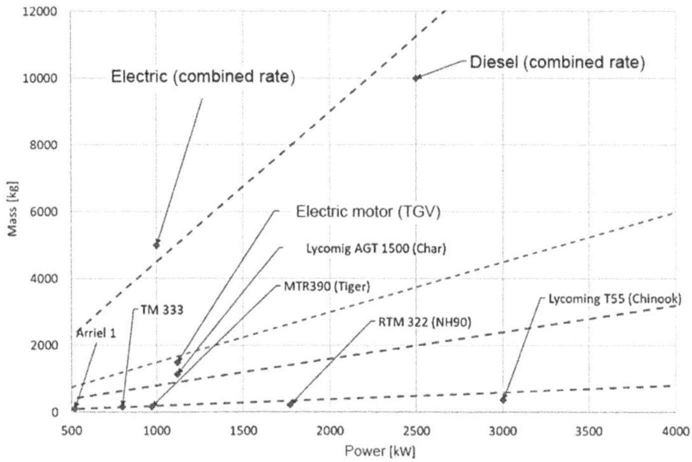

Figure 3.3. *Blade pitch angle of a conventional helicopter (source: Airbus Helicopters)*

These allow the pilot to control the variation in altitude, velocity and/or direction. The collective pitch lever (2) (Figure 3.4) controls the main rotor lift by means of collective variation of the pitch.

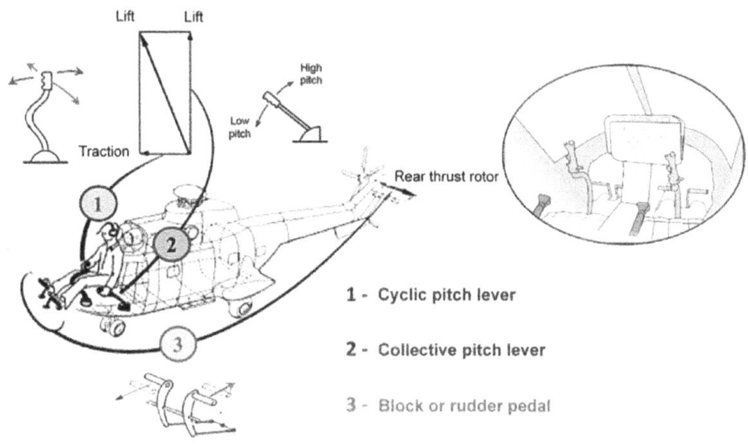

Figure 3.4. *Pitch control levers on a helicopter (source: Airbus Helicopters). For a color version of this figure, see www.iste.co.uk/krysinski/automotive.zip*

This control is coupled with engine velocity control to anticipate drops in power (Figure 3.5).

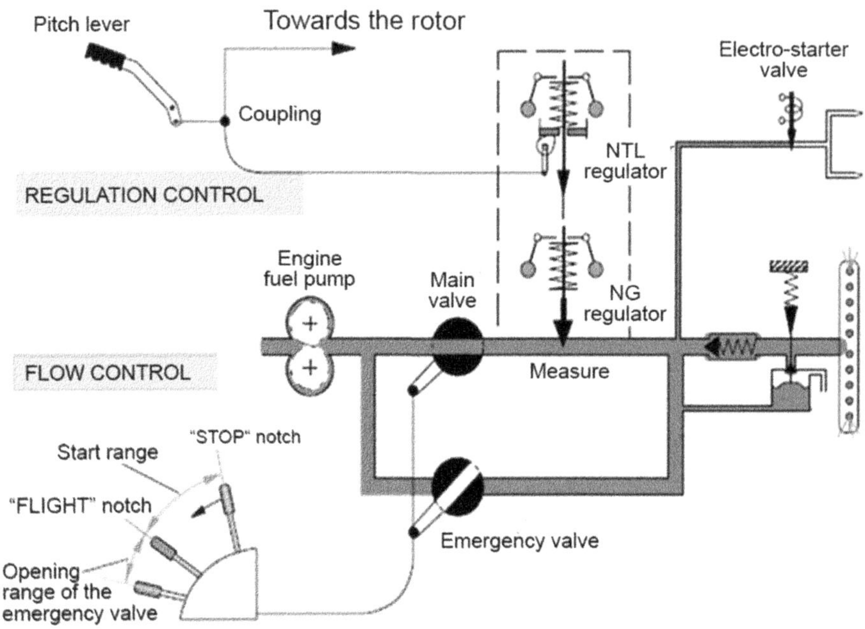

Figure 3.5. *Engine control anticipation and coupling system [KRY 11]*

When the engine's rotational speed – and therefore that of the rotor – is constant, the reduction ratio is constant.

A helicopter rotor rotates between 200 rpm and 500 rpm depending on the tonnage of the aircraft; turbomachines rotate between 40,000 rpm and 60,000 rpm. The overall reduction ratio is thus close to 200 and generally consists of a sequence of conventional gears, conical gears due to the relative positions of the rotor and motor and epicyclic gears allowing reductions while maintaining input-output coaxiality (Figure 3.6).

Since the turbine rotates very quickly, it is necessary to associate a first gearbox with a reduction ratio of about 20 to transport energy at reasonable speeds of about 5,000 to 6,000 rpm (Figure 3.7).

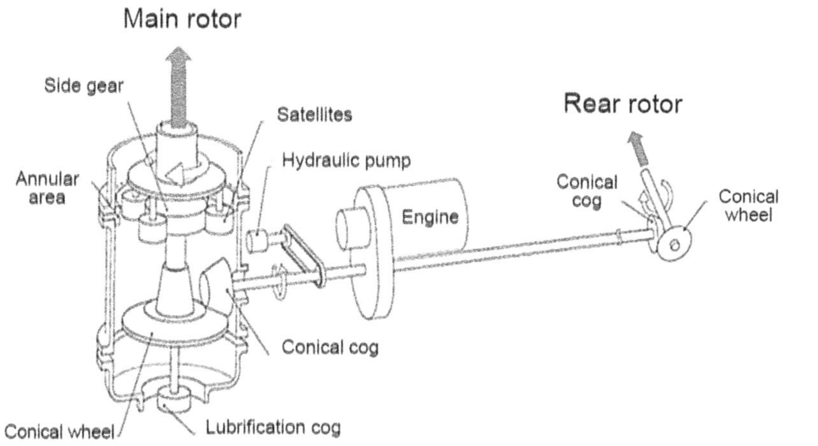

Figure 3.6. *Example of helicopter power transmission (source: Airbus Helicopters)*

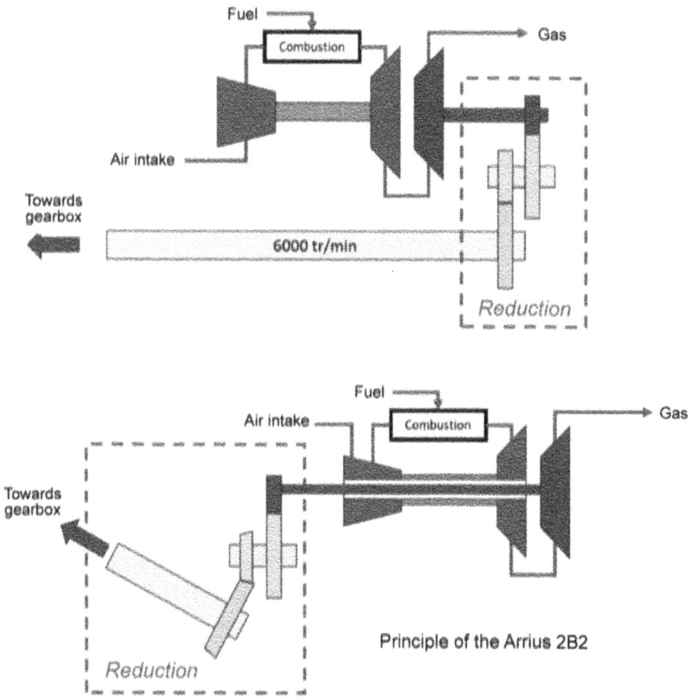

Figure 3.7. *Examples of engine output reduction. For a color version of this figure, see www.iste.co.uk/krysinski/automotive.zip*

The main gearbox integrates the angle gear between the engine's take-off power, the rotor shaft and the last reduction before the rotor. This part also includes the energy supply for the accessories (hydraulic pump, alternator, rotor brake, etc.) (Figure 3.8).

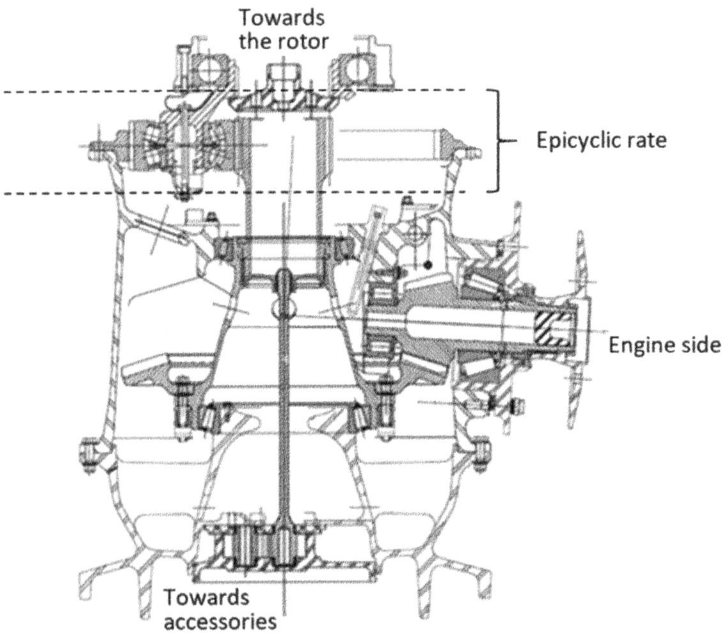

Figure 3.8. *Example of a helicopter's main gearbox (source: Airbus Helicopters)*

In practice, there are several operating rotational velocities, corresponding to different operating phases: nominal (rotor regime or NR), ground idle (close to 60% of the NR) and the take-off phase (close to 110% of the NR). It must be verified that at these velocities, dynamic couplings or resonance phenomena are not created [KRY 11]. This problem is accentuated on modern helicopters, where it is increasingly accepted to use velocity variations of the order of magnitude of 20% of rotor rpm to optimize engine performance (noise, consumption, etc.) according to flight phases or flight conditions (resources, cruise, etc.). Verification of non-coincidence of the system's natural frequencies and excitation frequencies is necessary. Most often, the problem lies in the b Ω which is the product of the number of blades and the rotational speed. This analysis is generally done using the Campbell diagram (Figure 3.9).

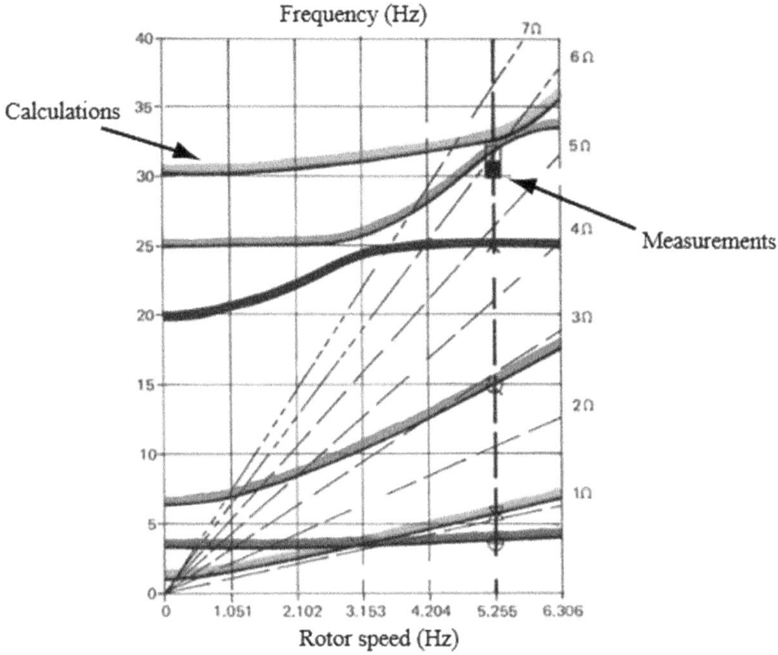

Figure 3.9. *Campbell diagram [KRY 11]*

3.1.2. *The case of multi-rotor structures*

A number of multi-rotor rotating wing projects are being developed. Designed for low to medium payloads, they are equipped with electric motors without gearboxes. Examples of this include the E-Hang project and the Volocopter project, designed to transport two or four people over short periods of time. The multitude of small rotors (8 for the E-Hang and 18 for the Volocopter) makes it possible for the rotors to turn much faster and thus to have a direct connection between the electric motor and the rotor.

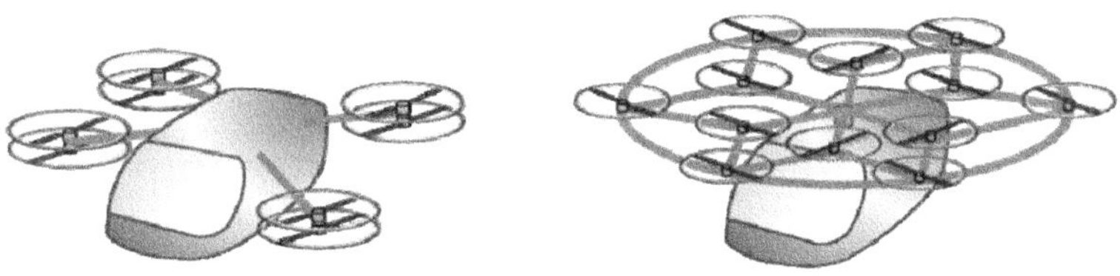

Figure 3.10. *Examples of multi-rotor rotary wings with electric propulsion*

3.2. Transmission system for aircraft

3.2.1. *Propeller aircraft cases*

For propeller aircraft, whether powered by a piston engine or gas turbines (turboprop), it is necessary to maintain propeller performance wherever possible, regardless of operating conditions (take-off, economic cruise or fast flight). To do this, the propellers must be variable-pitch propellers. A regulation system, often hydraulic, maintains a constant rotation speed.

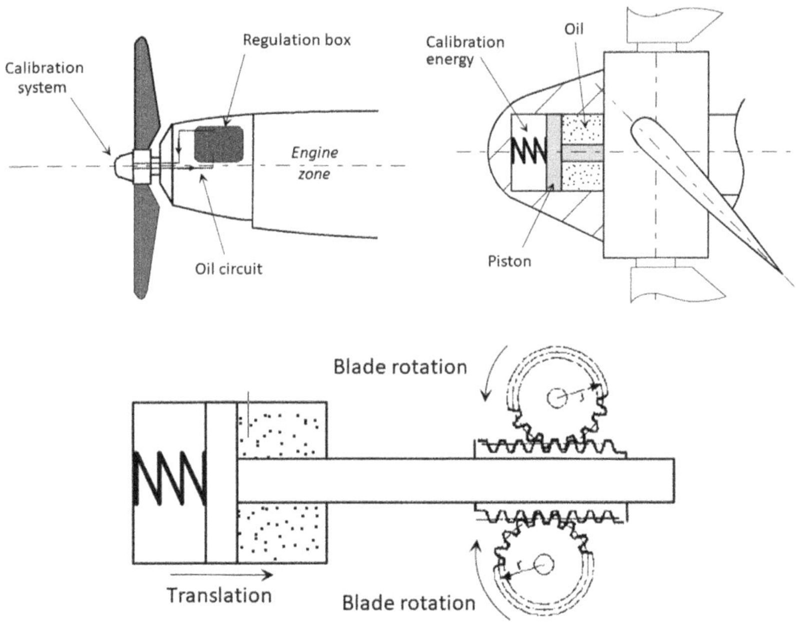

Figure 3.11. *Propeller control system [KRY 11]*

When superimposed on the N speed control, the pilot is then able to regulate the Q fuel flow rate to adapt the power requirement (Figure 3.12).

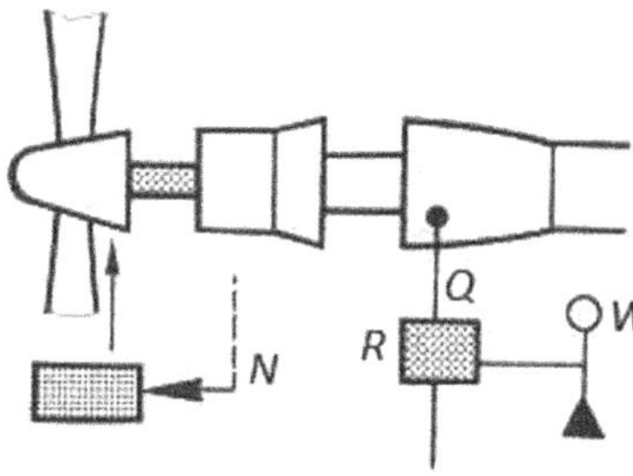

Figure 3.12. *Fuel flow control on a turboprop engine*

In the case of a turboprop, the turbine velocity is higher than the permissible propeller speeds. It is then necessary to introduce a reducer, which can be composed of conventional gear rates or epicyclic gear rates. These reducers can maintain coaxiality between the turbine axis and the propeller, or, on the contrary, shift this axis upwards, making it possible to have propellers that are larger in diameter for a given ground clearance (Figure 3.13).

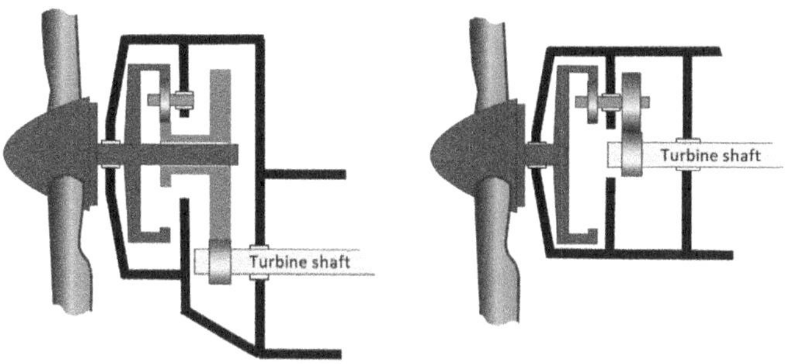

Figure 3.13. *Examples of turboprop reducer architectures. For a color version of this figure, see www.iste.co.uk/krysinski/automotive.zip*

3.2.2. *Turbojet aircraft*

For dual-flow turbojets, an additional flow rate flows in a bypass around the gas generator to improve performance. The proportion of air constituting the cold flow, which varies according to the engines, is expressed by the ratio of the secondary mass flow (or cold flow) to the primary mass flow (or hot flow). This ratio is called the dilution rate. On a passenger aircraft, this ratio is increasingly high, reaching values of 5 to 12. These large ratios require one to limit the rotation speed of cold flow fans. It is then necessary to introduce a reduction ratio (order of magnitude 3:1) by using epicyclic gearing to maintain coaxiality. This technical choice tends to increase the system's mass and the technical complexity of the engine, which affects manufacturing costs and system reliability. The decrease in speed makes it possible to obtain quieter engines.

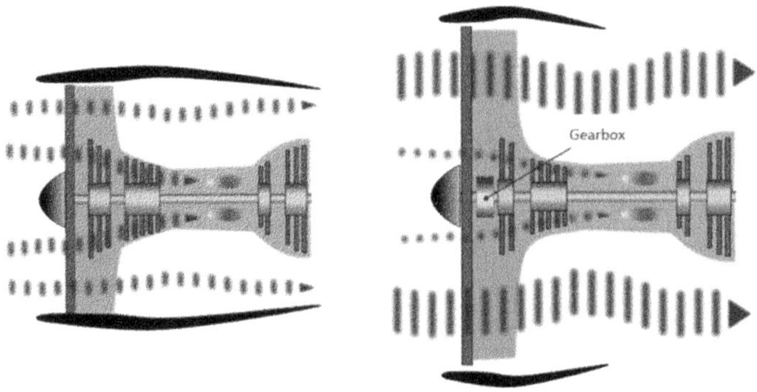

Figure 3.14. *Architectures without and with a double flow turbojet gearbox. For a color version of this figure, see www.iste.co.uk/krysinski/automotive.zip*

3.3. Transmission system for the automotive industry

3.3.1. *Gasoline or diesel internal combustion engines*

For piston combustion engines, it is necessary to have a variable reduction ratio. Technical solutions can be divided into two categories:

– sequential transmissions (multiple gears) with epicyclic or conventional gear rates;

– continuously variable transmissions (CVT) with a metal belt in an oil bath (e.g. Nissan Micra K11), or even infinitely variable transmissions (IVT) with a toroidal system (e.g. Extroid mounted on Nissan Cedric and Gloria).

3.3.1.1. *The need for a gearbox*

In the case of internal combustion engines, we observe that the torque curve is relatively "flat". As a result, the power increases almost linearly up to a certain point.

To try to get closer to the theoretical ideal curve, a multi-speed gearbox (manual or automatic) is used to increase or decrease the torque available to the wheel

(Figure 3.15). Thus, with this transmission, a curve that is closer to what is considered as ideal can be obtained.

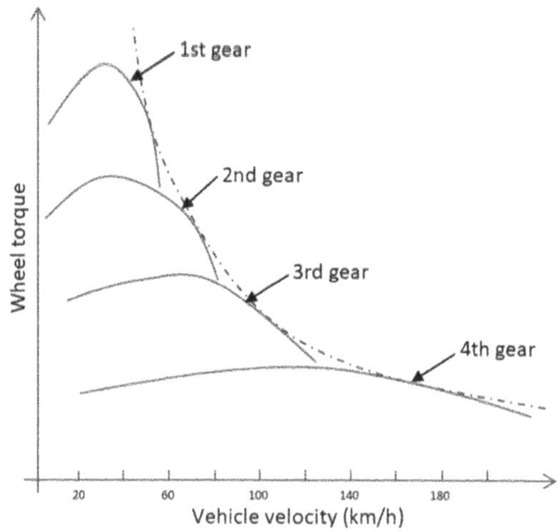

Figure 3.15. *Wheel torque according to 4 gear ratio velocity. For a color version of this figure, see www.iste.co.uk/krysinski/automotive.zip*

3.3.1.2. *Tiering rules for automotive gearboxes*

The choice and calculation of the number of ratios and the value of the reduction ratios are primarily determined by the type of vehicle, its unladen and laden mass and the intended use: passenger car, family car, sports car, road freight transport utility or urban use. Other cases may be considered for road use in mountainous, uneven or off-road areas.

To analyze the gearbox tiering, it is necessary to introduce the reduction coefficient between the engine and wheel velocities defined by:

$$k = \frac{\omega_{wheel}}{\omega_{engine}} \tag{3.1}$$

Similarly, the relationship between wheel velocity and vehicle velocity via wheel diameter is:

$$V_{veh} = \frac{D_{wheel}}{2} \omega_{wheel} \tag{3.2}$$

NOTE.– The designation of a tire gives as reference a formula of the type 195/65 R 15 91 H (e.g.):

– 195 is the width of the inflated tire, measured from one sidewall to the other (in millimeters);

– 65 is the "series" (sidewall height relative to tire width or h/l ratio) expressed as a percentage (here 65%, or 126.75 mm);

– R indicates the radial type;

– 15 is the rim diameter in inches (1 inch corresponding to 2.54 cm);

– 91 is the load capacity index (91 = 615 kg – see standards 31 of the European Tyre and Rim Technical Organisation (ETRTO);

– H is the velocity code (H = 210 km/h).

The diameter of the wheel in this configuration is then:

$$\begin{aligned} D_{\text{wheel}} &= [2 \times 195 \text{ mm} \times 65\%] + [25.4 \text{ mm / inch} \times 15 \text{ inches}] \\ &= 634.5 \text{ mm} \end{aligned} \quad [3.3]$$

Overall, the relationship is then:

$$V_{\text{veh}} = k \frac{D_{\text{wheel}}}{2} \omega_{\text{engine}} \quad [3.4]$$

For example, if $k = 0.07$ and $D = 0.6345$ m, the engine drives the vehicle at 8.4 km/h at 1,000 rpm and 54.4 km/h at 6,500 rpm.

It should be noted that a single coefficient k is not sufficient. By means of a gearbox, several values of k(i) are actually introduced which allow the engine to be adapted to the vehicle's velocity, both in terms of performance and fuel consumption.

The opening of a gearbox is defined as the ratio of its maximum reduction to its minimum reduction:

$$\text{opening} = \frac{k_{\text{max}}}{k_{\text{min}}} \quad [3.5]$$

The opening must be wide enough to ensure both sufficient hillside power and that the engine velocity is not too high when cruising. For example, for a 205 GTI, the opening is 3.759 and for a 205 BE1/5 diesel transmission, the opening is 4.371.

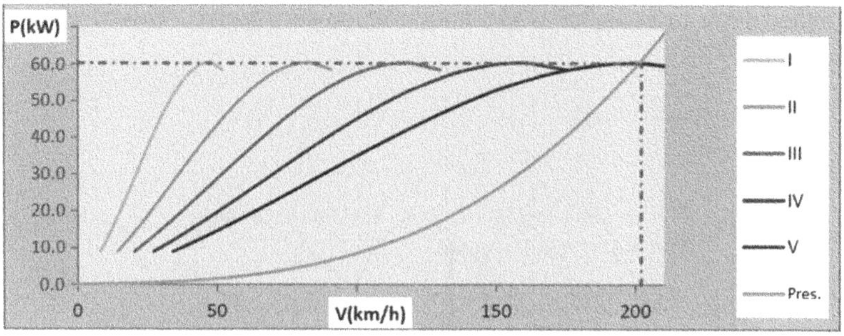

Figure 3.16. *Example of gearbox tiering. For a color version of this figure, see www.iste.co.uk/krysinski/automotive.zip*

A perfect tiering of a gearbox provides a considerable advantage for the vehicle's performance, especially if the engine has a limited or upwardly displaced operating range (in the case of 16-valve small-displacement engines), because it directly affects the efficiency of accelerations and throttle response.

In terms of torque, the ideal would be to have the maximum power at the wheels at all vehicle velocities (Figure 3.17, dotted line curve). If for a given power requirement (gradient and fixed weight, without acceleration) it happens that the maximum power is not available, the vehicle will be limited in velocity (point A). However, the addition of an intermediate ratio (dotted curve) allows a higher speed (point B). When tiering, it is essential to manage the notion of a power lag.

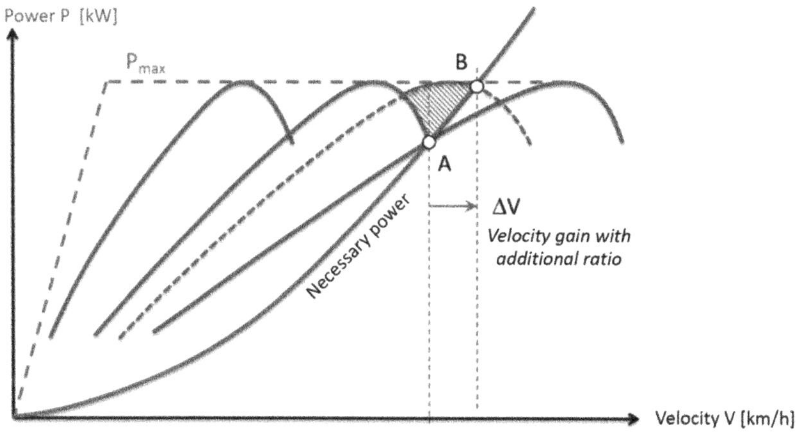

Figure 3.17. *Concept of power lag. Sequential gearbox. For a color version of this figure, see www.iste.co.uk/krysinski/automotive.zip*

Another point to consider is the notion of power reserve. This is illustrated in Figure 3.18.

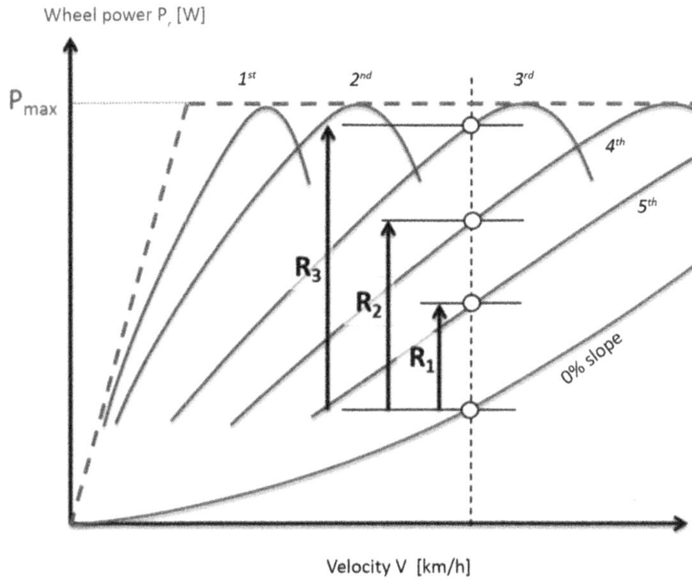

Figure 3.18. *Tiering of a gearbox. Concept of power reserve. For a color version of this figure, see www.iste.co.uk/krysinski/automotive.zip*

It can be seen that, for a desired velocity V and a necessary power configuration (0% gradient in the example), the driver can have the 3rd, 4th and 5th gear ratios at their disposal as a result of this step. The 1st and 2nd gears are inaccessible; this tends to cause the engine to run beyond its maximum velocity.

For the three possible gears, R_1, R_2 and R_3 measure the power available, by downshifting, to achieve acceleration. This tiering also has an impact on consumption via engine efficiency (Figure 3.19). We observe that for the same vehicle speed, the 5th gear ratio allows a lower consumption than the 4th or 3rd gear ratio. Everything is then a question of compromise.

With the widespread use of downsizing techniques, engines tend to deliver ever higher torque for the same displacement. This feature makes it possible to use gearboxes with higher gear ratios (increasingly "long" gearbox) and finally to reduce fuel consumption.

This notion can be measured by calculating the V1000.

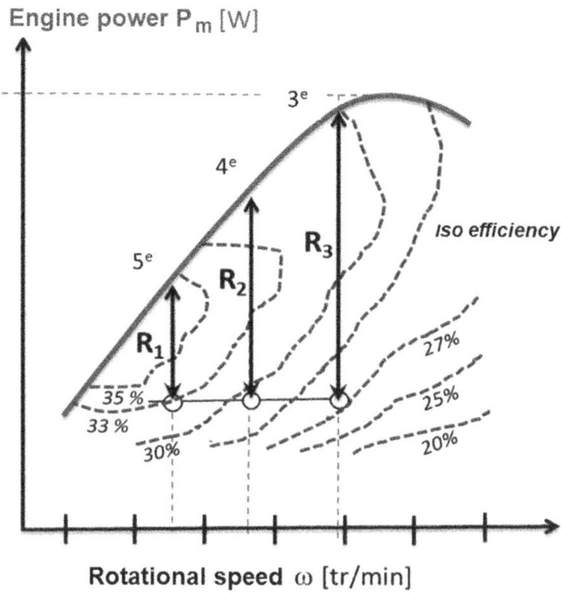

Figure 3.19. *Consumption of an engine according to the ratio used. For a color version of this figure, see www.iste.co.uk/krysinski/automotive.zip*

3.3.1.3. *Comparison of automotive gearboxes – the V1000 concept*

The V1000 velocity is the vehicle's reference velocity in km/h for a given gear when the engine is running at 1,000 rpm. Thus:

$$V_{veh}\left(\text{km/h}\right) = \text{V1000} \times \omega_{engine}\left(1000 \text{ tr/min}\right) \qquad [3.6]$$

By changing the unit of equation [3.6], we obtain the usual formula of V1000:

$$\text{V1000(i)} = \text{k(i)} \times 60 \times \pi \times D_{wheel} \qquad [3.7]$$

The V1000 is none other than the more intuitive expression of the k coefficient, both being equivalent (for the same wheel diameter). Figure 3.20 shows an example of the tiering of several boxes.

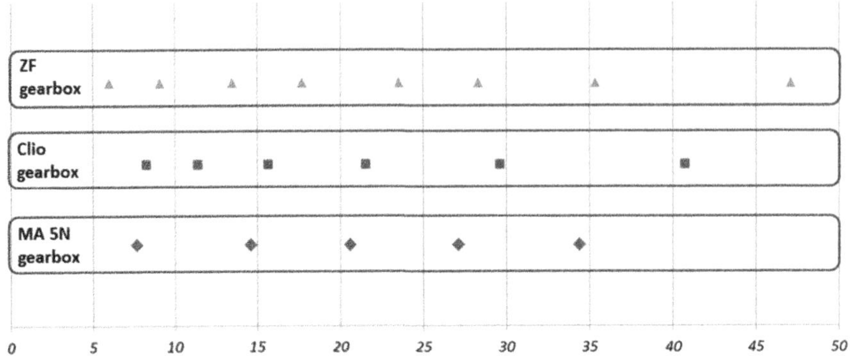

Figure 3.20. *MA5N power transmission stage (e.g. BV Citroën ZX 1.4 Petrol). For a color version of this figure, see www.iste.co.uk/krysinski/automotive.zip*

The vehicle's maximum velocity determines the ratio of the highest velocity (i.e. the lowest reduction ratio) and the previously defined maximum torque determines the lowest velocity ratio (highest reduction).

The number and ratios between the different speeds must be determined so that the previous curve "sticks" as closely as possible to the ideal curve.

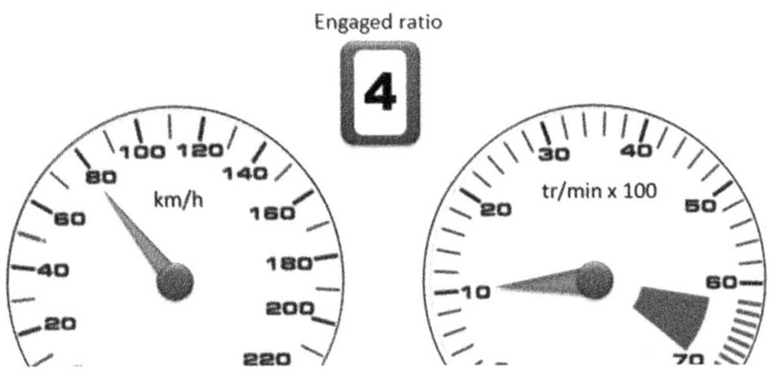

Figure 3.21. *Illustration of the V1000. For a color version of this figure, see www.iste.co.uk/krysinski/automotive.zip*

For long gearboxes, the gasoline-powered car with the longest gear is the (2010) Rolls-Royce Ghost: in the last gear, the speed is 72.9 km/h with an engine at 1,000 rpm. In terms of diesel, the Audi A8 4.2 TDI (2010) achieves an amazing 82.9 km/h at 1,000 rpm.

3.3.1.4. *Examples of gearbox technologies*

For electric cars, it was shown in Chapter 2 that the operating range of electric motors (from 0 to more than 10,000 rpm) only requires a constant reduction ratio in the order of 7 to 10 combined with a differential so that the wheels rotate at speeds allowing the usual speeds of passenger cars. Nevertheless, this situation could change in the coming years, especially in order to reduce electricity consumption at high velocity.

For gasoline or diesel cars, the operating range of piston engines (700 to 6,000 rpm) requires several gear ratios to allow the usual speeds of passenger cars. The gearboxes that allow this multitude of ratios can be classified according to the technology of their gearboxes and their control system: manual, controlled or automatic.

3.3.1.4.1. Manual gearboxes

Manual gearboxes are most common in European diesel or gasoline vehicles. They generally consist of two parallel shafts, the so-called primary input shaft – connected to the engine through the clutch – and the so-called secondary output shaft (Figure 3.22). The gearbox has several pairs of pinions, one fixed on the primary shaft and one idler on the secondary shaft. Synchronizers connected to the gear lever via a linkage make it possible to connect the idler sprocket to their carrier shaft according to the desired ratio.

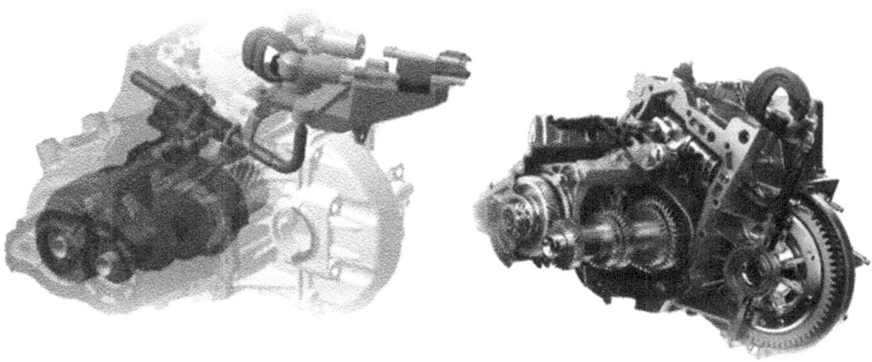

Figure 3.22. *6-speed manual transmission MB6 (source: PSA). For a color version of this figure, see www.iste.co.uk/krysinski/automotive.zip*

The driver selects the idler sprocket to be connected to the output shaft using a gear lever, a linkage assembly and idler sprocket forks.

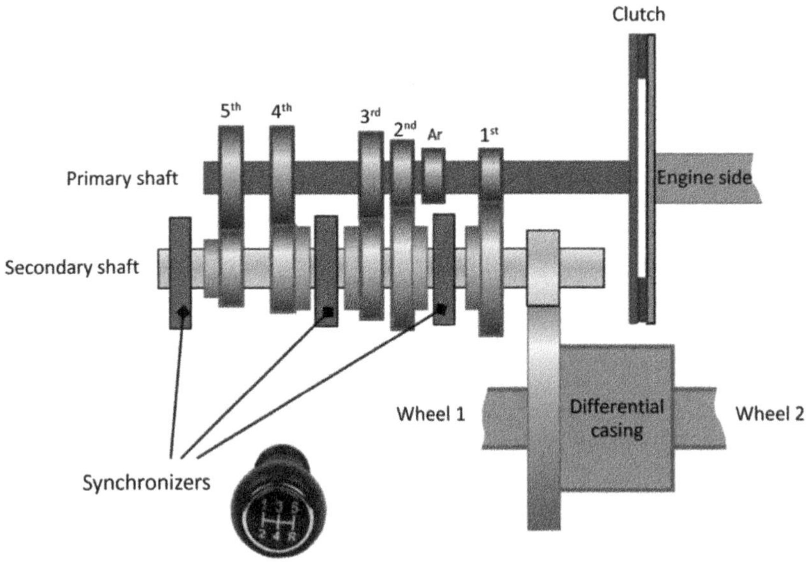

Figure 3.23. *Example of a 5-speed gearbox. For a color version of this figure, see www.iste.co.uk/krysinski/automotive.zip*

The tiering of connections is done according to the rules set out in the previous paragraphs.

3.3.1.4.2. Robotic or controlled gearboxes

For this type of gearbox, it is always the driver who controls the timing of the gear change via buttons, paddles or levers. However, it is an automated electrotechnical system that modifies the ratio under the condition that certain conditions are met (to avoid excess speeding or under-speeding).

Generally, the gearbox architecture remains the same as a manual gearbox. The control linkage is simply replaced by electric motors (Selespeed, Sensodrive PSA, Quickshift Renault, etc.) or hydraulic actuators (Sequentronic Mercedes, etc.). This system is advantageous for the manufacturer in that it is not necessary to change the gearbox structure or design new gearbox models.

Figure 3.24. *Robotic gearboxes (SensoDrive – source: PSA). For a color version of this figure, see www.iste.co.uk/krysinski/automotive.zip*

3.3.1.4.3. Automatic gearbox

Automatic transmission is beginning to become popular across the European continent after having largely conquered the American market. For this type of gearbox, it is the system that manages the most appropriate time for the gear change and selects the most appropriate transmission ratio for a particular situation. Generally, there are several operating modes (sport, city, etc.).

This type of gearbox consists of epicyclic trains. The modification of the gears is performed by closing and opening brakes to block the rotation of one of the components or by using a clutch to synchronize two components. By combining different speeds, manufacturers are able to offer gearboxes with eight to nine gears.

The traditional clutch is replaced by a torque converter. It allows gears to be shifted without torque breakage and dampens any vibrations transmitted by the crankshaft or wheels.

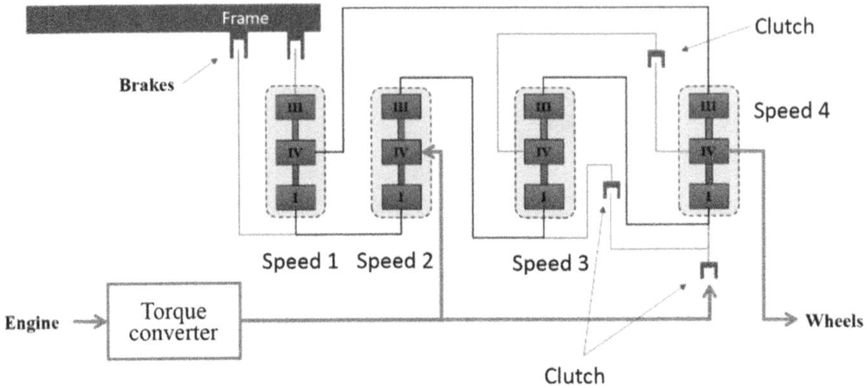

Figure 3.25. *Composition of an eight-speed automatic gearbox based on 4 epicyclic gearboxes. For a color version of this figure, see www.iste.co.uk/krysinski/automotive.zip*

Thanks to the possibility of having a greater number of gears and an optimized use of the torque converter, a smoother operation is obtained. Nevertheless, automatic transmission is still much more expensive than a manual gearbox, but it is not necessarily significantly more expensive than a dual-clutch robotic gearbox.

3.3.1.4.4. Dual-clutch transmission (DCT)

There are also dual-clutch gearboxes available to provide an engaged gear in anticipation of the next change. They are composed of two classic half boxes placed in parallel. One is made up of even connections and the other of odd connections. This type of system makes it possible to reduce the time between the shift from one gear to another, thus avoiding torque interruptions that generate jolts typical for conventional gearboxes.

Several systems have been developed by car manufacturers. The Porsche DoppelKupplung (PDK) system was developed by Porsche in the 1980s and used on competition cars (Porsche 962 or Audi S1). Volkswagen took inspiration from this to develop the direct-shift gearbox (DSG) system, which can be found in Volkswagen, Audi or Skoda models. Renault has developed the EDC system, which can be found in several models (Megane, Captur, Kadjar, Clio or Twingo) in six- or seven-speed versions.

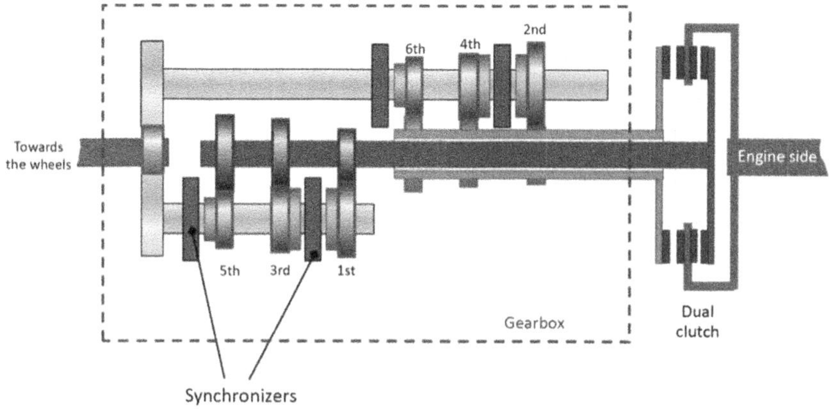

Figure 3.26. *Diagram of a dual-clutch gearbox (DC). For a color version of this figure, see www.iste.co.uk/krysinski/automotive.zip*

3.3.1.4.5. Continuously variable transmission

Continuously variable transmission (CVT) allows the transmission ratio to be continuously varied over a specified range. Using this solution, it is possible to let the engine work at the most favorable speed for the best possible acceleration or the lowest possible fuel consumption.

Speed variation is achieved by means of two conical flange pulleys, the spacing of which is controlled and can be continuously varied. The power transmission is achieved by means of a belt or chain.

One of the precursors was the DAF manufacturer, then the process was later adopted on carts and by Volvo which used belts.

Other more recent solutions use a metal belt or chain in an oil bath. This type of drive may be found in the case of a larger number of manufacturers: Ford (Fiesta and Focus), Fiat (Uno and Punto), Audi (Multitronic), Mercedes, Honda (Jazz and Civic Hybrid, Insight), Nissan (Micra K11, Murano, Altima, Juke, Tiida) and Dodge (Caliber).

These metal belt systems have the advantage of a long service life and are much better able to withstand the high tensile forces delivered by modern automotive engines, most of which produce high torque.

For this type of component, the clutch is necessary for low speeds and starting the vehicle, since the variation does not pass 0.

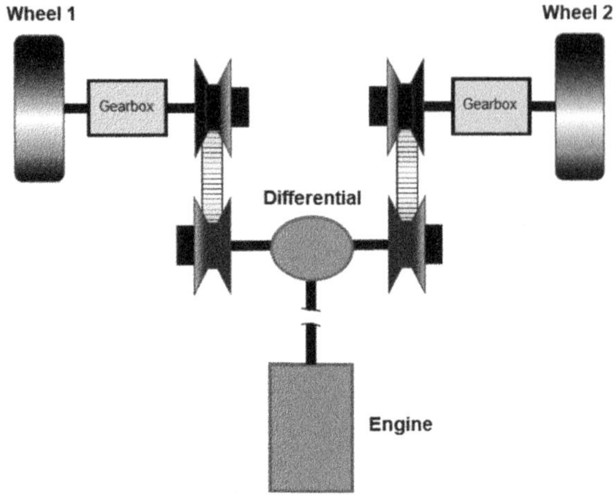

Figure 3.27. *Example of a transmission principle with belt drive*

A development without the need for a clutch of this type has been developed to enable passage through 0. This solution is called the infinitely variable transmission (IVT). In this case, the transmission is capable of shifting from forward to reverse through neutral gear and continuously. Technically, the solutions use adhesion drives through mobile rollers based on toroidal shapes.

These solutions require a high degree of contact control to ensure the passage of high torque while having a good efficiency. To achieve this, several rollers are used in parallel. As the system is not sufficient to ensure the transmission ratio between the engine and the wheels in itself, it is associated with one or more reducers in a series. The Torotrack system, widely used on trucks, buses and tractors, is the most common example (Figure 3.28).

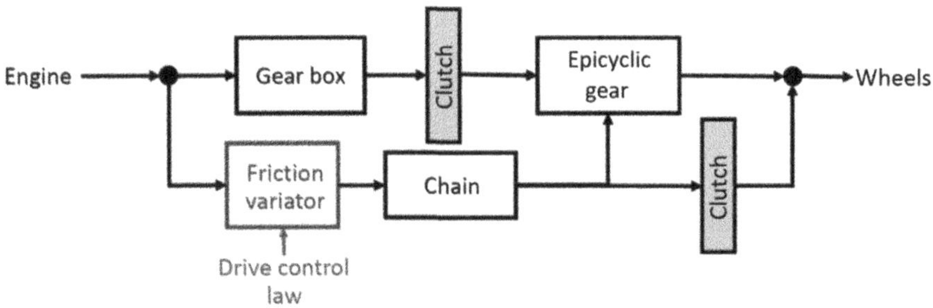

Figure 3.28. *Principle of a Torotrack infinite ratio transmission (IVT). For a color version of this figure, see www.iste.co.uk/krysinski/automotive.zip*

In comparison, Figure 3.29 shows and compares the accessible range of vehicle speed according to engine speed.

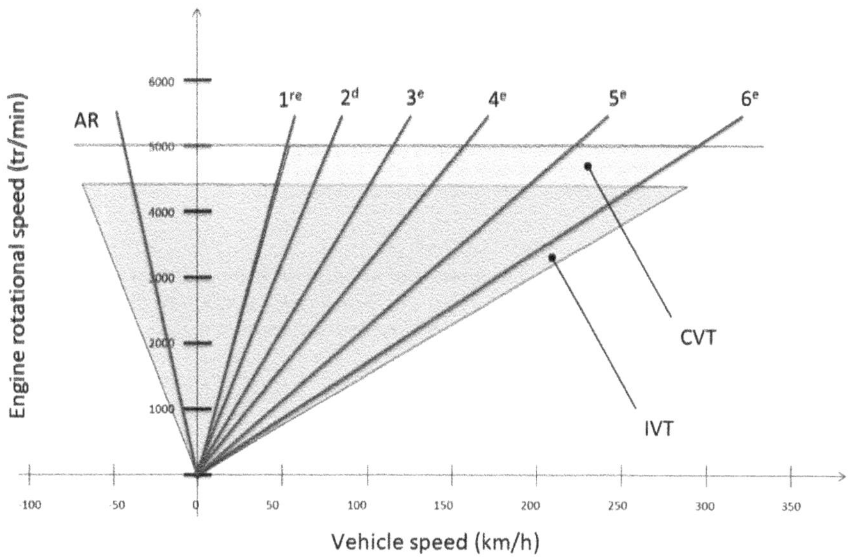

Figure 3.29. *Comparison of gearbox opening. For a color version of this figure, see www.iste.co.uk/krysinski/automotive.zip*

3.3.2. *The case of electric motors*

When it comes to cars, the characteristics of an electric motor – which provides torque at very low speed and can rotate very quickly, up to 10,000–15,000 rpm – allows a fixed reduction ratio between the engine and the wheels (in the order of 7 to 10) to achieve the desired vehicle performance.

Thus, Renault's ZOE proposes a reduction of 9.32, and Tesla a reduction of 9.73.

Very rarely used in electric vehicles, the multi-speed transmission is beginning to be developed on the market. Having several gears would allow high speed to operate in areas of higher efficiency. Figure 3.30 shows the impact of second gear, the first gear (in blue) allowing torque at low speed (in town) but not good efficiency at high speed (on the road). The solution of a continuous variation due to the low ratio variation can be noteworthy [RUA 18].

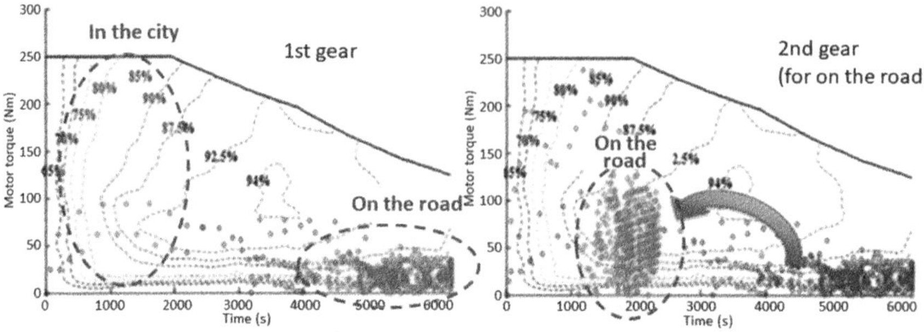

Figure 3.30. *Impact of the number of speeds on the operating area of the high-speed motor [RUA 18]. For a color version of this figure, see www.iste.co.uk/krysinski/automotive.zip*

3.4. Conclusion

It has been shown that for medium- and high-power aeronautics, it is more advantageous to run the turbines at optimal speeds – generally high – for efficiency reasons. These motors are then regulated in speed. It is necessary to have fixed gearboxes in the power chain to drive either a rotor in the case of helicopters or propellers for turboprop aircraft. These ratios can be high and add significant mass to the system. For example, for large helicopter turbines, the mass of the transmission box corresponds to the mass of the two turbines that generate the power.

For turbojet engines in large aircraft, due to the high dilution rate, reducers are in place to slowly rotate the fans.

For the smallest aeronautical powers – requiring piston engines – the choice of a fixed gear is always preferable. However, the ratio is lower because their optimal operating speed is lower. There is a factor of 10 when compared to the turbines.

For electric motors used in aeronautics – if multi-rotor systems are used – which rotate quickly, it is possible to drive the propellers directly without a reduction system.

For cars, given their limited speed range, internal combustion engines require the transmission ratio to be varied – if necessary – to adapt the engine torque to the torque at the wheel required to drive the vehicle. This can be done either sequentially through manual or automatic gearboxes or through continuous drives. For low speeds and to carry out transient phases, it is necessary to connect one or more clutches or a converter. All these components add mass to the structure.

For electric car engines, due to their wide speed range, manufacturers currently only offer a fixed gear ratio. To make the engine work better in maximum efficiency or torque capacities, it is conceivable to use continuous or sequential gearboxes to adapt the engine to an optimal operating point.

4

Energy Storage

4.1. Classification of energy sources

4.1.1. *Primary energy sources*

There are several energy sources that can be used for a land vehicle or for aircraft. Using one of these sources rather than another can be beneficial and judicious. Similarly, it may be possible to consider exchanges or complementary uses between these different forms of energy.

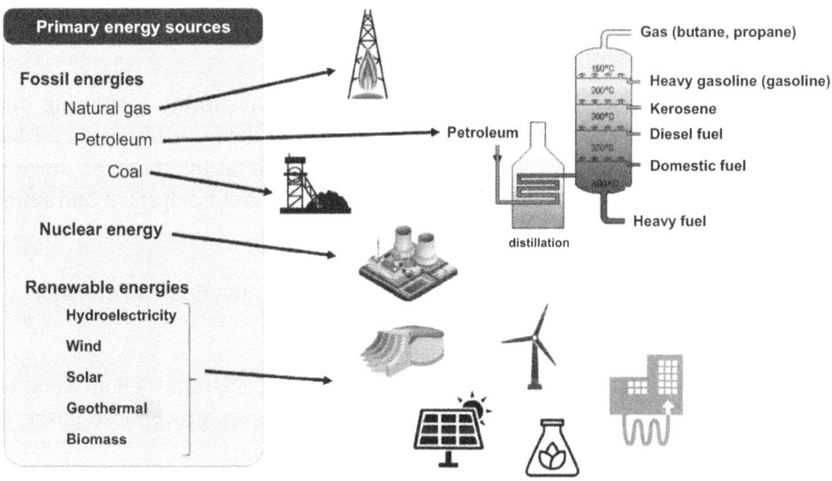

Figure 4.1. *Different primary energy resources. For a color version of this figure, see www.iste.co.uk/krysinski/automotive.zip*

Figure 4.1 shows the classification of different energy sources that can be found to date. Raw materials and natural phenomena are the main sources of energy. The former provide the so-called fossil and nuclear energies, the others the so-called renewable energies. They cannot be directly used in the forms in which they are found.

Fossil fuels come from the combustion of raw materials such as coal, oil or natural gas. These fuels have been created over hundreds of millions of years. Raw material reserves are abundant but unfortunately not renewable on our time scale.

Nuclear energy also comes from a raw material, uranium. It is often considered as an alternative to other fossil fuels because it reduces greenhouse gas emissions and offers a certain level of energy independence, even if it raises issues about safety and, especially, the storage of radioactive waste.

Renewable energies, as their name suggests, are not exhaustible, unlike fossil fuels. They are also known as clean energy because they come from natural phenomena (wind, solar radiation) and their use is booming: they do not yet replace other energy sources but offer the possibility of significantly reducing the use of fossil fuels.

The electricity supplied by hydropower represents about 5% of the total electricity produced in France. Hydropower plants convert energy from rivers, waterfalls and even tides or waves and are an inexhaustible and relatively clean source of energy.

Wind energy has a bright future because of its inexhaustibility (capturing one thousandth of the wind energy available on Earth would theoretically meet all the world's electricity needs). However, it is still unpredictable at the moment, since it depends on wind strength and causes problems with the ground surface and noise and visual pollution.

The sun's light energy is gathered on solar panels and converted into electrical or thermal energy.

Biomass energy is produced by direct combustion of vegetation such as wood or by its conversion into biofuels. Even if combustion produces greenhouse gases, it remains the main green energy.

4.1.2. *Energy carrier concept*

Primary energy forms (crude oil, coal, biomass, nuclear energy, natural gas and renewable energy) are not considered to be directly usable in vehicles. It is therefore necessary to introduce the energy carrier notion (Figure 4.2).

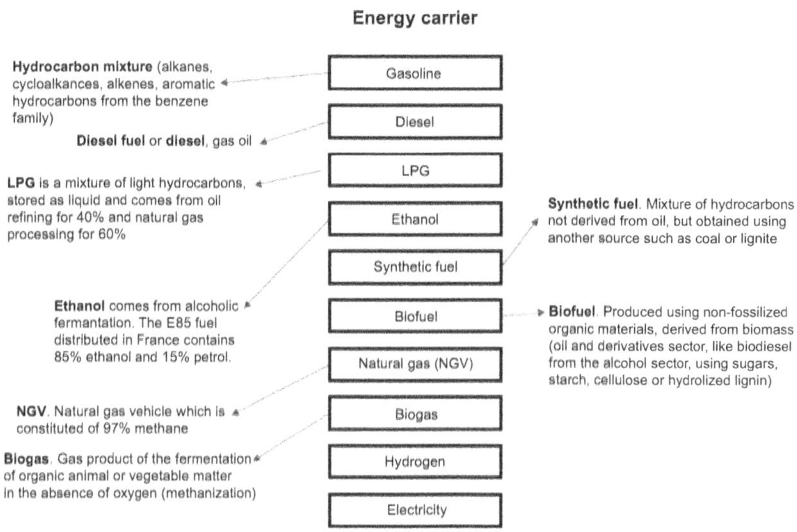

Figure 4.2. *Energy carrier concept*

The energy carrier is by definition the energy form used by the vehicle's transportation powertrain. Fuels are the historically used energy carriers. As its French name "*carburant*" suggests, a fuel contains carbon. Fuel is often a liquid, diesel or gasoline, and sometimes a gas. With a view to sustainable development with a reduction in net greenhouse gas emissions, other carriers are developing rapidly: electricity, the most widely used, hydrogen, compressed air, etc.

Hydrogen is considered here as an energy carrier and not as an energy source because it is practically non-existent in its natural state. In nature, it is found mainly in combination with oxygen (in water) or long carbon chains (in hydrocarbons). It is therefore necessary to first spend energy to produce dihydrogen from water or hydrocarbons before recovering the energy, through combustion in a fuel cell or engine, for example. The great advantage of this energy carrier is that it could be stored more easily than electricity.

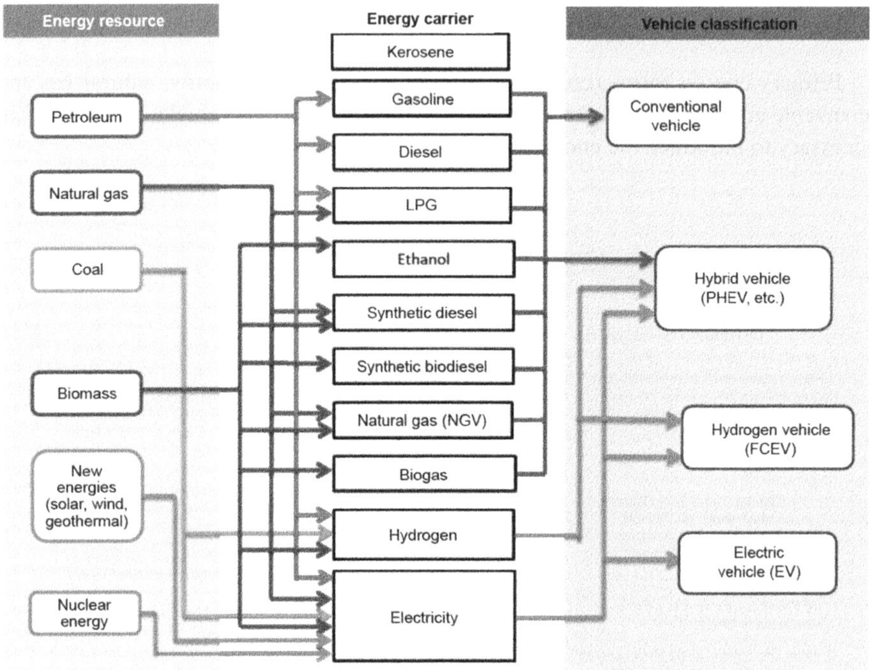

Figure 4.3. *Vehicle energy classification. For a color version of this figure, see www.iste.co.uk/krysinski/automotive.zip*

4.1.3. *Use of different energy sources in automotive and aeronautical transport*

In the context of automobile or aeronautical transport, it is interesting to analyze the different energy carriers used. To date, not all vectors have the same technological maturity. Figure 4.3 shows the relationship between the classification of motor vehicles, the base resource and the energy carrier. Apart from kerosene, which is not used for conventional motor vehicles, there is a great diversity of energy carriers. After a long period of supremacy of fossil energy, and associated energy carriers such as diesel, gasoline or liquefied petroleum gases (LPG), the automotive industry is looking for innovations to meet the ecological and economic pressures to become independent of fossil fuels.

The desire to rely more on new energies, such as biomass or even nuclear energy, is leading to the emergence of new or updated energy carriers (ethanol, biogas, hydrogen or electrical solutions). These solutions can lead to hybrid (multi-vector) or 100% electric solutions. Figure 4.4 shows the aircraft classification and energy carriers. For this type of transport, depending on the size and power of the aircraft, kerosene, gasoline and diesel from fossil fuels have strongly predominated. Since 2010, many projects (e-fan, Volocopter, HY4, ERAole, X-57) have been proposing prototypes to possibly consider alternatives to fossil energy by clean energy, through hybrid or 100% electric solutions. The maturities of these solutions are not yet at the same level as those for land transport, particularly because of the problem of storage and the weight generated, which is fundamental for aircraft, as shown in Chapter 5: Hybridization.

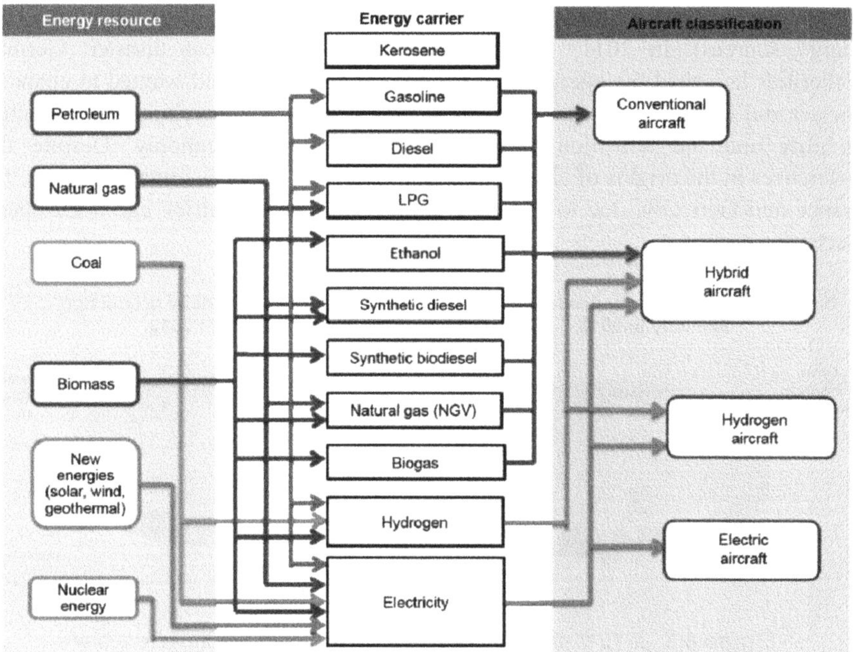

Figure 4.4. *Aircraft energy classification. For a color version of this figure, see www.iste.co.uk/krysinski/automotive.zip*

As this change in energy carrier is linked to economic (depletion of resources) and ecological (CO_2 emissions) considerations, it is interesting to demonstrate the link between a vehicle's CO_2 emissions per kilometer of use and the type of energy resource. This calculation is done for a complete analysis of the emission produced from well to wheel (WTW) as opposed to the principle of the tank to wheel. The mere consideration of the energy carrier (diesel, petrol or electric) is not sufficient as such. It is also necessary to analyze the origin of the energy resource (fossil, coal, nuclear or new energies).

As a result, an electric car will not have the same carbon impact in France, where, for electricity production, the share of nuclear power is high and less carbon intensive, as in other countries (China or Germany) where the share of coal is much higher. Figure 4.5 shows, for example, the distribution of energy resources in Germany and France in 2013. It can be seen that energy production sources are historically diversified in Germany (coal, renewable energy, nuclear energy, gas), while they are quite concentrated in France (mainly on nuclear and renewable energy sources). In 2011, following the Fukushima nuclear disaster, German authorities launched the energy transition or *Energiewende* and wanted to abandon nuclear and fossil fuels in favor of renewable energies, the development of electric mobility and the transition to a non-carbon emitting economy. Despite the differences in the origins of electricity production, the problems remain the same for France and Germany, due to the integration of the two countries into a European electricity system.

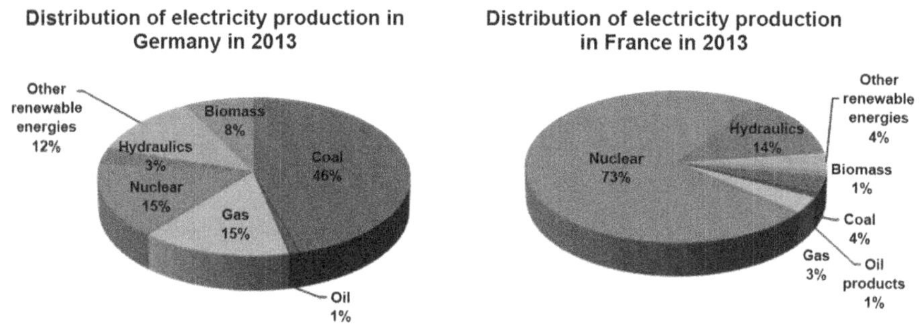

Figure 4.5. *2013 distribution of energy resources. France–Germany comparison (source: Observatoire de l'industrie électrique). For a color version of this figure, see www.iste.co.uk/krysinski/automotive.zip*

To have an order of magnitude, in France, a gasoline vehicle of 1,200 kg with a consumption close to 6 l/100 km had an emission of 160 g/km in 2010 (source: EDF). Figure 4.6 compares CO_2 emissions according to the origin of the energy resource in France, Germany or on average in Europe. It can be seen that the "all fuel" or "all coal" origins for electricity production are not a good alternative to fossil fuels in thermal vehicles. Mixing energies for electric cars is a good alternative from the point of view of pollution only if the proportion of polluting energies (coal, fuel oil) is low (France vs. Germany).

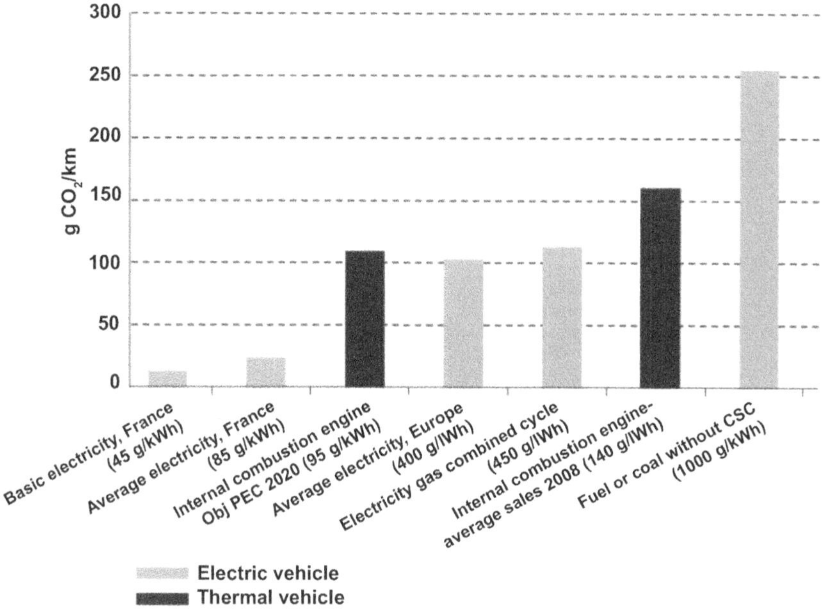

Figure 4.6. *CO_2 emissions per 1 km traveled depending on the energy resource (source: ADEME). For a color version of this figure, see www.iste.co.uk/krysinski/automotive.zip*

The socio-political vision must also be analyzed. Tensions observed in some countries in recent decades have led to high volatility in the price of crude oil (Figure 4.7). Since 2002, we have observed that demand has been growing, particularly in emerging countries, especially China and India. In addition, the reduction in production capacity caused by climate phenomena (hurricanes Katrina and Rita) and geopolitical crises (Iraq, Iran, Nigeria, Libya, Venezuela, etc.) have fueled the rise in crude oil prices. Finally, global inflation has also contributed to sustaining the rise in oil prices, which has become a safe haven.

This variability is accentuated by speculation, particularly from hedge funds. In comparison, the price of electricity is much more stable. For European countries, it is interesting, in the long term, to no longer be dependent on these uncontrollable variabilities.

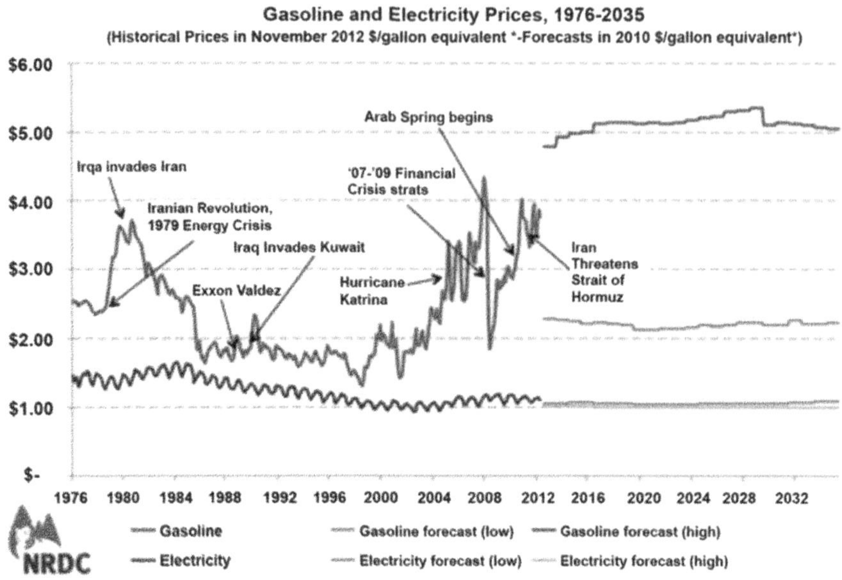

Figure 4.7. *Evolution of oil and electricity prices (source: NRDC association). For a color version of this figure, see www.iste.co.uk/krysinski/automotive.zip*

4.2. Energy storage for transport

From a technological point of view, the ideal storage system for transport would be having a high energy density in a limited volume and for a limited weight, while having the possibility of using this energy easily and quickly without too much intermediate loss.

In some cases – such as the automobile – energy can be stored again during braking phases. In this case, storage must be carried out over the same period of time as braking with traditional brakes, not require overly complex mechanical/electrical conversion elements and have a high efficiency.

4.2.1. *Different forms of energy storage*

Storage methods depend on the type of energy. On traditional combustion engine vehicles, liquid hydrocarbons are currently the dominant form of energy storage in terms of volume. Fuels come from fossil fuels and have an efficiency of 75% "from source to pump". Biofuels are made from biomass, with an efficiency of 70% "from biomass to pump".

The objective is to analyze storage technologies and performance as the system's only energy source but also as the second energy source for hybrid vehicles. These energies can be stored in electrochemical (batteries), electrical (supercapacitor), thermal or mechanical (kinetic energy or potential energy) form. In terms of the last two possibilities, the following focus can be made.

Currently, thermal storage is little used and difficult to consider in terms of the automobile. A distinction can be made between storage by sensitive heat and storage by latent heat. Raising the temperature of a material allows energy to be stored. The principle by sensitive heat is the same as that of solar water heaters: they recover the heat during the day and then release it, with an average efficiency of around 40% for the most recent systems. The preferred materials are water, synthetic oil, rock or concrete. Latent heat storage is based on the energy used when a material changes state (e.g. solid–liquid). Reverse transformation releases the energy stored in the form of heat with an efficiency of about 60%. This technique is applied in buildings using phase change materials. Incorporated into the walls, they serve as a thermal regulator, according to the heat provided by the sun.

Mechanical storage includes storage paths using potential energy (compressed air or springs) or kinetic energy (flywheel).

Storage using gas compression is currently a development focus. For the automotive sector, PSA's Hybrid Air system or the ETH Institute's Hybrid Pneumatics Engine project (thermal) are representative of the potential of this type of storage. Energy can also be stored as kinetic energy in a "flywheel", a wheel-shaped device rotating around its central axis. An electric machine provides it with kinetic energy (motor operation) and recovers it as needed (generator operation), resulting in a decrease in the flywheel's rotation speed. This system makes it possible to restore more than 80% of the accumulated energy, but for a limited storage time. In practice, the flywheel is used for very short-term smoothing of the energy supply in production equipment. This is particularly the case for internal combustion engines and especially diesel engines.

4.2.2. *Different energy storage technologies*

4.2.2.1. *Electrochemical battery technologies*

By definition, a battery stores energy in a chemical form and allows it to be released in electrical form. In transport, it is important that the technology allows charging and discharging cycles, so they are accumulators rather than batteries. A redox type chemical reaction can result from electricity: this is the charging phase of the battery. Depending on demand, the chemical reaction can be reversed to produce electricity and thus discharge the system. The oxidation-reduction torques vary from one model to another, giving the possibility of having different types of batteries (lead-acid, lithium-ion, nickel-metal hydride, etc.) with different performances.

In practice, a battery pack is constituted of an assembly of modules made up of elementary cells. Each cell is voltage limited according to the type of oxidation-reduction.

In order to provide the voltage necessary for the vehicle to operate, it is then necessary to make serial or parallel associations of these cells and modules (Figure 4.8).

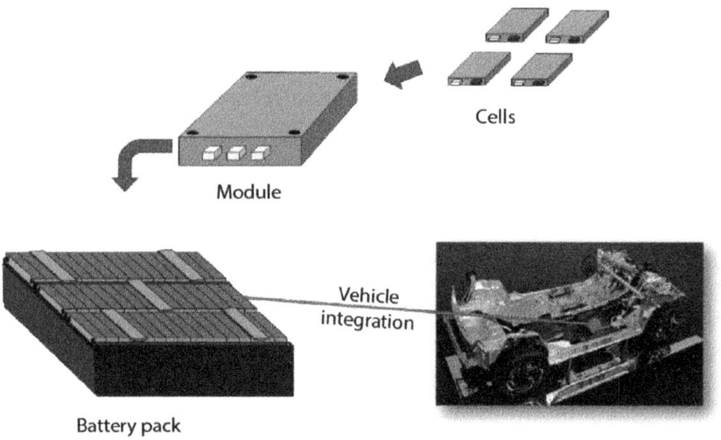

Figure 4.8. *Battery pack consisting of modules and cells. For a color version of this figure, see www.iste.co.uk/krysinski/automotive.zip*

The objective for each cell is to limit the voltage to values between 2 and 5 V. For example, the first version of the Li-ion battery in Renault's ZOE (22 kWh) consisted of 12 modules each with 16 cells or 192 cells. Each cell had a nominal voltage of 3.7 V at a current of 400 mA. The combination made it possible to have a 400 V battery for a weight of 290 kg, a capacity of 22 kWh (or 76 Wh/kg) and therefore 55 Ah.

In the 2000s, the most common batteries were nickel and cadmium (CadNi) or lead (Pb) based. In order to protect against the harmful effects of heavy metals, regulations, including a European directive, banned this type of technology. Lithium-based batteries then developed significantly.

Lithium remains in the ionic state through the use of an insertion compound at both the negative and positive electrodes, hence the name lithium-ion.

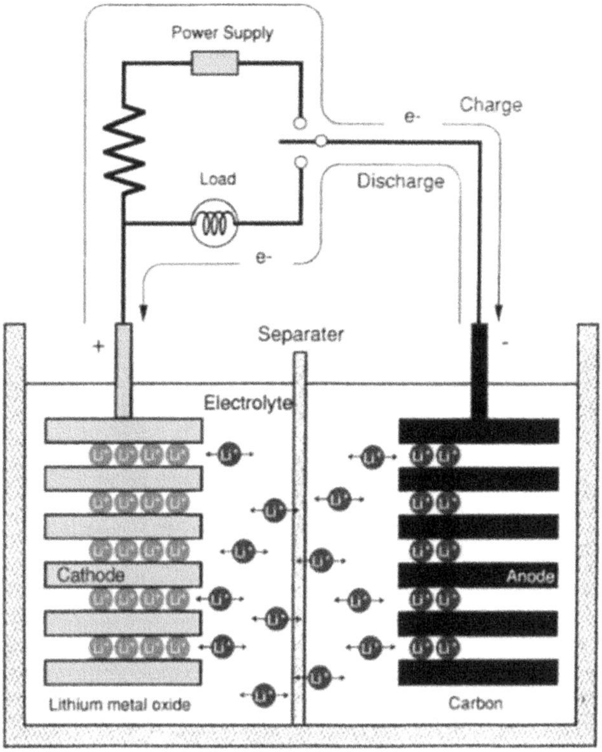

Figure 4.9. *Operating principle of a Li-ion battery [WAK 01]*

Lithium-based technology has several advantages:

– as lithium is a light metal, the energy density of this type of technology is improved;

– lithium does not have a memory effect that requires a battery to be completely discharged before being recharged on other material pairs;

– it is also recognized as having low self-discharge, which avoids regular maintenance and is an asset for transport applications;

– the voltage for this type of battery is almost constant throughout the discharge phase.

It also has some disadvantages:

– shallow depth of discharge. As its lifetime is associated with partial rather than total recharging, manufacturers recommend avoiding complete discharge and therefore not using the battery completely. The state of charge (SoC) in % is the equivalent of the fuel gauge to measure the amount of energy available in a battery;

– danger of overheating.

Depending on the use of the battery, 100 % electric, hybrid (rechargeable or not), the battery size and the strategy of state of charge management to preserve its life span will change (Figure 4.10).

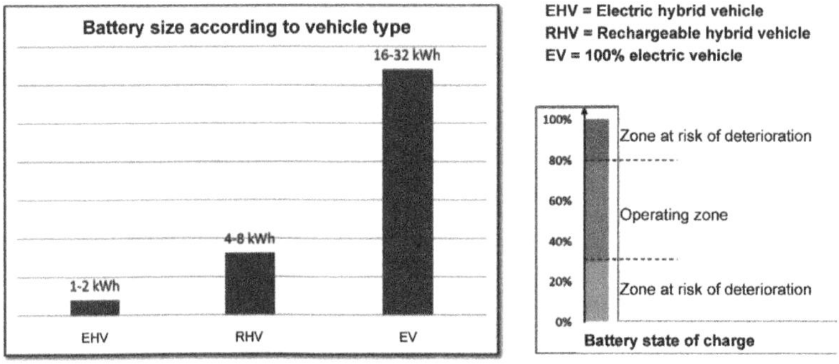

Figure 4.10. *Battery size and strategy for using the state of charge (SoC). For a color version of this figure, see www.iste.co.uk/krysinski/automotive.zip*

The batteries in these vehicles require both high power charging capabilities when recovering braking energy and high power discharging capabilities for start-up or overboost assistance. For this reason, their batteries must be maintained at a state of charge that can provide the required power while having sufficient margin to accept the necessary regeneration power without risking overloading the cells. Full charging of the battery reduces the charging capacity during braking and therefore the braking efficiency. The lower limit is set to avoid excessive discharge, which could reduce the battery life.

There are several lithium battery technologies available.

The lithium-cobalt battery ($LiCoO_2$) now has the highest energy density of any lithium technology. Among its advantages, we can mention the absence of a memory effect and a very low self-discharge, which, in theory, allows it to be stored for a long time. It has conquered the world of laptops and telephones, and also equips the Boeing 787 Dreamliner. It requires great care when charging. In case of incorrect balancing of the cells, the battery overheats or even explodes. After two aircraft fires – fortunately without human loss – Boeing and its partners have completely redesigned the BMS (Battery Monitoring System) which manages the balancing of the battery cells, monitors the temperature and disconnects if the voltage is too low. Mechanical protection is also necessary to avoid shocks. A damaged internal structure or a perforation can cause a fire.

The lithium-cobalt-aluminum battery (LiCoAl), which is similar to the lithium-cobalt battery in terms of performance, is used in Tesla electric vehicles. With its low internal impedance, it allows high discharges, necessary for the impressive accelerations of these cars. Similarly, it allows recharging at high currents, limiting the duration of recharging. Cases of fire in these cars have been reported, but it appears that all three known cases are due to an accidental short circuit of the battery terminals. A battery stores energy and – regardless of its technology (lithium, lead or other) – short-circuiting the terminals causes a sudden release of heat by the Joule effect.

Lithium-manganese cobalt (LiMnCo) technology, still used in electric vehicles but also in some portable tools, is chosen for its long life and greater stability than the previous battery. General Motors has selected it for its range of electric vehicles, but it is a battery that does not withstand temperatures above 50 °C. Tesla therefore preferred to confine this technology to stationary storage batteries for private individuals, as it takes up more volume for the same capacity. These batteries still present the risk of fire or even explosion that lithium-cobalt batteries are accused of.

The lithium-metal-polymer (LMP) battery equips the Bolloré Bluecar. It does not present an explosion risk but has a relatively modest capacity/weight ratio of 110 Wh/kg, which is still nearly three times higher than the lead battery. It is a "hot" battery that requires a temperature of 80 °C to operate.

Finally, the Lithium-iron-phosphate (LiFePO4) battery has been widely adopted by the industry. Although this technology does not achieve the capacity/weight ratios of cobalt technologies (120 Wh/kg compared to 200 Wh/kg for lithium-cobalt), it is very stable, displaying no fire or explosion risks, even with an elementary BMS. This battery is also characterized by a large number of permissible charge-discharge cycles, about 3,000 discharges compared to a few hundred for most commercial batteries (lithium or lead). In addition, it has a particular characteristic which is a kind of reverse memory effect: the less significant the discharge, the more cycles there are, which makes it particularly insensitive to microcycling. Finally, it is quite resistant to extreme ambient temperatures, from 0 to 60 °C. With a quality BMS, it is a very reliable battery, capable of providing service for many years.

The most promising technologies currently under development are:

– lithium-sulfur battery (Li-S) technology, which uses very cheap materials, increasing the energy density of cells by four times compared to current lithium cells;

– magnesium-ion battery technology (developed by Toyota, LG, Samsung and Hitachi). Magnesium ions carry a charge twice as high as lithium ions. Moreover, magnesium is an abundant material, so the technology makes it possible to build batteries that are denser in energy and cheaper;

– lithium-air battery technology. Consisting of an oxidation of lithium with oxygen absorbed from the air, these "breathing" batteries have a huge weight advantage over others in that they do not have to carry one of their main ingredients (air). A lithium-oxygen (Li-air) battery can theoretically store 10 times more energy than conventional batteries. This technology now allows a range of 400 kilometers to be achieved. Many manufacturers such as Volkswagen, BMW and Toyota are working on the development of this technology which does not yet have a possible application. It seems to be too expensive to achieve in the short or medium term;

– sodium-air battery technology. A battery cheaper than Li-air technology is being studied on the basis of sodium, a metal that is both available in large quantities and inexpensive, replacing lithium. These batteries could provide five times more energy than Li-ion batteries.

4.2.2.2. *Supercapacitor technologies*

Supercapacitors consist of two electrodes separated by an electrolyte. The equivalent scheme of a supercapacitor is given by the series assembly of two capacitors with a resistance associated with an electrolyte.

These are rechargeable systems. The system releases the accumulated energy with excellent efficiency and over a very large number of cycles.

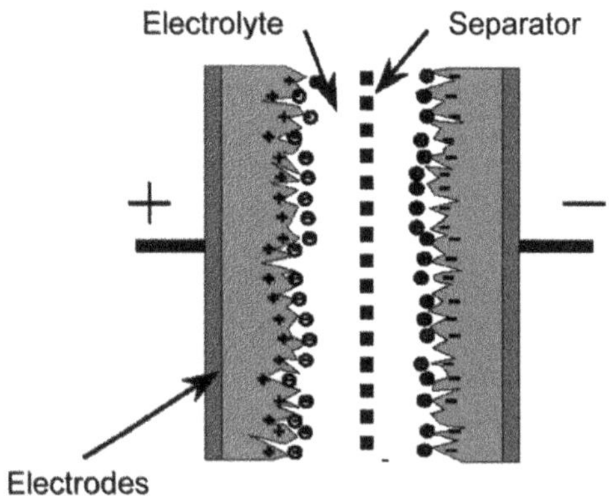

Figure 4.11. *Operating principle of supercapacitors [GUA 04]. For a color version of this figure, see www.iste.co.uk/krysinski/automotive.zip*

The internal resistance is very low, which allows charging or discharging with high currents. As a result, the charging time can be in the order of a few seconds. The charging and discharging times are from 1 to 30 sec while for batteries it is necessary to have one to five hours, depending on the technology.

They store less energy, 20 to 50 times less, at the same volume as a battery. For vehicles, they can be auxiliaries to the usual storage systems. For example, on some hybrid vehicles, electrochemical batteries provide traction for the vehicle, while supercapacitors provide energy for start-up, acceleration and other specific operations. The supercapacitors can then be recharged during braking.

The energy density of this type of component is 1 to 10 Wh/kg, compared to 20 to 200 Wh/kg for conventional batteries, and its power density is 1 to 3 kW/kg, compared to 0.05 to 0.200 kW/kg for conventional batteries.

The number of charge-discharge cycles that a supercapacitor can withstand is about 1,000 times higher than a battery.

4.2.2.3. *Hydrogen storage technologies*

Another solution for storing energy is to use hydrogen, or, more precisely, dihydrogen (H_2), a light gas that is very abundant on Earth. Its oxidation with oxygen produces electricity through a fuel cell. For transport, this could be an alternative solution to fossil fuels such as gasoline, diesel or kerosene, provided that its overall environmental impact is beneficial.

Hydrogen can be produced from hydrocarbons or water and a primary energy source. Thus, natural gas is both an energy and a hydrogen source. In this case, the most common production technique is the reforming of natural gas with water vapor. In the case of hydrogen production by electrolysis of water, the hydrogen source is water and the energy source is electricity produced from renewable or nuclear energy sources. The sector is not yet sufficiently developed and economically profitable; in 2018, 96% of hydrogen was produced from fossil energy (oil, natural gas and coal), the most profitable solution (source: IFPEN).

For renewable sources there are no greenhouse gas emissions or pollutants. For hydrogen produced from fossil fuels, carbon dioxide (CO_2) is released in significant quantities during its production and the utility can then be discussed. Techniques are being developed to capture this emission in refineries; however, this solution does not solve the problem of energy dependence on oil-producing countries or the announced end of fossil fuels. In addition, the energy efficiency required for hydrogen extraction and storage must also be considered.

Hydrogen has a very high mass energy density (120 MJ/kg), but since it is a very light gas, its density of energy by volume is low and therefore requires a large tank, and thus its storage and transport are real challenges. At room temperature and atmospheric pressure, 1 kg of gas has a volume of 11 m^3 (11,000 l). The purpose of hydrogen storage technologies is essentially to reduce the volume naturally occupied by hydrogen in its thermodynamically stable state under ambient conditions. Thus, hydrogen can be physically stored in different forms: gaseous, liquid or solid (Figure 4.12).

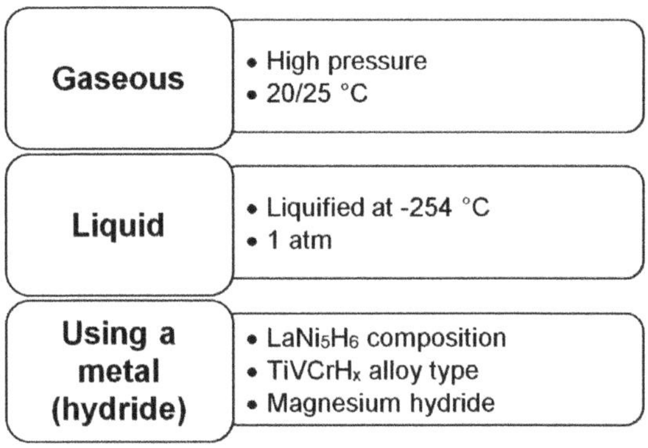

Figure 4.12. *Forms of hydrogen storage*

4.3. Forms of hydrogen storage

All solutions are still being developed and improved. For the aeronautics and automotive sectors, the objective is to obtain efficient storage facilities that take into account the different constraints to be respected: tank weight and volume, cost, life cycle, energy efficiency of filling, filling time, operating temperature and leaks.

The Züttel diagram compares the density of hydrogen according to the percentage of hydrogen mass stored on the mass of the solid tank (mass density) for different forms of storage (Figure 4.13). For example, at 700 bar for a composite tank, hydrogen gas represents 12% of the full tank. There are three areas representing the three possible storage modalities of hydrogen: gaseous, liquid (cryogenic) and solid (conventional hydrides). For liquid storage, the volume density of hydrogen is independent of the tank. For gas storage, the volume density is dependent on the storage pressure and the tank technology. Thus, two types of tank with different potentials are shown, steel and composite. For this type of storage, high pressures allow a higher volume density (smaller tank) but the mass density decreases and the tanks can then be very heavy. It can be seen that storage in solid form of complex hydrides has a high potential. This can be detailed as follows:

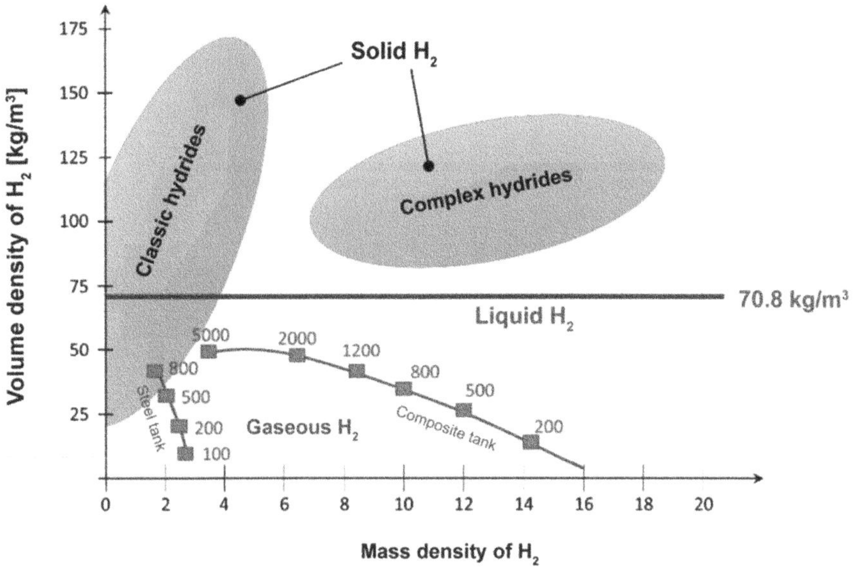

Figure 4.13. *Volume density of hydrogen storage in its different forms. For a color version of this figure, see www.iste.co.uk/krysinski/automotive.zip*

4.3.1. *Storage in gaseous form*

For gas storage, the simplest form, the higher the pressure, the greater the amount of hydrogen stored. Static storage tanks (e.g. transport bottles) operate at pressures of 300–350 bar; the volume density achieved is not sufficient for automotive application. For on-board automotive tanks, a high pressure storage, 700 bar, is used. At this pressure, 1 kg of hydrogen can be stored in a 25-liter tank and contains about 33 kWh. Toyota's Mirai has two tanks of 5 kg each, that is 2 x 125 l, allowing a range of around 200 km (Figure 4.14). In addition, the weight of the tank itself is a disadvantage, given that it must withstand very high pressures. It takes about 10 kg of full composite tank to store every 1 kg of hydrogen. In comparison, a gasoline tank for the same amount of energy is 10 times smaller and 6 times lighter.

The same applies to compression efficiency. Thus, in practice, to carry out industrial electrolysis, it takes 1 liter of water and 5 kWh of electricity to produce a "normal" cubic meter of H_2 which contains 3 kWh of energy. Once compressed to 700 bar, only 2.55 kWh remains. The efficiency is then estimated at 60% and then 85%.

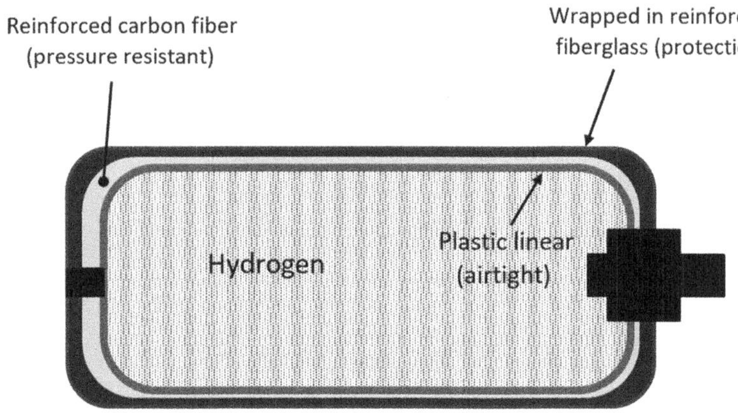

Figure 4.14. *Example of a gaseous hydrogen tank for cars (source: Ube Industries). For a color version of this figure, see www.iste.co.uk/krysinski/automotive.zip*

4.3.2. *Storage in liquid form*

Liquid storage involves several techniques. The conventional method consists of cooling hydrogen below its dew point, that is 20.28 °K (−252.87 °C) at atmospheric pressure (101,325 Pa). The liquid then has a density of nearly 71 kg/m^3. Under these conditions, 1 kg of hydrogen can be stored in a 15-liter tank containing 33 kWh.

To remain liquid, hydrogen must be maintained at a temperature of −252.8 °C. The tank must have secondary systems that maintain hydrogen at a low temperature and at a pressure at least slightly above atmospheric pressure. The liquefaction of hydrogen and its temperature maintenance have a high energy cost. This system is instead reserved for space launchers, even if it is considered for aeronautics.

If hydrogen is liquefied from the "normal" state of the cubic meter of H_2, which contains 3 kWh of energy obtained from the electrolysis of one liter of water, only 2 kWh will remain, the efficiency of the liquefaction operation being close to 35%.

As the pressures involved are lower than for high gas pressure tanks, the weight problem is less critical.

For this type of storage at −253 °C, over time it is the liquid hydrogen that poses a problem. Despite the tank's isothermal jacket, it evaporates continuously as it boils. For a car where the volume and insulation of the tank are necessarily reduced compared to industrial solutions, the losses are estimated at between 1 and 5% per day. In two weeks, half of the tank may have evaporated.

4.3.3. *Storage in solid form*

A third solution is to store hydrogen in a solid state by taking advantage of the property of some compounds to absorb hydrogen reversibly when exposed to hydrogen pressure. Metal hydrides formed by gas-solid reactions from certain metals are the most conventional ($LaNi_5H_6$ compounds, $TiVCrH_x$ type alloys, magnesium hydride). To recover hydrogen, it is necessary to provide heat, which can come from the electric motor or fuel cell for the car. Nevertheless, these solutions are still at the research stage. It is even possible to apply this process to mobile units, when weight is not essential (large vehicle, train, ship, submarine, etc.) and under conditions where cost constraints remain marginal.

4.3.3.1. *Fuel cell*

The first way of using hydrogen energy would be to burn the gas through a combustion engine, as proposed by BMW on a 7 series in the mid-2000s. These developable solutions to decarbonize existing vehicles (truck, car or plane) have the advantage of being cheap but do not have good efficiency. In the case of aeronautical or automotive transport, the use of a fuel cell is envisaged to convert stored energy into electricity.

In this case, the generation of the electrical voltage is achieved by oxidation on one electrode of a reducing fuel, such as hydrogen, coupled with the reduction on the other electrode of an oxidant, such as oxygen from the air (Figure 4.15). The operation of a dihydrogen-dioxygen battery is particularly clean since it only produces water and consumes gas. The advantage of the battery, in addition to the non-emission of CO_2, is that it eliminates all other pollutants.

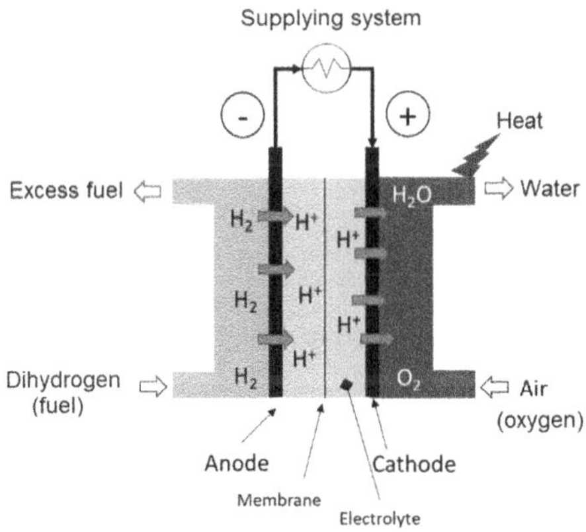

Figure 4.15. *Principle of the membrane fuel cell. For a color version of this figure, see www.iste.co.uk/krysinski/automotive.zip*

There are mainly two types of fuel cells: the Proton Exchange Membrane Fuel Cell (PEMFC) for conventional transport (automotive, air transport) and the Solid Oxid Fuel Cell (SOFC) for very high power transport (submarines, ships, etc.).

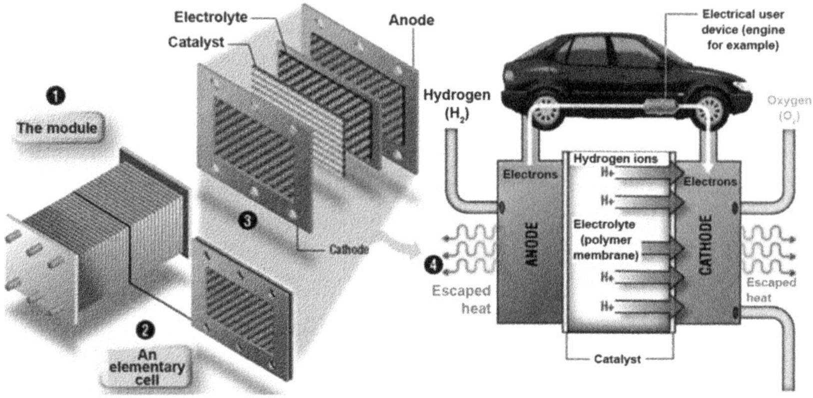

Figure 4.16. *Operating principle of a GENEPAC fuel cell electric generator (source: CEA and PSA) . For a color version of this figure, see www.iste.co.uk/krysinski/automotive.zip*

PEMFC technology has the best performance and its operating temperature is 20 to 80 °C. SOFC technology operates at a higher temperature of about 800 to 1,000 °C which is necessary to have sufficient conductivity of the solid electrolyte (thin ceramic based on zirconium oxide).

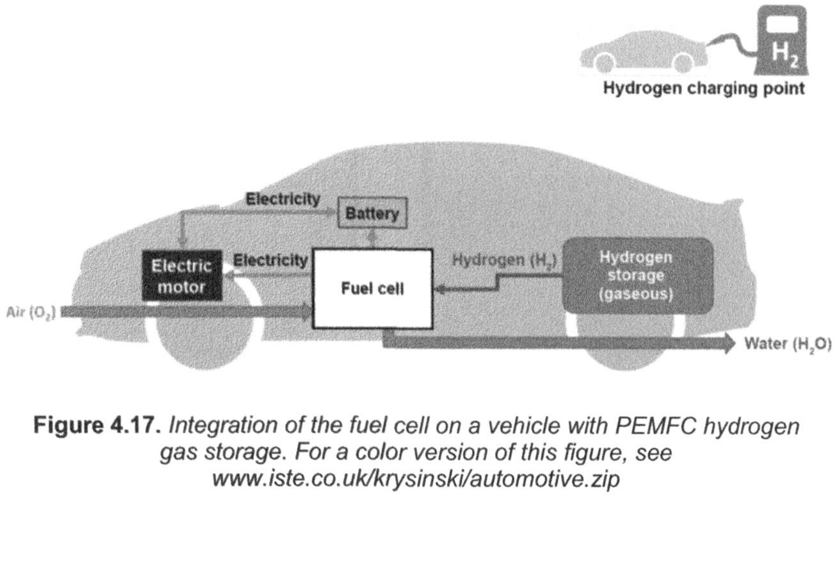

Figure 4.17. *Integration of the fuel cell on a vehicle with PEMFC hydrogen gas storage. For a color version of this figure, see www.iste.co.uk/krysinski/automotive.zip*

Another technology called the Direct Methanol Fuel Cell (DMFC) uses methanol or ethanol (DEFC) injected directly into the fuel cell as fuel (Figure 4.18). This solution allows easier storage than hydrogen, requiring neither high pressures nor low temperatures, as it is liquid at room temperature. The volume energy density of methanol is much higher than that of hydrogen, even very highly compressed.

The efficiency is slightly lower and the obstacle to using this type of battery is mainly related to the difficulty of transporting methanol cartridges (air transport is prohibited) and the low power that can be used technologically.

The interest in using electricity in terms of the automobile is also linked to the ability of engines to work in generators to recover energy during braking. In fact, hydrogen storage is systematically associated with electrochemical storage (batteries) to enable this function. The combination can be in series, with hydrogen then used to increase the battery's autonomy (range extender) or in parallel (the battery used as storage for braking, as a complementary source for a power boost or even as the only energy source in 100 % electric mode) (Figure 4.20).

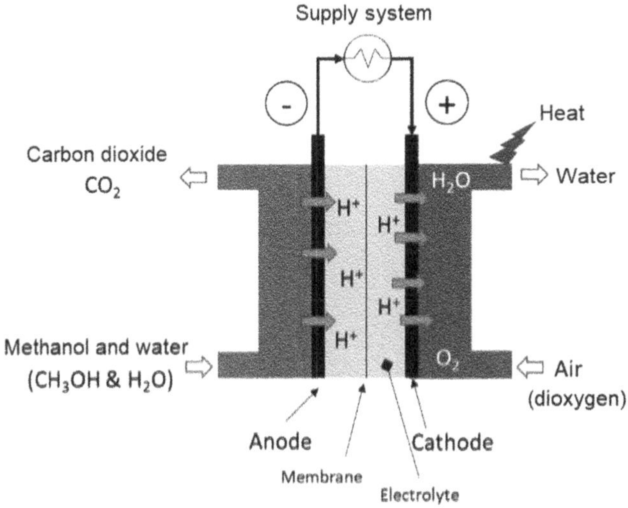

Figure 4.18. *Direct Methanol Fuel Cell (DMFC) or ethanol (DEFC) fuel cell. For a color version of this figure, see www.iste.co.uk/krysinski/automotive.zip*

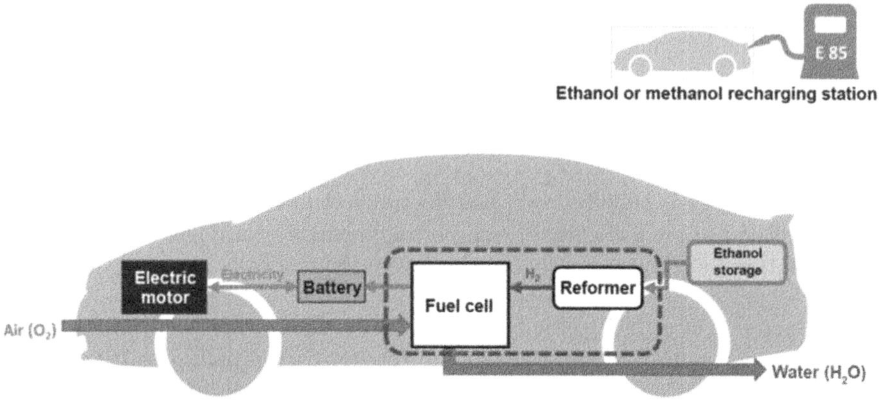

Figure 4.19. *Integration of the SOFC system on a DEFC vehicle. For a color version of this figure, see www.iste.co.uk/krysinski/automotive.zip*

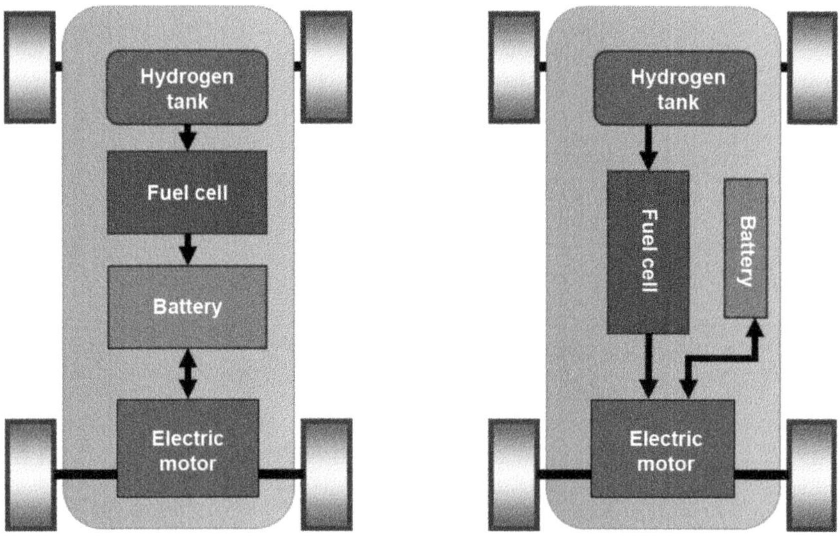

Figure 4.20. *Hydrogen fuel cell operating modes. For a color version of this figure, see www.iste.co.uk/krysinski/automotive.zip*

The fuel cell requires an environment to be integrated into an automobile or aircraft, known as a "fuel cell" system.

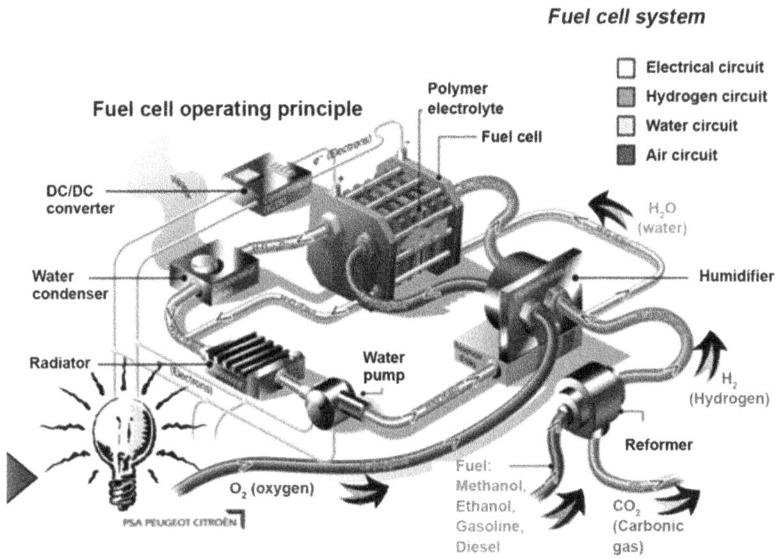

Figure 4.21. *"Fuel cell" system with reformer (source: PSA Peugeot Citroën). For a color version of this figure, see www.iste.co.uk/krysinski/automotive.zip*

The use of fuel cells leads to energy losses compared to the compression of hydrogen, but especially to the chemical reaction producing electricity. Overall, the efficiency is close to a configuration with an internal combustion engine.

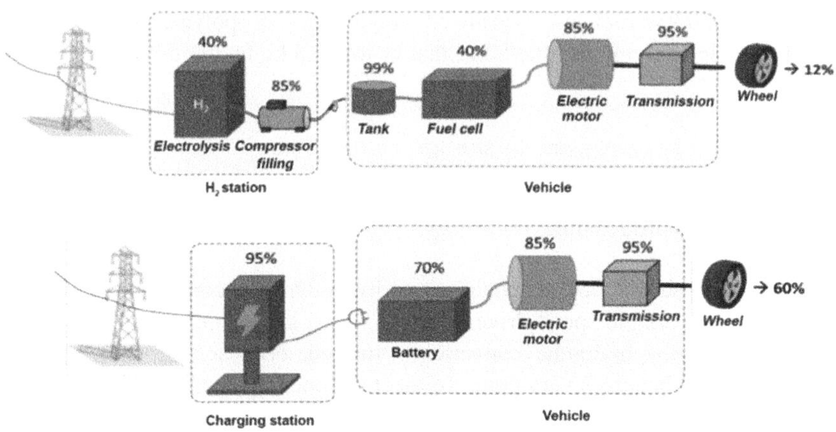

Figure 4.22. *Comparison of the performance of the "fuel cell" system and electric transportation powertrain. For a color version of this figure, see www.iste.co.uk/krysinski/automotive.zip*

Since 2015 Toyota has been marketing the Mirai vehicle, with a possible range of 650 km for an NDEC cycle. This vehicle is composed of two energy sources: a 114 kW (155 hp) fuel cell combined with two 5 kg hydrogen capacity tanks (330 kWh in total) and a 1.6 Wh Ni-Mh battery that recovers energy during the braking and deceleration phases and provides additional power during the acceleration phase.

In the field of aeronautics, fuel cells are being considered for the development of the "more electric aircraft" by supplying power to a number of essential aircraft functions (taxiing, wing de-icing, engine starting, etc.). For example, in 2011, the DLR (*Deutsches Zentrum für Luft- und Raumfahrt*) experimented on an Airbus A320 ATRA with a front axle equipped with an electric motor powered by a fuel cell, allowing the aircraft to move on the ground without the help of the engines. Such a system would reduce ground emissions by 20%, save 200 to 400 liters of kerosene per day, reduce reactor operating time by 1,200 h/year (and thus reduce maintenance costs) and reduce noise by 100%. This work, which began in 2008 as part of the German LuFo IV project, is being carried out in collaboration with Airbus and Lufthansa.

For aircraft propulsion, various projects identified concern propeller aircraft powered by hydrogen, which powers either a turbojet, a combustion engine or a fuel cell and electric motor assembly.

Used in a number of drone projects (unmanned flight), the fuel cell is being tested on small manned aircraft. Thus, the European project called "Enfica-FC", started in October 2006 and finished in 2011, aimed to develop a two-seater aircraft powered by a 20 kW fuel cell, hybridized with a 20 kW Li-polymer battery for take-off and capable of flying for one hour, thanks to a 1.5 kg hydrogen reserve stored at 350 bars.

4.3.3.2. *Gas or compressed air storage technologies*

For the automotive industry, there are two types of hydraulic transmissions: hydrostatic and hydrodynamic.

Figure 4.23a shows a power chain with a hydrodynamic converter. This element is a continuous variable speed drive that can give a kinematic opening of about 2. The opening of the hydraulic converter is limited, because, for high values, its energy efficiency becomes very poor. To meet the opening needs of the power chain it must be combined with a mechanical gearbox with a certain number of gears and a reverse gear.

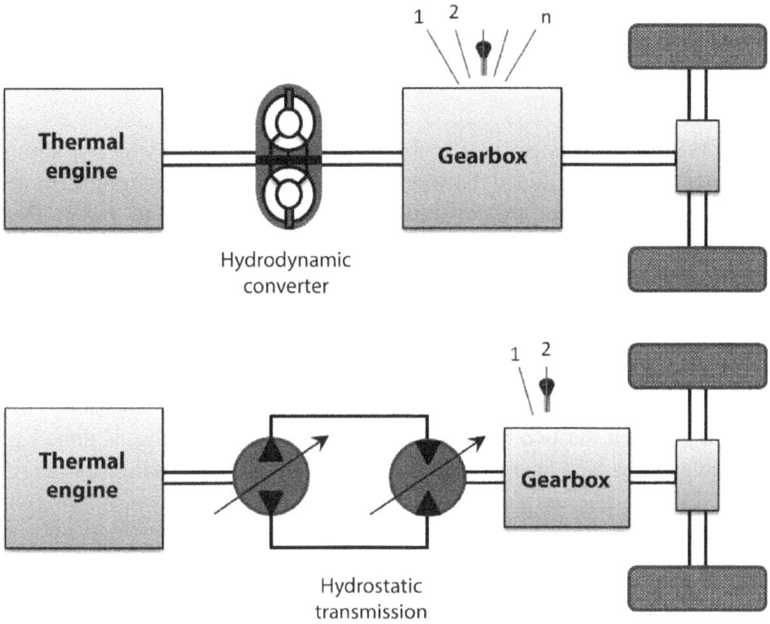

Figure 4.23. *Comparison of a hydrostatic transmission and a hydrodynamic transmission*

Figure 4.23b illustrates the principle of hydrostatic transmission. The combustion engine drives a hydraulic pump, which converts mechanical energy into hydraulic energy. The power of a pump is generally given by:

$$P_h = \Delta p \cdot Q \qquad [4.1]$$

with:

– Δp the pressure difference (expressed in Pascal) between intake and discharge;

– Q the volume flow rate (expressed in m^3/s);

– P_h is the hydraulic power W (expressed in watts).

The hydraulic flow is transferred through the pipes to the hydraulic motor, which again converts the hydraulic energy into mechanical energy. The kinematic opening of the variable displacement pump or variable displacement hydraulic motor ensures that only a minimum number of gears are available on the gearbox. A hydrostatic transmission allows a continuous variation of ratios with the variation of the displacement. This transmission therefore provides IVT (Infinite Variable Transmission) services.

4.3.3.3. *Hybrid Air system*

PSA's Hybrid Air system is a hydrostatic transmission to which a storage unit (accumulator) has been added between the pump and the hydraulic motor. The combustion engine drives the hydraulic variable displacement pump through a continuously variable transmission. This pump can drive the hydraulic motor to pass part of the torque to the wheels or store energy in the high-pressure accumulator. These two actions are possible at the same time, depending on the need for overall energy optimization of the power chain. The low-pressure accumulator serves as an expansion vessel for the oil. Another part of the power goes directly to the wheels through a parallel mechanical connection of the hydrostatic transmission. The power distribution between the two branches, mechanical and hydraulic, is done by means of epicyclic gearing. The energy is then collected by conventional gears.

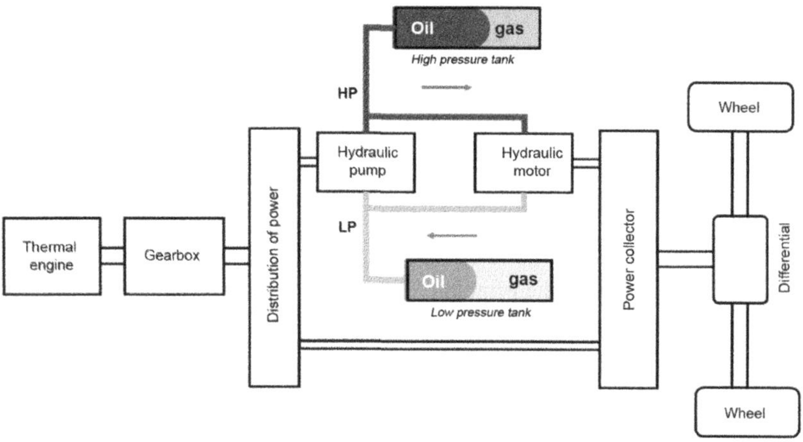

Figure 4.24. *Diagram of PSA's Hybrid Air system. For a color version of this figure, see www.iste.co.uk/krysinski/automotive.zip*

By definition, pneumatic energy is the energy stored in a gas that has been compressed. In the context of vehicle hybridization, the idea is to use compressed gas as a means of energy storage, at least temporarily, with gas being used as a resource. The energy storage is carried out using an accumulator consisting of an enclosure and a deformable element (gas). Resource accumulators are only rarely used. Several principles are then possible. Gas accumulators without separating elements are rarely used, as for these, the gradual dissolution of gas in the oil inevitably leads to operational disruptions. These are due to the gradual decrease in the quantity of gas, which causes a drop in pressure, as well as an increase in the compressibility of the oil. Depending on the nature of the separation surface between the gas and the hydraulic fluid, there are essentially two types of hydropneumatic accumulators, bladder and membrane.

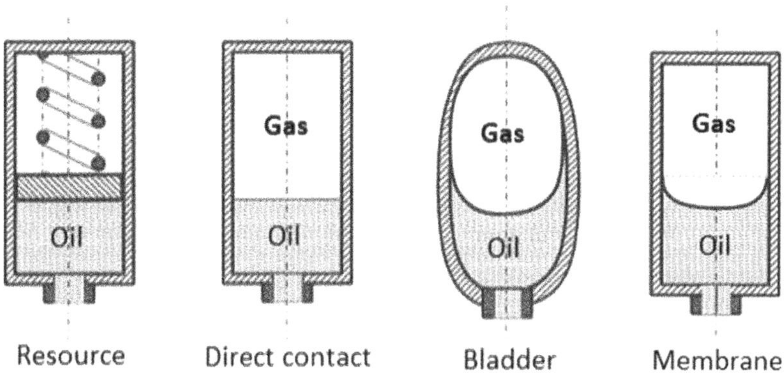

Figure 4.25. *Schematic views of a pressure storage unit*

In order to understand the design elements of pneumatic hybridization, it is useful to analyze the pneumatic accumulator's behavior.

Suppose that gas compression is rapid and the transformations adiabatic.

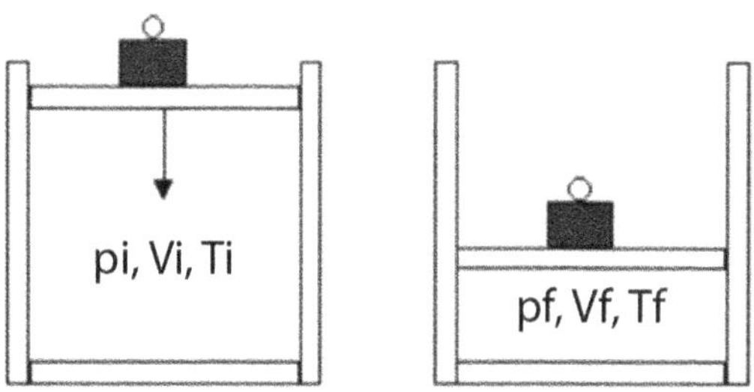

Figure 4.26. *Diagram of the modification of the gas state*

$$V = \left(\frac{Cste}{P}\right)^{\frac{1}{\gamma}} \qquad [4.2]$$

Let:

$$\mathrm{dV} = -\frac{1}{\gamma}\frac{\mathrm{Cste}}{\mathrm{P}^{\frac{1}{\gamma}+1}}\mathrm{dP} \qquad [4.3]$$

Which gives us:

$$V = \left(\frac{Cste}{P}\right)^{\frac{1}{\gamma}} \Rightarrow dV = -\frac{1}{\gamma}\frac{Cste}{P^{\frac{1}{\gamma}+1}}dP \qquad [4.4]$$

By integrating the energy this gives:

$$W = \int \frac{Cte\ P}{\gamma\ P^{\frac{1}{\gamma}+1}}dP = Cte\left[\frac{P^{1-\frac{1}{\gamma}}}{\gamma-1}\right]_{P_1}^{P_2} \qquad [4.5]$$

We know that:

$$\mathrm{Cte} = \mathrm{P_1 V_1^{\gamma}} = \mathrm{P_2 V_2^{\gamma}} \qquad [4.6]$$

And on the other hand, considering that it is the initial volume V_1 that is fixed (volume of the tank, etc.), we can thus express the storable energy:

$$W = \frac{P_1 V_1}{\gamma-1}\cdot\left(\left(\frac{P_2}{P_1}\right)^{1-\frac{1}{\gamma}} - 1\right) \qquad [4.7]$$

We observe that the energy is a function of low pressure P_1. By setting the maximum high pressure threshold P_2 (e.g. 350 bar, technological limit of hydraulic components), the change in energy can be traced according to P_1. We observe that there is an optimal pressure value (which can be obtained from the derivative of the function) allowing for a given gas (γ), for an initial volume V_1 of gas and a pressure P_2 (limited by technology), to optimize the quantity of storable energy.

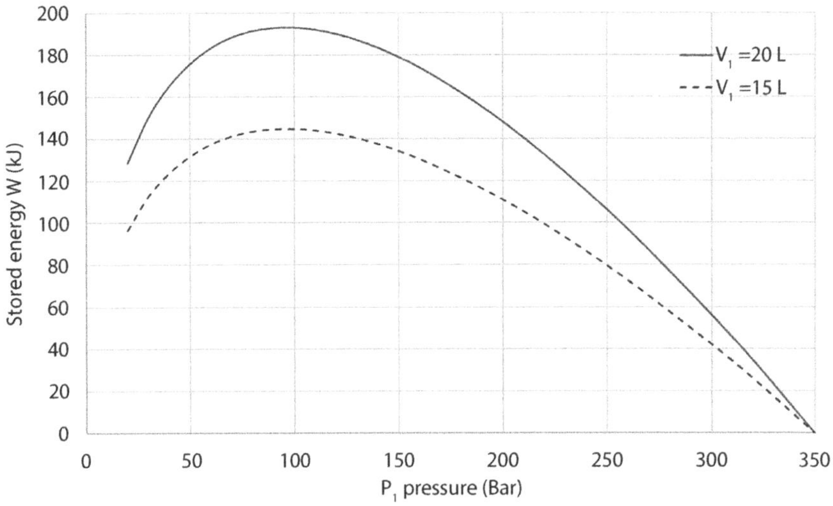

Figure 4.27. *Evolution of storable energy according to the initial pressure for nitrogen, $\gamma = 1.7$, $P_2 = 350$ bar and different values of V_1*

The value ensuring an optimum energy storage would be about 100 bars.

The torque to be supplied to the pump, for a mechanical/hydraulic efficiency η, is given by the formula:

$$C_{pump} = \frac{1}{\eta} c_y \Delta P \qquad [4.8]$$

We observe that from this point of view it is interesting to have the lowest pressure drop, so that P_1 is close to 350 bars. The best compromise between stored energy and power to be supplied is close to 200 bars.

The advantage of this type of system is that it can recover energy during braking, which can be used for acceleration at low speeds, where the combustion engine has the lowest efficiency.

For a vehicle of 1,250 kg launched at 60 km/h for a constant deceleration from 1 m/s² to 0 km/h, the energy W to be expended is 175 kJ in 16.7 s. During this phase, the instantaneous braking power is 10 kW:

$$W = \left|(E_c)_{final} - (E_c)_{initial}\right| = \frac{1}{2} M V^2 \qquad [4.9]$$

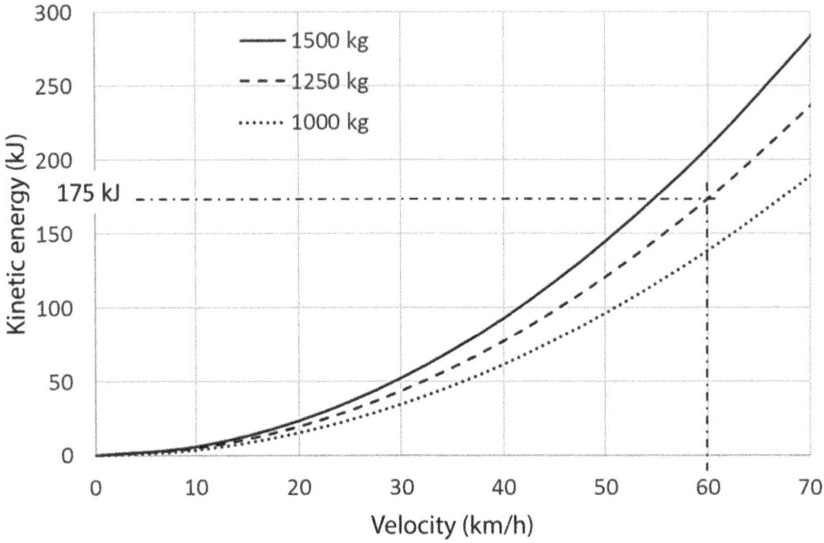

Figure 4.28. *Evolution of kinetic energy according to velocity*

Assuming that we can store 100% of this energy and using the relationship [4.7], for a pressure P_2 of 350 bars, a pressure P_1 of 200 bars, for a nitrogen gas, we obtain a volume of 20 l:

$$V = \frac{(\gamma-1)}{P_1\left(\left(\frac{P_2}{P_1}\right)^{1-\frac{1}{\gamma}}-1\right)}W \qquad [4.10]$$

It can be seen that the volume of the tank is proportional to the energy to be stored and therefore to the vehicle's mass. This approach is very simplified. The advantage of such a system is to use it in the context of hybridization, a case study presented in Chapter 5. Hybridization allows the downsizing of the combustion engine, possible thanks to the contribution of the hydraulic motor, and repositions the operating point in urban areas towards optimal efficiency.

Hydraulic machines have quite different powers and operating ranges. For a 1,250 kg vehicle, the hydraulic motor must have a torque of 180 Nm at low speed to ensure the vehicle's "take-off" and around 50 kW of power at high speed to obtain the excess power from the combustion engine in mixed operation. To ensure storage, it is necessary to have a 20 kW power pump.

The total hydraulic mass, engine, pump and tanks and accessories amounts to around 100 kg.

The 175 kJ corresponds to 48 Wh or a density of 0.48 Wh/kg. The calorific value of gasoline is 35,475 kJ/l, which is equivalent to 4.9 ml of gasoline.

These values are very small compared to the capacities of a battery, but this technology has a very high power density, which gives an advantage in both quick storage and return time. It can correspond to urban travel applications with frequent stops and starts.

4.3.3.4. *Compressed air system*

Other solutions aim to use compressed air storage, ambient air being the cheapest and most available.

Among the possible configurations, storage systems using only volumetric compression without additional energy input were studied. These systems have the advantage of being able to store energy over a long term – without significant losses in the stationary charged state – and make it possible to define an energy efficiency of the round-trip cycle similar to that of an electrochemical accumulator. This pressure energy storage system exists in a classic way on heavy trucks and large mining machines.

It is proposed that we analyze the case of a reversible compressed air motor that can compress ambient air to a pressure of about 300 bar and use the expansion of this compressed air to produce mechanical work to set a vehicle in motion. Can such systems be developed for motor vehicles and aircraft?

If we want to compress a gas, considering, at first, an ideal gas, to pass from a state [P_0, V_0] to a state [P_1, V_1], assuming a polytropic (reversible adiabatic) transformation of polytropic coefficient k, the work W to provide is given by:

$$W = \frac{P_0\, V_0}{k-1}\left(\left(\frac{V_0}{V_1}\right)^{k-1} - 1\right) \qquad [4.11]$$

For an initial pressure of 1 bar, an initial volume of 350 l with a compression ratio of 15.6, that is, a compressed volume of 22.46 l, and a polytropic coefficient k of 1.4, a work of 175 kJ is obtained. It can be seen that this system requires a 350-liter tank, much larger than the 20-liter Hybrid Air system for the same energy stored. This energy corresponds to the energy that can be recovered when braking a vehicle of 1,250 kg from 60 km/h to 0 km/h. The size of this tank is unacceptable for passenger vehicles.

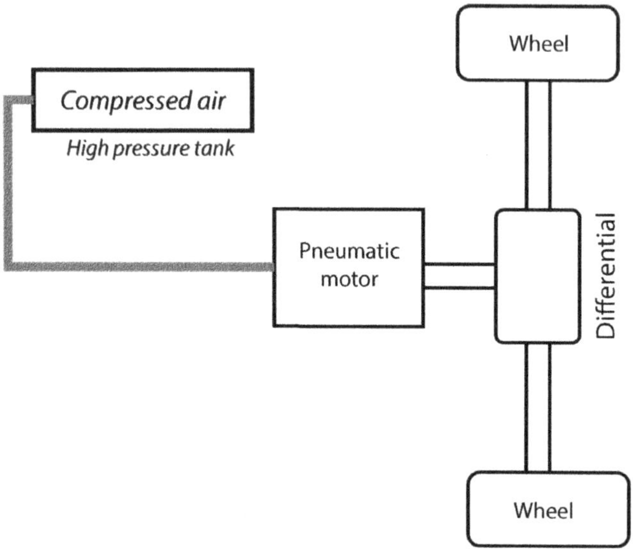

Figure 4.29. *Diagram of a compressed air system*

Another problem may be caused by the temperature increase due to compression. The ideal gas law shows that for a polytropic transformation, the temperature increase is given by:

$$T_1 = T_0 \left(\frac{V_0}{V_1} \right)^{k-1} \qquad [4.12]$$

Thus, in the proposed case, for an initial temperature T_0 of 20 °C, a temperature T_1 of about 615 °C is obtained. If the energy is reused quickly, the air will not have time to cool down and the energy can be returned almost entirely. If the gas is completely cooled to room temperature, the resulting expansion of the constant volume gas must be taken into account. The conservation of the mass in the tank shows:

$$PV = nRT \quad \Rightarrow \quad P_{cooled} = P_1 \frac{T_{cooled}}{T_1} \qquad [4.13]$$

Thus, the state of the usable gas after cooling to 20 °C is about 16 bar. From this state, with a compressed volume of 22.46 l and a pressure of 16 bar at room temperature, the reusable energy supplied to the wheel by an expansion is 47 kW. The gas temperature is then −139.3 °C and the gas volume is 160 l.

$$W = \frac{P_{cooled}\ V_{cooled}}{k-1}\left(\left(\frac{V_{cooled}}{V_{final}}\right)^{k-1} - 1\right) \qquad [4.14]$$

The limitations of such a purely compressed air system are clear here. Storage in the form of compressed gas could be used to hybridize a combustion engine: the exhaust gases would maintain the temperature of the compressed gases, the cold recovered during the expansion would help the vehicle's air conditioning. It is possible to consider using the combustion engine to provide energy for gas compression by making it work at its optimal efficiency point during certain driving phases.

Some manufacturers, such as Guy Nègre, have sought to use compressed air as an energy carrier with an auxiliary via a burner. Patents have been filed in this regard, but no vehicle is currently on the market with this operating principle. The proposed system uses a compressed gas (here ambient air) as an energy source to produce, by expanding this gas, mechanical energy that drives a crankshaft, such as an internal combustion engine. The temperature of the air expelled by the exhaust pipe is between 0 and 15 °C, which could be used for indoor air conditioning in hot countries. The initial theoretical energy content is 14.3 kWh (300 l at 300 bar). First, the loss that occurs due to the 300 to 20 bar pressure regulator is removed (unused technical work), and then the operation of the motor itself in polytropic process is removed (Figure 4.30). In fact, the energy available to the wheels for propulsion does not exceed 2–3 kWh for the proposed configuration. Excluding the mass of the tank, the proposed system has a density of 135 Wh/kg. It can be seen that in this case, the available energy is limited by the maximum possible pressure of the tank (300 bar).

The engineer Louis Mékarski of Polish origin is the inventor of the compressed air tramway that bears his name. In 1875, the compressed air traction system was considered a possible variant of the Francq locomotive steam engine process for the Paris metropolitan railway. Both technologies were quickly replaced by electric traction. For transport it is necessary to have a great expansion. To make the system viable, Louis Mékarski imagined using, instead of dry air, a mixture of compressed air and water vapor during expansion to limit the temperature drop. This Mekarski-type compressed air self-propelled rail system was once favored by many urban tramway networks (Paris, Nantes, Saint-Quentin, etc.).

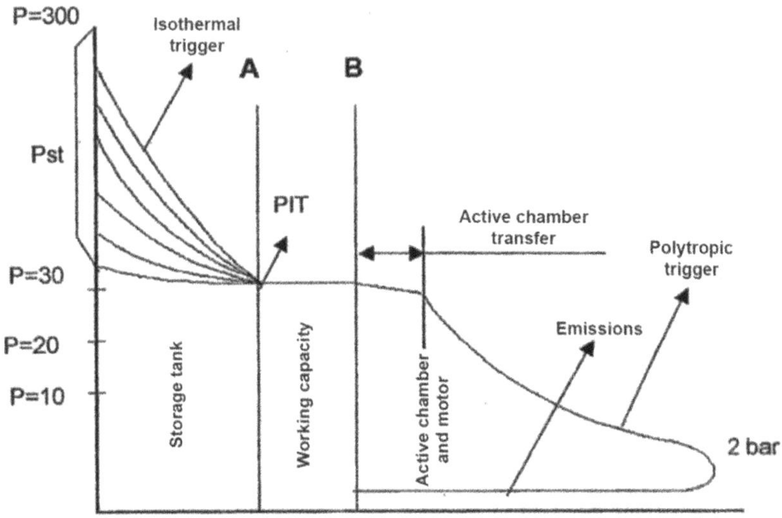

Figure 4.30. *Thermodynamic cycle of the compressed air engine cycle. Extract from the patent [NEG 08]*

Figure 4.31. *Mekarski tramway at the Gare de l'Est (Paris) in 1900*

Compressed air can be used for larger scale storage by electricity market players and particularly those associated with renewable energy. The CAES (Compressed Air Energy Storage) system uses larger natural cavities, such as caverns, or artificial cavities, such as former mines (Figure 4.32). The energy we want to store is used to compress the air, which is free and available, to pressures limited by the storage medium (strength and tightness). On the other hand, compression must be adiabatic to be effective. This constraint is technically difficult to achieve. One solution is to store the heat produced during the compression of the air in order to reuse it elsewhere through storage in a fluid, such as hot oil. In fact, the order of magnitude of the pressures is between 70 and 100 bars for very large cavities of several hundred thousand m^3. To recover energy, simply expanding the compressed air in a turbine driving an alternator suffices. It is then necessary to heat the air at the outlet. Overall, the efficiency of such a system is less than 50%.

Although this type of system has the advantage of not emitting CO_2 and having large storage and power capacities, for the time being it has not become established due to its complexity and low efficiency.

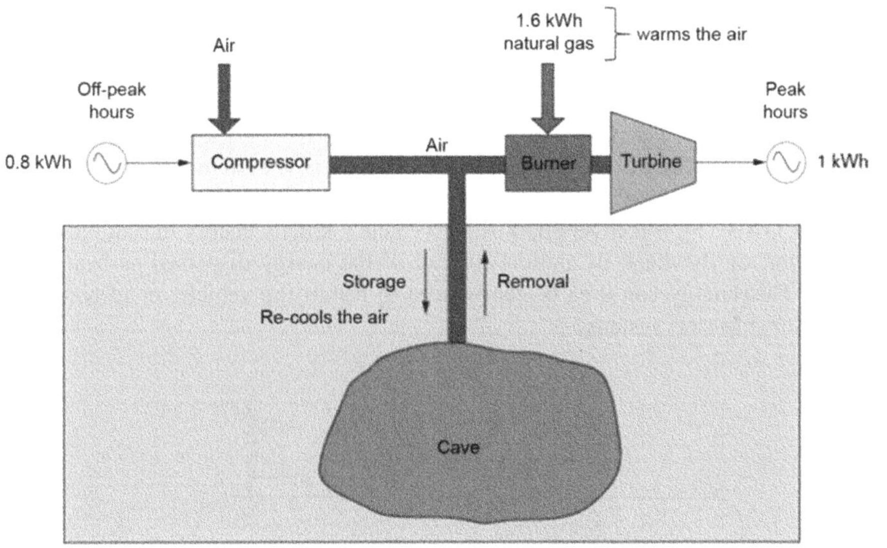

Figure 4.32. *Large scale compressed air storage system (CAES). For a color version of this figure, see www.iste.co.uk/krysinski/automotive.zip*

As shown in Figure 4.33, such a system regulates the availability of electricity by storing energy during the day when it is produced but not used and then returning several hundred MWatt for two to three hours.

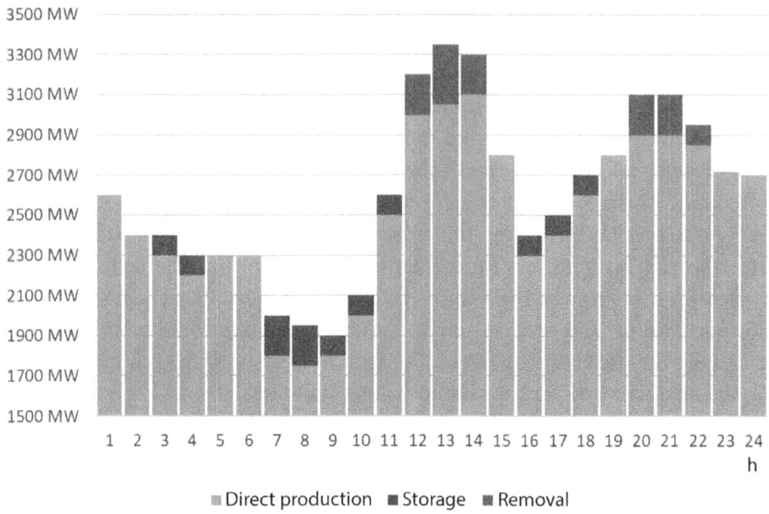

Figure 4.33. *Example of using a CAES over a 24-hour period. For a color version of this figure, see www.iste.co.uk/krysinski/automotive.zip*

4.3.3.5. *Flywheel system technologies*

The KERS (Kinetic Energy Recovery System) is a system that stores energy in kinematic form. Used in land transport such as buses [DEA 16] or hybrid cars [TIE 13, GAB 15], it recovers part of the vehicle's kinetic energy during braking to set a rotor, or flywheel, in motion, instead of the energy dispersed as heat in the brakes. This energy can then be reproduced to restart the vehicle, or to restart the engine for stop-start systems.

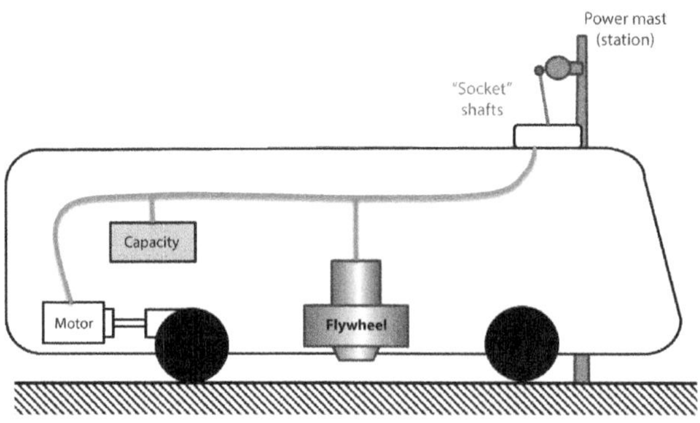

Figure 4.34. *Example of the Gyro-Bus [TRI 12]. For a color version of this figure, see www.iste.co.uk/krysinski/automotive.zip*

This type of storage is also used for sustainable power generation systems such as wind or solar power, as overproduction of electricity can be stored by this type of rotor instead of the more commonly used batteries [LEQ 04].

The main qualities of such a system are the speed of energy recovery, a number of use cycles that is much greater than batteries and an absence of toxic substances. It should be noted that weather conditions do not affect the performance of the steering wheel. Its high weight nevertheless limits it in its applications.

For a flywheel, the energy stored at time t is associated with its rotational speed and inertia. The latter is related to the density of the material and its geometry.

As a reminder, the kinematic energy of a system rotating at time t is defined in Joule by:

$$E_c\left(S/R\right) = \frac{1}{2} J\, \Omega_{S/R}^2 \qquad [4.15]$$

Thus, the energy stored by a flywheel is related to its geometry, its material (inertia J) and the rotational speed Ω reached at the moment.

For the automobile, the objective is to transfer the energy that the vehicle has at a speed V to KERS to slow down or even stop the vehicle. However, the brakes still contribute to the braking. The same applies to the rest of the power chain elements (measurable via efficiency) as well as rolling resistance or aerodynamic forces. In fact, only a fraction of the vehicle's energy is transferred to KERS during braking.

For a vehicle that changes from an initial speed V_i to a final speed V_f, the quantity stored for a steering wheel that changes from 0 to Ω is given by:

$$E_c\left(S/R\right) = \frac{1}{2} J\, \Omega_{S/R}^2 = x\%\ \frac{1}{2} M\left(V_i^2 - V_f^2\right) \qquad [4.16]$$

The choice of material, geometry and speed of rotation define the energy that can be stored. We can observe that, to store a large quantity of energy, it is in our interest to have a high rotational speed and a high inertia.

It is also necessary to take into account the other sizing criteria which are:

– avoiding the risk of the disc bursting under centrifugal force;

– avoiding the effects induced by a high peripheral rotation speed (noise, air heating, friction losses);

– finding bearing technologies that can withstand high rotational speeds.

The choice of material and geometry also has consequences on mass and size, which are also criteria to be taken into account.

The risk of explosion is the first criterion to be analyzed. For this purpose, a first simplified study makes it possible to evaluate the maximum speed not to be exceeded. The study is based on a cylinder the stress state of which is studied when it is rotating at constant speed.

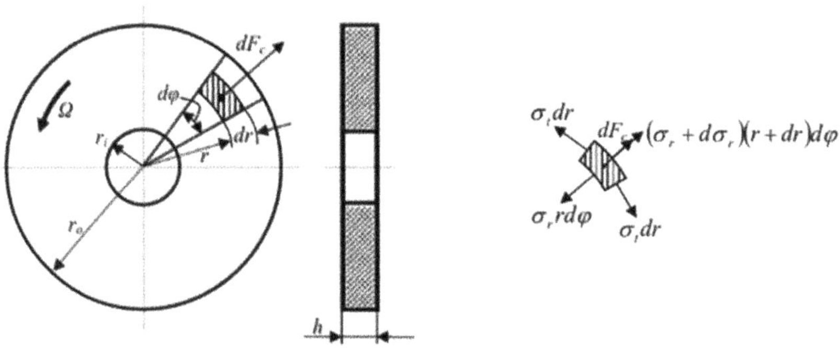

Figure 4.35. *Modeling the constraints of a rotating disc*

The stress status for a point M is then given by:

$$\sigma_{ri} = \sigma_{ro} = 0$$

$$\sigma_{ti} = \frac{3+\mu}{4} \cdot \rho \cdot \Omega^2 \left(\frac{1-\mu}{3+\mu} r_i^2 + r_o^2 \right)$$

$$\sigma_{to} = \frac{3+\mu}{4} \cdot \rho \cdot \Omega^2 \left(r_i^2 + \frac{1-\mu}{3+\mu} r_o^2 \right)$$

σ_{ri}, σ_{ro} are the normal stresses on the radius contours r_i, respectively r_o.

σ_{ti}, σ_{to} are the normal stresses on the radius contours r_i, respectively r_o.

Assuming that the sizing criterion is to avoid the disc bursting, we obtain:

$$\sigma_{max} \leq \sigma_{adm} \quad \rightarrow \quad \Omega \leq \frac{1}{r_0} \sqrt{\frac{\sigma_{adm}}{\rho}} \qquad [4.17]$$

For an imposed external radius, for example, by space requirements, it is the strength-density limit ratio that is important. The advantage is to have a high ratio such that it can be obtained for Kevlar or carbon fiber and resin composites, compared to the usual metallic materials.

Material	Kevlar	Carbon fiber and resin	Fiberglass and resin	Titanium alloy	High strength steels	Aluminum alloy
Density ρ (kg/m³)	1800	1500	2000	4500	7800	2700
Resistant to breaking (MN/m²)	4800	2400	1600	1215	1300-2100	594
$\sqrt{\frac{\sigma_{adm}}{\rho}}$	1.63	1.26	0.90	0.51	0.51	0.47

Table 4.1. *Comparison of the strength-density ratio of different materials*

It is necessary to control the energy transfer from the vehicle to the flywheel during braking, and the energy transfer from the flywheel to the vehicle during use as an energy source.

In the phases excluding energy storage and removal (driving or stopping the vehicle), it is necessary to have clutches to decouple the two systems. When storing energy, the flywheel is stationary or at a low speed and the wheels rotate at a given speed; at the end of the process, the flywheel rotates at high speeds while the wheels are at a low or zero speed. In fact, it is necessary to have a mechanism between the two systems that allows them to act as a reducer at the start of braking and as a multiplier at the end of storage. It is the control of this transmission ratio that controls the energy transfer, usually via a continuously variable transmission (CVT). To isolate the flywheel as well as possible in the phases of non-use of stored energy, it is preferable to insert a clutch between the CVT and the flywheel. Figure 4.36 shows the architecture used on the TRIVIC system [DHA 13].

This type of storage is useful in hybridization with the possibility of recovering energy during the braking phases.

For air transport, as braking phases do not exist, the use of flywheels is not suitable. It should be noted that, for rotary wings the rotor of which is sized for autorotation, the principle is similar. When the aircraft drops, the rotor is driven and stabilized at a certain rotational speed. The pilot then has this stored energy at his disposal to brake the aircraft near the ground and land while ensuring the safety of the occupants.

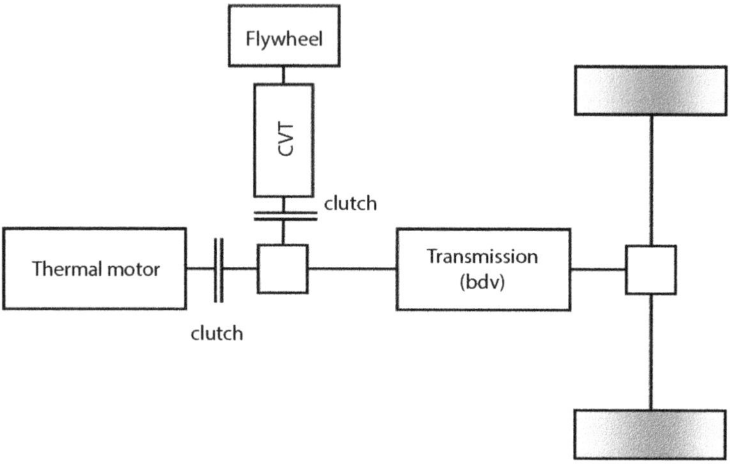

Figure 4.36. *Diagram of a possible architecture of the storage system by flywheel*

We can consider that if the vehicles are on the same network, there is a pooling of the energy recovered during braking. During periods of heavy traffic, the storage system requires high power due to traffic and very short discharge times. Flywheel storage allows this operation at high speeds for countless cycles, replacing conventional battery or supercapacitor technologies.

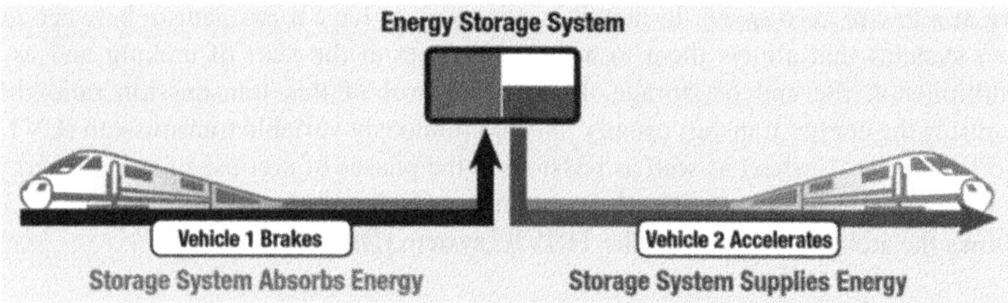

Figure 4.37. *Flywheel storage system for a metro network (source: CALNETIX Technologies)*

4.3.3.6. *Storage of energy in thermal form*

Any material has the ability to release or store heat through thermal transfer. This transfer can be achieved:

– by sensitive heat, that is, by changing the material's temperature;

– by latent heat, that is, by phase change of the material, generally solid/liquid change of a material for which the volume variation is low.

The thermal battery principle presented here and associated with the latent heat principle is a physical structure for storing thermal energy and releasing electrical energy.

The technology of these batteries uses an electrolyte of the molten salt type, which is solid and inactive at room temperature and standard pressure (0 °C, 1 bar). Its function is thus to separate the anode and cathode. To trigger the release of energy, each cell making up the battery has a pyrotechnic source (hot source) to bring the temperature to 400–550 °C. At these temperatures, the electrolyte melts, so there can be ion exchange, releasing a high power, related to the ionic conductivity of the molten salt which is higher than the sulfuric acid of conventional batteries (lead acid or other).

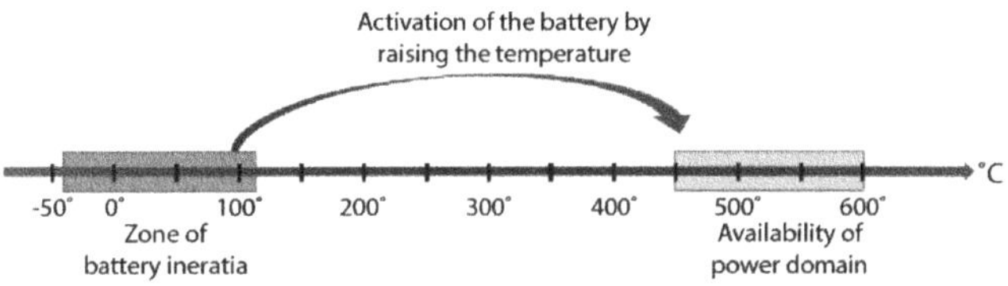

Figure 4.38. *Activation of the thermal battery. For a color version of this figure, see www.iste.co.uk/krysinski/automotive.zip*

As long as the electrolyte is solid, the battery is inert and remains inactive. This inactivated storage property has the double advantage of avoiding the deterioration of active materials during storage and at the same time eliminating the loss of capacity due to self-discharge until the battery is put into service (several years' latency). There are no special storage conditions and the batteries can be stored "ready for use".

However, once activated, the thermal battery provides energy for a very short period of time, from several seconds to more than one hour, depending on its size. Current technologies are non-rechargeable. The traditional application remains in the field of missiles and space due to the storage conditions that can be the most extreme and the unique use of this type of system.

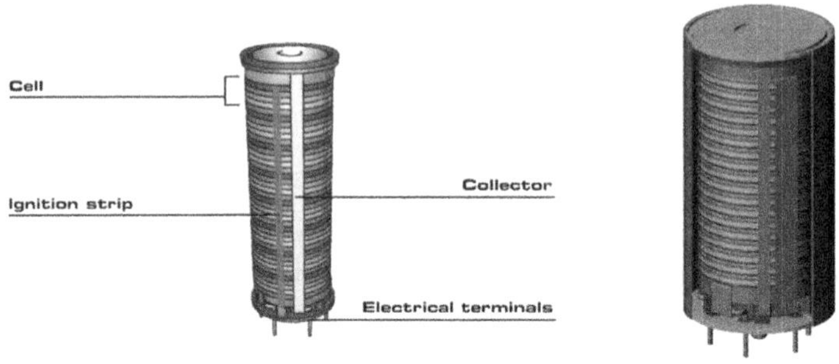

Figure 4.39. *Arrangement of a thermal battery (source: ASB Group). For a color version of this figure, see www.iste.co.uk/krysinski/automotive.zip*

This type of battery has a very high energy density (kg/Wh) and power density (kg/kW).

This technology is not yet mature enough to be used in automotive or air transport such as airplanes or rotary wings, due to the short service life, single-use nature, high surface temperatures and, to a lesser extent, the non-linearity of the output voltage. The development of these technologies requires the search for new electrolytes and other combinations of anodes and cathodes. It can be considered for emergency power supplies (failure of hydraulic generation on aircraft, engine failure on helicopters), or for use in ejector seats or canopy ejection.

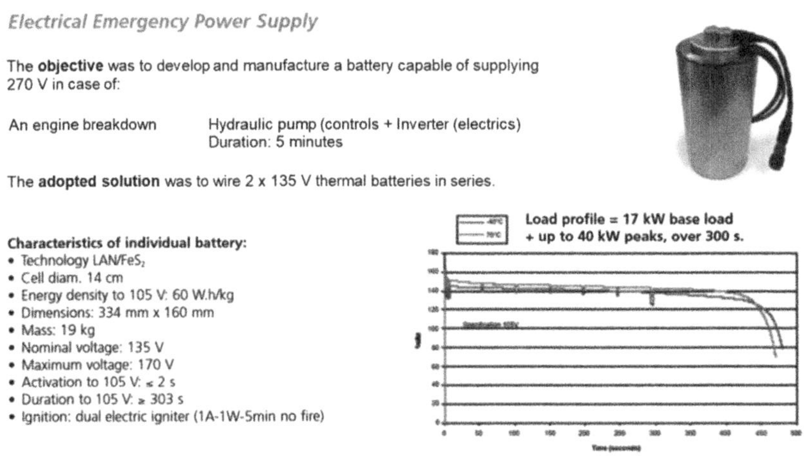

Figure 4.40. *Application example proposed by the company ASB Group (source: ASB Group). For a color version of this figure, see www.iste.co.uk/krysinski/automotive.zip*

4.3.4. *Comparison of diesel fuel tanks and automotive batteries*

It is interesting to directly compare a gasoline tank for a combustion engine and a battery for an electric motor in the context of automotive use.

The calorific value of gasoline is 35 475 kJ/l (9.85 kWh/l) and that of diesel is 38 080 kJ/l (10.6 kWh/l). The simplified thermal traction chain, shown in Figure 4.41, is considered.

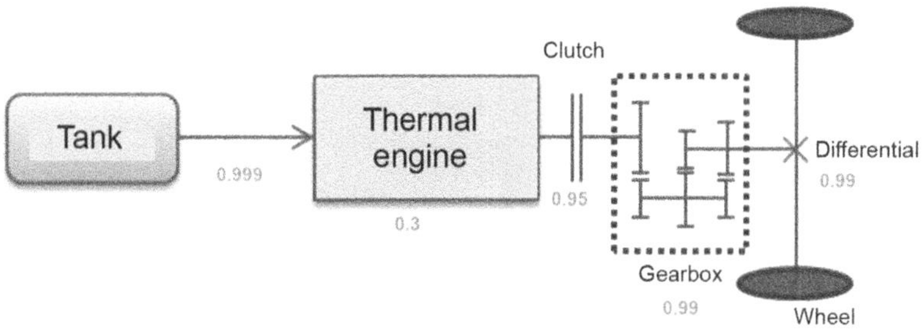

Figure 4.41. *Simplified diagram of a thermal traction chain with the main efficiencies. For a color version of this figure, see www.iste.co.uk/krysinski/automotive.zip*

Taking into account all the efficiencies that occur between the tank and the wheel, this gives an overall efficiency of 28%, for example, for a segment B vehicle. This finally results in 2.75 kWh/l of useful energy produced per liter of gasoline, the rest being converted essentially into heat. It should be noted that the density of gasoline is 0.702 kg/l and the price of a liter is around €1.55 (2014).

Now let's look at what this energy stored in a battery represents. The simplified electric traction chain, in Figure 4.42, is considered.

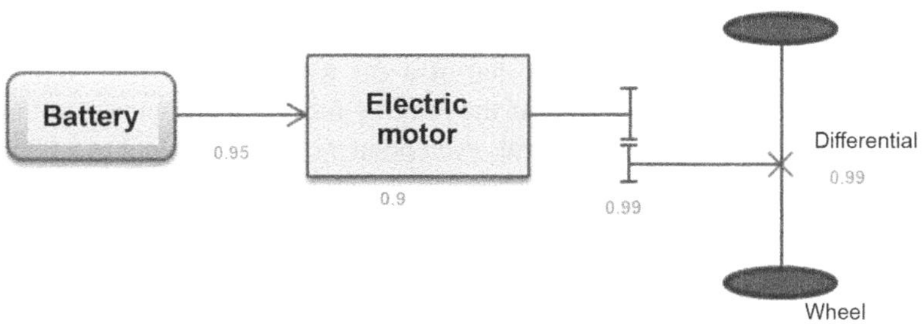

Figure 4.42. *Simplified diagram of a thermal traction chain with the main efficiencies*

Taking into account all the efficiencies that occur between the battery and the wheel, this gives an overall efficiency of 83.8%, and we finally obtain that 2.75 kWh of useful energy (equivalent to one liter of gasoline) corresponds to 3.28 kWh of battery. The properties of the batteries mean that they cannot be fully recharged or discharged, which means that, in practice, only 80% of the energy available in the battery is used over a cycle: a total of 4.10 kWh of battery to be charged is therefore required. The energy density of a lithium-ion battery is about 150 Wh/kg and the energy density by volume is about 300 Wh/l (typical data in 2014). The equivalence, illustrated by Figure 4.43, can therefore be established.

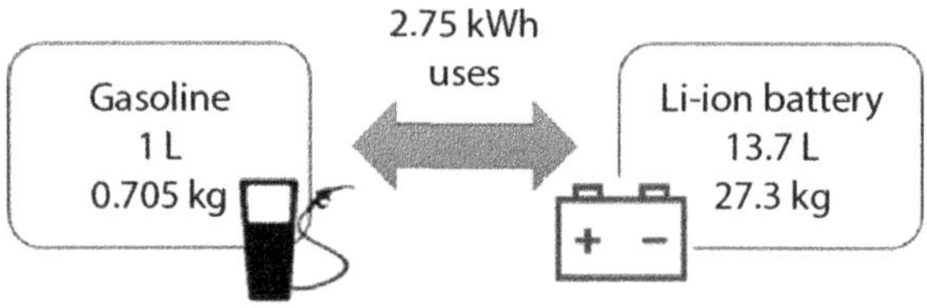

Figure 4.43. *Gasoline-lithium-ion battery equivalence*

The mass ratio is therefore 38.7 for the same useful energy.

The energy cost of these two configurations can also be compared. Based on a fuel consumption of 6 l/100 km and a fuel price of €1.40 per liter, the price of gasoline is €8.40/100 km. The price of electricity depends on charging in peak hours (12.75 ct/kWh) or off-peak hours (8.64 ct/kWh). For an electric vehicle consuming 13.6 kWh per 100 km, the cost is then €1.73/100 km or €1.18/100 km. The cost of renting the battery is about €80/month and the cost of routine maintenance is €400/year in gasoline and €200/year in electricity. It can be seen that the cost of use is not very noticeable compared to the price of electricity.

In terms of the energy price, in 2018, the cost of using an electric car (off-peak OP or peak P) becomes lower than that of a car with a combustion engine from 10,000 km/year. On the other hand, if the price of fuel were to increase to the level of €2.06 liter of diesel, the annual break-even point would be lowered to 5,000 km.

The previous comparison is based only on the available energies. However, we can also look at the distances that both vehicles can travel with this useful energy available to the wheels. The addition of a large and heavy battery will tend to increase power demands, and therefore energy requirements, compared to a conventional thermal vehicle. However, a major advantage of the electric vehicle is that it can recover a large part of the energy lost during braking. In overall terms, the vehicle is often less energy consuming and can therefore travel longer distances with the same amount of energy.

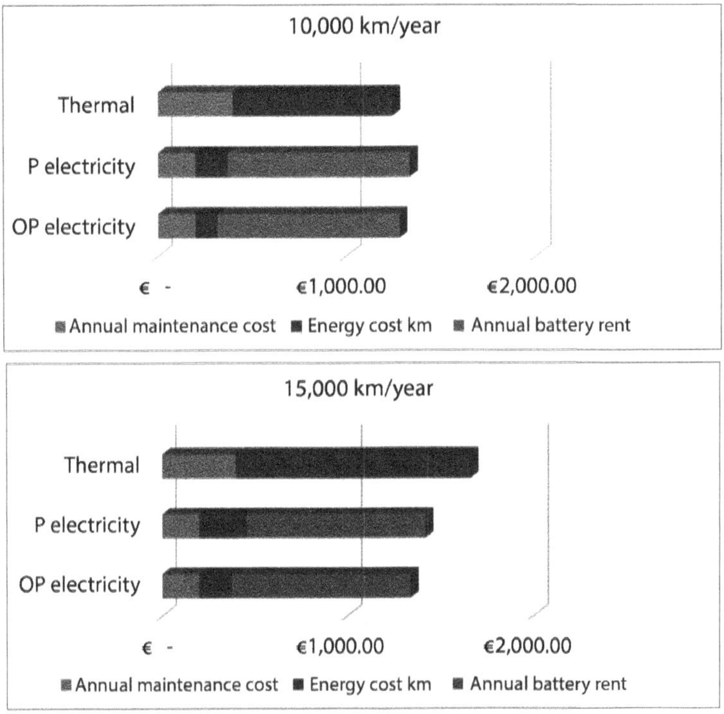

Figure 4.44. *Comparison of the cost of using gasoline and electricity. For a color version of this figure, see www.iste.co.uk/krysinski/automotive.zip*

If we consider a 1,000 kg thermal car (Crr = 0.008, SCx=0.61 m^2) and the same electric vehicle of the same mass, we can estimate the distance that can be covered with these 2.75 kWh of useful energy as defined in Table 4.2.

2.75 useful kWh	Gasoline	Electricity	Saving
Urban	16.36 km	20.66 km	26 %
Roads	26.48 km	32.5 km	22.7 %
Highways	18.75 km	20.29 km	8.2 %

Table 4.2. *Comparison of the distances that can be covered (with or without energy recovery) on the Artemis urban/road/highway cycle*

It is therefore clear that the ability to recover and store energy plays an essential role. However, it should be kept in mind that the addition of a battery significantly increases the vehicle's mass, which results in a lower saving on out-of-town trips. With a wheel energy of 2.75 kWh (corresponding to 27.3 kg of a lithium-ion battery) and a consumption of 150 Wh/km, the range is 18.3 km. With a 273 kg pack, you can travel 183 km. For example, after a first version, at 22 kWh for a weight of 290 kg and a range of 235 km, Renault proposed a second version, of a 305 kg battery for the ZOE car with an energy of 41 kWh [BER 17]. To increase the battery's energy performance, new cell materials were developed and the amount of active ingredient increased, while preserving space. By optimizing the cell design, 10% of active surfaces have been added. This allows a better exchange between anodes and cathodes. The manufacturer has a range of 400 km based on an NDEC cycle, and more than 300 km for real use in summer and 200 km in winter.

This simple calculation shows the current limit of electric vehicles related to the current battery capacities. Figure 4.45 shows the integration of the battery into PSA's electric vehicle platform. It will be difficult to integrate more batteries of the same technology without degrading the vehicle's habitability performance.

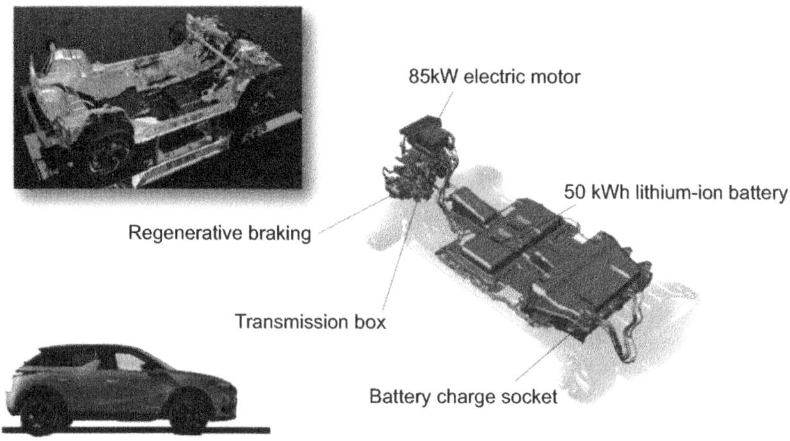

Figure 4.45. *Electric vehicle and its batteries (source: PSA). For a color version of this figure, see www.iste.co.uk/krysinski/automotive.zip*

This simple reasoning means that, in a context where CO_2 emissions were not taken into account, the automotive industry has turned towards combustion engines since energy storage is much less voluminous and much cheaper. In the current state of battery technology, electric vehicles cannot replace combustion engine vehicles for conventional uses. This can only be considered for very limited uses.

4.4. Conclusion

After several decades of using fossil fuels, such as gasoline, diesel or kerosene, it is now necessary to find clean and sustainable alternative energies. In the field of transport, to ensure autonomy, it is necessary to store energy. Different solutions described in this chapter, such as electrochemical batteries, hydrogen, compressed air or flywheels, can be an alternative or complementary solution through hybridization to the use of fossil fuels.

Several criteria can be used to compare them. The Ragone diagram [CHA 13] allows the different technologies to be located with respect to the mass power (W/kg) as a function of the mass energy (Wh/kg) (Figure 4.46). This diagram shows the interest of developing supercapacitors and flywheels for their high mass power capacity despite their low stored energy quantity capacity. It can also be seen that due to their high mass energy capacity, fossil fuels or fuel cells are a solution adapted to the demand for large quantities of energy to be stored, as is the case for systems requiring a large autonomy.

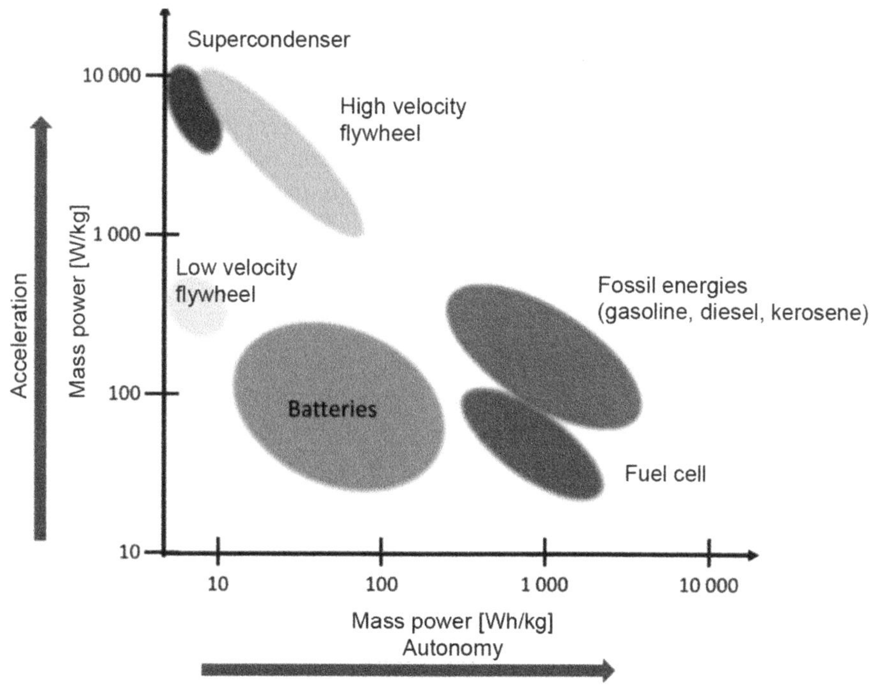

Figure 4.46. *Ragone diagram. For a color version of this figure, see www.iste.co.uk/krysinski/automotive.zip*

Compared to stationary storage systems that are dedicated to storage in support of renewable energy production sites, embedded storage systems (automotive, aeronautics, rail, etc.) are small capacity and low power, integrated into a mobile system. When selecting the type of storage, it is also necessary to use the time criterion associated with the storage system's charging and discharging time. For very short durations of a few milliseconds to a few seconds, supercapacitors or flywheels are well adapted, thus being able to bring an extra boost to another type of energy source (fossil or electrochemical).

For storage systems for mobility, in the short and medium term, electrochemical systems seem to be the most promising option for on-board storage in electric or plug-in hybrid vehicles. Lithium-ion technology is currently in the best position to meet this challenge; however, high temperature technologies may have an opportunity for captive fleets. In the longer term, magnesium and metal-air batteries could be technically and economically attractive. Similarly, research and development activities are pushing for the use of electric cars for their ability to store electrical energy, to regulate energy consumption and, thus, to smooth out peaks and troughs in production linked to renewable energies (wind, solar). The car network will thus contribute to better use of the electricity grid and reduce maintenance costs.

The use of hydrogen is in progress but it does not seem that it will be ready in the near future, due to the various technical barriers, linked to the volatility of hydrogen and its reactivity with oxygen, which require significant financial investment in development.

5

Hybridization

In the field of transport, hybrid propulsion is, by definition, a technology that uses several energy sources. The most classic historical combination is the use of a thermal energy source (gasoline, diesel or kerosene tank) with an electrical energy source (battery). However, other combinations can be used, such as human energy combined with electrical energy for electric bicycles or, thermal or electrical energy combined with mechanical energy (flywheel). These different sources can then be used separately or in combination.

The presence of several energy sources introduces additional technical complexity and often a higher cost, compared to the traditional single energy source solution. This complexity is offset by the fuel consumption or CO_2 emission savings that hybrid propulsion can bring. Hybridization can be seen either as a solution to increasing the initial energy system's autonomy or as a solution to improving the energy consumption of the means of transport. This can be done by making each source work at its optimal point of performance.

Autonomy increase is achieved by using a serial architecture of energy systems. This is the case – for example – for certain electric powered cars which, in addition to a battery, use an auxiliary system (range extender), the purpose of which is to extend the range by using another type of storage. There is no mechanical connection between the auxiliary system and the propulsion system (wheels or propeller). The most traditional is the use of a combustion engine combined with an electric generator and a fuel tank (Figure 5.1). The energy efficiency of the propulsion system is achieved when the combustion engine operates at low power. For submarines combining diesel and batteries, the 100% electric mode is used in submerged mode due to the lack of oxygen required for diesel engines. They then need to have periods when the engines can recharge the batteries using generators, coupled with diesel engines.

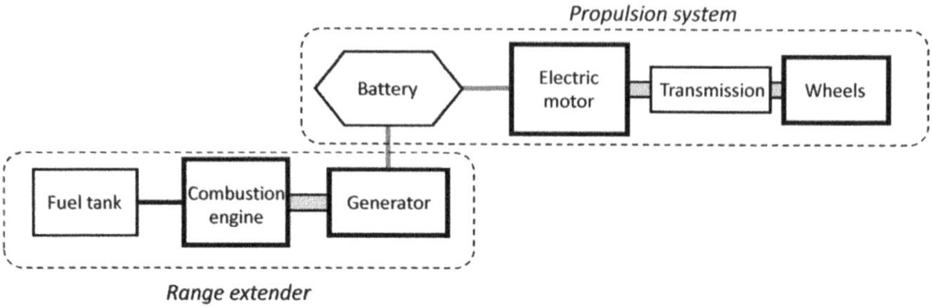

Figure 5.1. *Principle of the range extender. For a color version of this figure, see www.iste.co.uk/krysinski/automotive.zip*

The second interest in hybridization is the improvement it can make to energy efficiency by using a propulsion system distributed between several energy systems. The association can be parallel or, more complex, with power derivation. The most common association is the combination of a combustion engine and an electric motor. Energy efficiency is linked to the fact that, in mixed mode, the combustion engine works in the torque-speed ratio, where the efficiency is most advantageous.

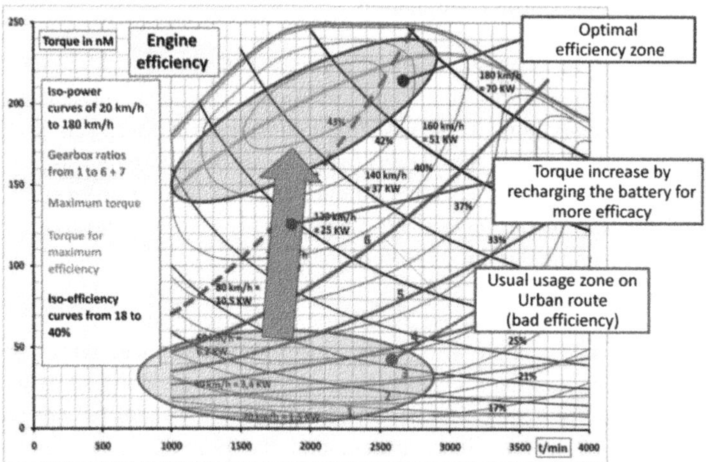

Figure 5.2. *Ideal operating mode of the combustion engine in hybrid mode*[1]*. For a color version of this figure, see www.iste.co.uk/krysinski/automotive.zip*

1 Source: http://www.8-e.fr.

For vehicles – particularly in urban cycles – or rail traction, the phase of recovering the vehicle's kinetic energy during braking is an additional advantage since it increases the vehicle's range without, oversized batteries (Figure 5.3).

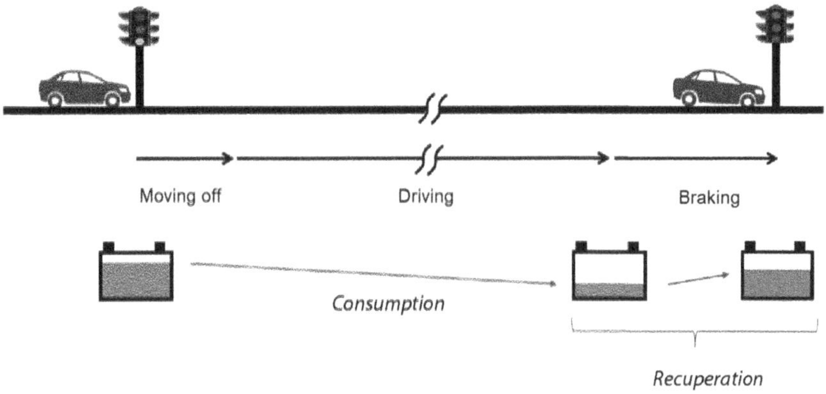

Figure 5.3. *Operating phase in urban areas*

For aircraft, hybridization can be useful for improving helicopter performance under emergency and transitional conditions, for example autorotation assistance and twin-engine take-off. The auxiliary source (electric) is then used for a very short time to supplement or complement the main energy source (combustion engine). The only way for an aircraft to recover energy may be during the landing phase. For example, the energy produced by an aircraft's braking system during landing – currently wasted in heat produced by the friction of the disc brakes – would be captured and transformed into electricity by motor-generators, integrated into the landing gear. It would participate in a partial recharge of the batteries. As these transitional phases are less frequent than for a car or train, these solutions remain less studied.

5.1. Hybridization of electric motors: range extender

Range extender systems on an electric vehicle consist of adding a second energy source (e.g. fuel energy) to the existing source (e.g. electrochemical energy contained in the battery). This will make it possible to:

– maintain the advantages of electric traction: the electric solution is known for its driving comfort. Indeed, whether in terms of silence or directly available torque, electric traction is perfectly in line with what manufacturers and drivers are looking for comfort of ride. It also has the considerable advantage of not producing any direct CO_2 emissions;

– limit disadvantages: the major problems linked to electric vehicles are the high price and the lack of range. However, these two parameters are directly related (but inversely) to the amount of batteries available on the car. Thus, a high-capacity battery enables a long range, but the battery is one of the most expensive components of a vehicle and therefore plays an important role in the price.

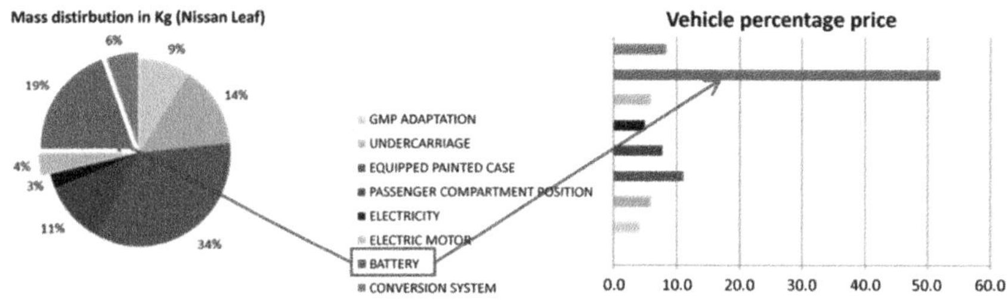

Figure 5.4. *Battery's share of the mass and price of the electric vehicle (source: Nissan Leaf 2012). For a color version of this figure, see www.iste.co.uk/krysinski/automotive.zip*

The addition of a range extender allows a significant reduction of the battery capacity, while maintaining the possibility of covering long distances and thus simultaneously reducing the cost, the mass and the risk of having an empty battery in the middle of nowhere.

It is important to keep in mind that a range extender is first and foremost an electric vehicle.

5.1.1. *Application examples for the automotive industry*

We propose here an analysis of the range extender's operating principle – and its efficiency – through the example of an electrically powered motor vehicle, each wheel powered independently (Figure 5.5). An architecture with range extender by adding a combustion engine then allows us to increase the autonomy without using an oversized battery.

It is worthwhile analyzing the performance of this architecture through the vehicle's consumption.

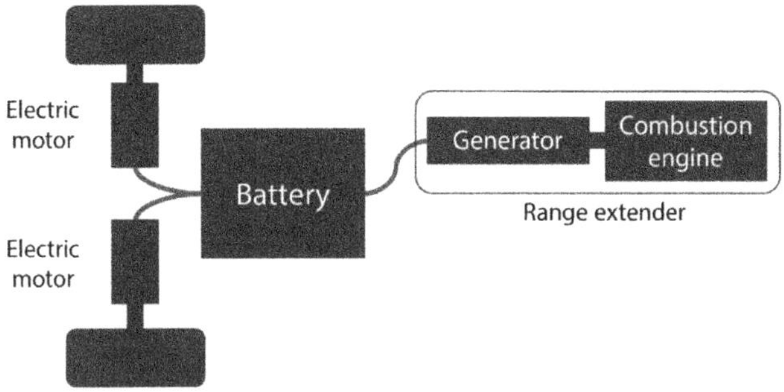

Figure 5.5. *Example of range extender application on a vehicle*

Two energy tanks are the battery and the fuel tank. In the braking or lowering phase (engine brake), the battery can be recharged.

In order to analyze a range extender solution, we propose studying several transmission architectures: 100% electric propulsion, 100% thermal propulsion and a thermal range extender to increase the autonomy of an electric propulsion.

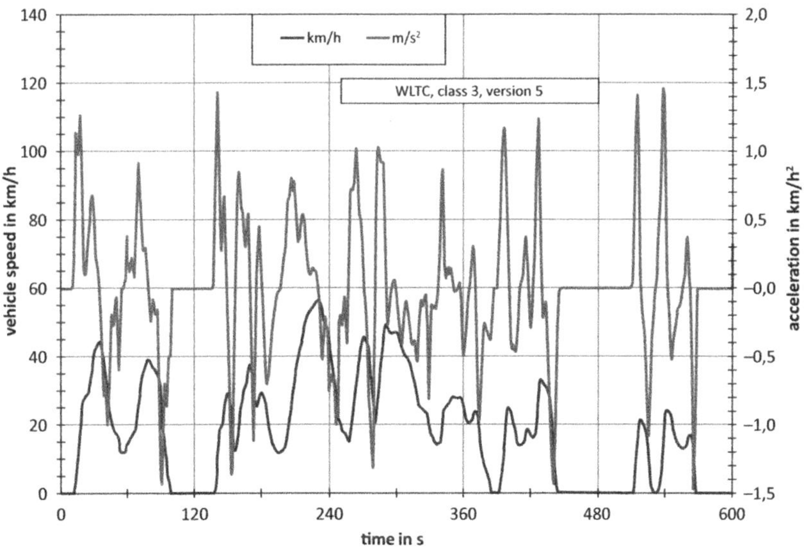

Figure 5.6. *WLTC approval cycle (phase 1). For a color version of this figure, see www.iste.co.uk/krysinski/automotive.zip*

The specifications for the use of the vehicle presented as an example are the same for all three cases: the vehicle has a mass of approximately 1,800 kg and the operating cycle used is based on phase 1 of the WLTC urban cycle. It represents a basic cycle of 600 s which corresponds to a traveled distance of 3 km. Figure 5.6 shows the change in speed and acceleration imposed on the vehicle over the considered cycle.

During this phase, it is assumed that the journey is made on a flat road. The power required is then related to the aerodynamic resistance, rolling resistance related to tires and acceleration effects. The useful power *P(t)* is defined from the speed of the vehicle V(t) and the tractive effort F_T(t) required for drag:

$$P(t) = F_{\mathrm{T}}(\mathrm{t}) \times \mathrm{V}(\mathrm{t}) \tag{5.1}$$

If we isolate the vehicle, the fundamental principle of dynamics makes it possible to express the tractive effort by:

$$F_{\mathrm{T}}(\mathrm{t}) = M\frac{dV}{dt} + F_{\mathrm{tires}} + F_{aero} + \mathrm{F}_{\mathrm{gravity}} \tag{5.2}$$

Each of these efforts can be expressed by:

$$\begin{cases} \mathrm{F}_{\mathrm{tires}} = \mathrm{C}_{\mathrm{rr}}\ \mathrm{M}\,\mathrm{g} \\ \mathrm{F}_{\mathrm{aero}} = \dfrac{1}{2}\rho\,\mathrm{S}\,\mathrm{C}_{\mathrm{x}}\ \mathrm{V}^2 \\ \mathrm{F}_{\mathrm{gravity}} = 0 \end{cases} \tag{5.3}$$

In the proposed configuration, a rolling resistance Crr of 0.001 (dimensionless) and a penetration coefficient SCx of 0.75 m^2 will be assumed.

Assuming an overall efficiency η and a kinematic ratio of K (including gearbox, differential and wheel geometry), the engine torque C_m and the engine speed Ω_m can be expressed as:

$$\begin{aligned} Cm(t) &= \frac{1}{\eta K} F_{\mathrm{T}}(\mathrm{t}) \\ \Omega_{\mathrm{m}} &= \mathrm{K}\ \mathrm{V}(\mathrm{t}) \end{aligned} \tag{5.4}$$

5.1.1.1. *Case of 100% electric propulsion*

The proposed example uses a synchronous electric motor with 64 kW of power (88 hp) and a maximum torque of 220 Nm. This type of motor has a speed range from 0 to 11,000 rpm. Consequently, a fixed transmission ratio is sufficient to cover the passenger vehicle's change in speed. The transmission ratio, including gearbox and differential, is 9.32. The wheels used are 185/65/15 in size. It is assumed that the transmission has an overall efficiency of 0.95. The use of the WLTC cycle shown in Figure 5.6 and the configuration studied, make it possible to analyze the electric motor's areas of use on the torque/speed graph.

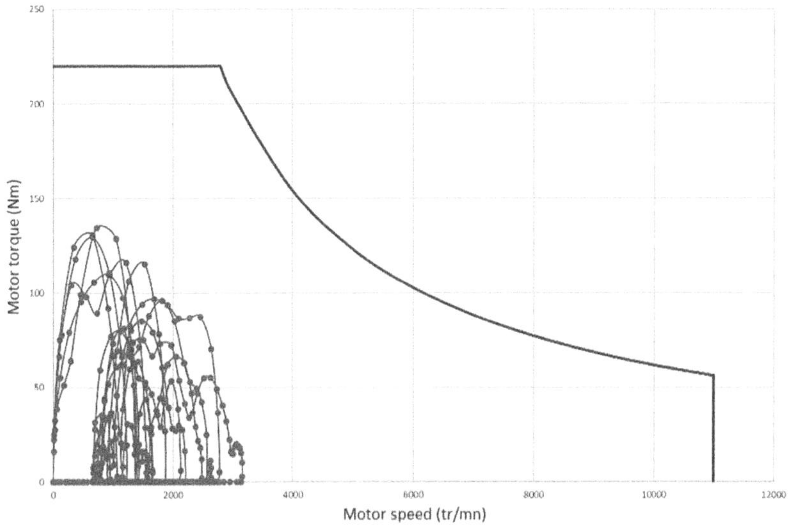

Figure 5.7. *Use of an electric motor during the WLTC cycle (phase 1). For a color version of this figure, see www.iste.co.uk/krysinski/automotive.zip*

By using the formula [5.1], it is possible to trace the change in motor power used during this cycle (Figure 5.8). The integration of this power over the cycle time then makes it possible to define the energy consumed during a cycle, which corresponds to 0.346 kWh. A 41 kWh battery, which would correspond to a weight of 305 kg, would make it possible to cover approximately 366 km (118 cycles).

The advantage of the battery is that it is rechargeable. It is then possible to evaluate the power lost during the braking phases of the WLTC cycle and to quantify, taking into account the efficiency, a fraction of the energy that can be stored to increase the overall range of the vehicle. The recoverable energy on the

WLTC cycle (phase 1) with an efficiency of 0.3, leads us to restore 0.088 kWh, and to have a range of 492 km, an increase of 34%.

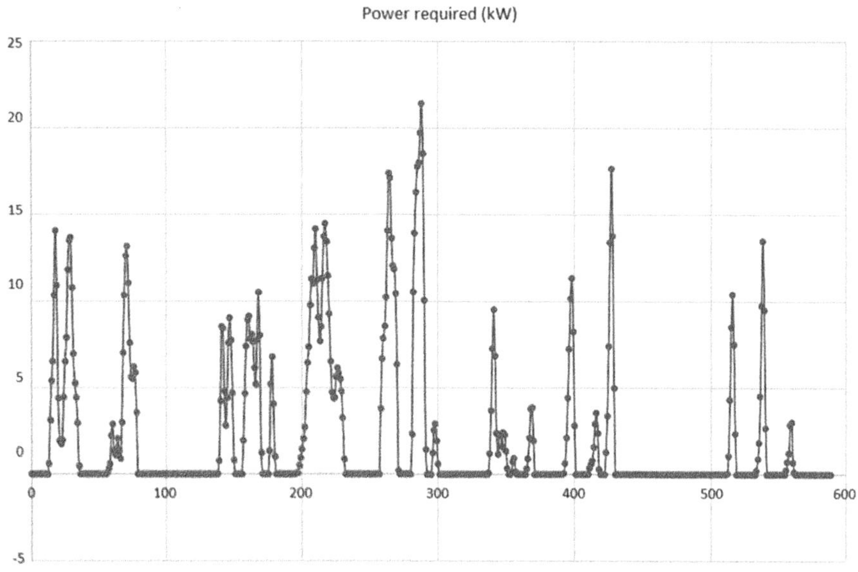

Figure 5.8. *Evolution of motor power during the WLTC cycle (phase 1)*

5.1.1.2. *Case of a 100% thermal propulsion*

For example, a 65 kW (89 hp) combustion engine is used. It has a maximum torque of 175 Nm at 2,500 rpm. The engine is associated with a gearbox and a differential. The gear ratios are respectively 3.73 in 1st, 2.05 in 2nd and 1.32 in 3rd. The differential ratio is 5.18. It is assumed that the wheels are identical to those in the previous case (185/65/15) and that the transmission has an overall efficiency of 0.95. The use of the WLTC cycle shown in Figure 5.6 and the configuration studied, make it possible to analyze the areas of use of the combustion engine on the torque/speed graph, taking into account changes in speed ratio so that, over the cycle, the engine speed remains below 3,500 rpm.

By using the formula [5.1], it is possible to trace the change in the transportation powertrain used during this cycle (Figure 5.8). The integration of this power over the cycle time then makes it possible to define the energy consumed over a cycle that corresponds to 0.346 kWh. It can be seen that the engine is working for this cycle, mainly in an area where performance is poor. For a fuel energy density of 34 MJ/l, the 42-liter tank contains 1,428 MJ. For an average engine and transmission efficiency of about 24%, the energy available to the wheel is 342 MJ. This WLTC cycle (phase 1) requires, from a theoretical point of view, 1.25 MJ for a distance traveled of 3 km. The 42-liter tank would therefore allow 822 km, which would correspond to a consumption of 5.1 l/100 km.

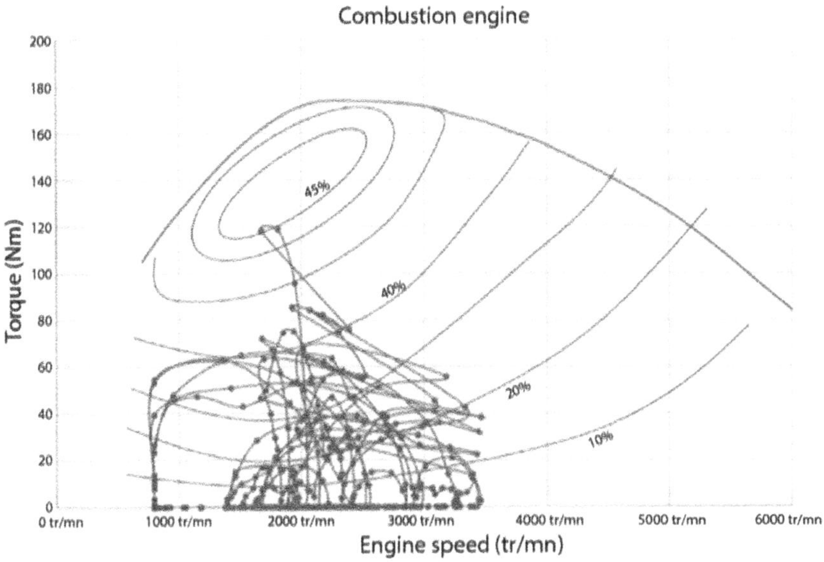

Figure 5.9. *Use of a combustion engine during the WLTC cycle (phase 1). For a color version of this figure, see www.iste.co.uk/krysinski/automotive.zip*

5.1.1.3. *Case of an electric propulsion system with range extender*

As a first step, it is possible to imagine using the previous combustion engine to increase the range of the electrically propelled vehicle (see Figure 5.1). The advantage is then to use the combustion engine at an operating point where the efficiency is the best.

Assuming an efficiency of 40%, including engine efficiency and a generator, a 10-liter tank can store 37.8 kWh in the battery. Taking the same characteristics of the electric motor form of engine power previously proposed, the additional range by the range extender is 337 km, which can be added to the 366 km range provided by the initial 41 kWh of a full battery.

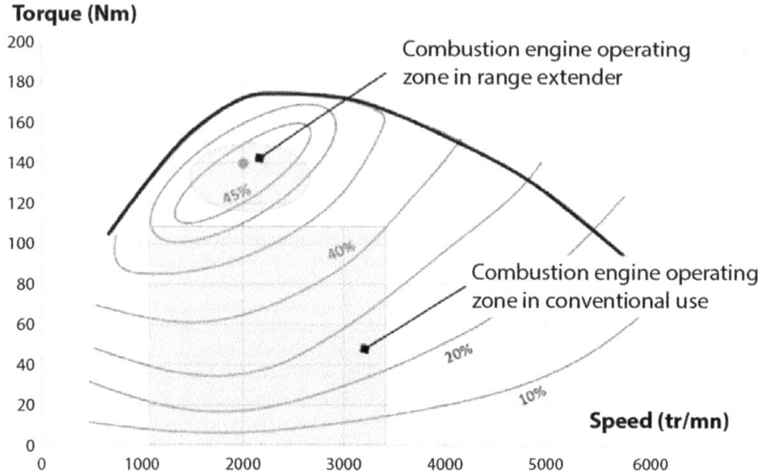

Figure 5.10. *Operating area of the combustion engine in range extender compared to conventional use*

This solution would add a significant mass to the basic transmission. With a density of 0.135 kWh/kg for a battery, 4 kW/kg for the electric motor, the electric part can be estimated at 320 kg. With a density of 0.702 kg/l for fuel and 1.5 kW/kg for the combustion engine, the thermal part can be estimated at 30 kg. A total of 350 kg.

Since the battery corresponds to 86% of the weight of the transportation powertrain in this configuration, another one must be considered. The system can offer a 4 kWh battery for a range of 80 km in optimization, with the possibility of recovery during braking phases. The combustion engine may be undersized in power since it is not used directly for propulsion. A 35 kW motor can be envisaged, the remaining tank 42 l. In this configuration, the system's mass would be 46 kg for the electrical part and 53 kg for the thermal part, forming a total of 98 kg.

In the city, the range of 80 km is more than enough for 100% electric use. If a longer range is required, the range extender takes over to use the 822 km range provided by the 42 l tank.

5.1.1.4. *Overview*

The previous results can be summarized in Table 5.1.

		Range extender 1	Range extender 2	100% electric	100% thermal
Electrical	Battery	41 kWh	4 kWh	41 kWh	
	Engine	64 kW	64 kW	64 kW	
	Electrical autonomy	492 km	80 km	492 km	
Thermal	Tank	10 l	42 l		42 l
	Combustion engine	65 kW	35 kW		65 kW
	Thermal autonomy	337 km	822 km		822 km
Balance	Total autonomy	829 km	902 km	492 km	822 km
	Total weight	393 kg	98 kg	320 kg	73 kg

Table 5.1. *Comparison between the range extender and 100% electric or 100% thermal system*

This mode of operation is offered by some car manufacturers and is directly integrated into the vehicle. This is the case with the Chevrolet Volt, or some versions of the BMW i3, which can have an optional range extender. The latter offers a 22 kWh battery and a 170 hp electric motor, with the option of integrating a 647 cm^3 two-cylinder engine with 38 hp (in kW) and a 9 l tank to double the range.

Some manufacturers propose to position the range extender in a trailer to extend the range of any electric car. EP Tender offers an auxiliary system (APU or Auxiliary Power Unit) mounted on a small 200 kg trailer with a 35 kW motor and a 35 l tank for a possible range of 600 km.

5.1.2. *Application examples for aeronautics*

The range extender concept can be extended to aeronautics.

For an aircraft, it is possible to consider electric propulsion associated with a propeller. Either the battery life would be increased by an auxiliary source using a combustion engine, or the generator would directly power the electric motor.

For example, at the 2011 Paris Air Show, a hybrid-powered aircraft was presented, consisting of a 70 kW electric motor that alone propelled the aircraft, and an auxiliary system consisting of a 30 kW Wankel engine supplying electricity to the engine or recharging the batteries (Figure 5.11). Since aircraft do not have usable energy recovery phases, the battery is necessary for aircraft accessories (pump, etc.) and for power peak calls, such as during take-off phases.

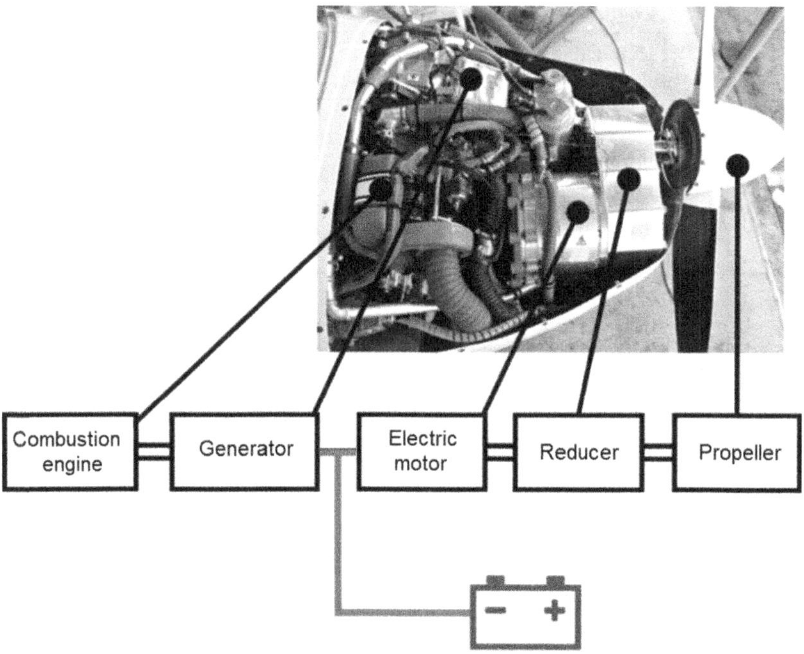

Figure 5.11. *Hybrid powered aircraft, EADS-Siemens project (source: Siemens). For a color version of this figure, see www.iste.co.uk/krysinski/automotive.zip*

During the take-off or landing phases, the required power is higher (Figure 5.12); the Wankel engine creates electricity directly from the battery's energy. In flight, the engine supplies the propeller directly ("generator set" mode). In flight phases where the energy required for the propeller is less, the combustion engine recharges the battery.

It may be interesting to compare the range extender architecture for this type of product on a one-hour mission, with 100% electric or 100% combustion formulas. A 100% electric formula would require, for this type of aircraft, 80 kW of power and a battery with a capacity of 60 kWh. With characteristics of 4 kW/kg the engine would be about 20 kg and the battery for 135 kWh/kg would be about 450 kg. The total weight of the 470 kg transportation powertrain is then very significant.

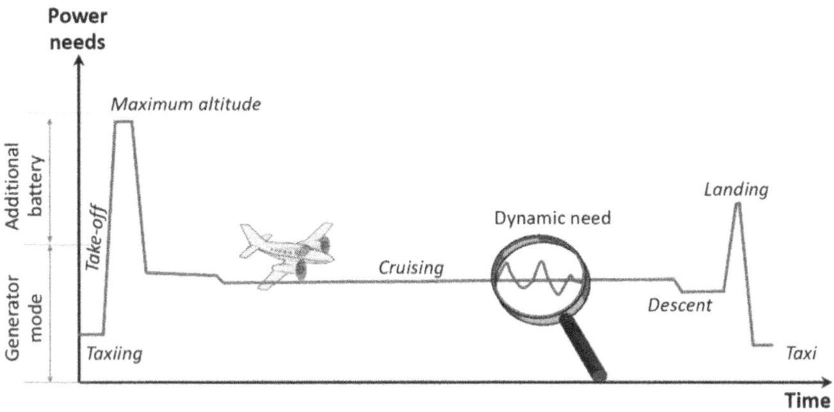

Figure 5.12. *Evolution of power requirement during an aircraft mission*

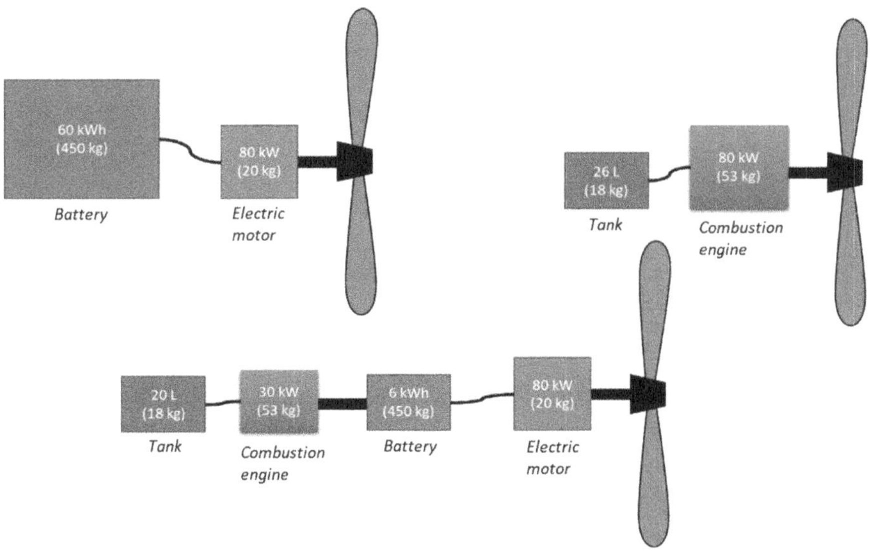

Figure 5.13. *Comparison of 100% electric, kerosene and range extender solutions. For a color version of this figure, see www.iste.co.uk/krysinski/automotive.zip*

For a 100% combustion formula and for the same power, the engine would be 53 kg (1.5 kW/kg). For the same autonomy, it would be necessary to have a 26-liter tank weighing 18 kg (0.700 kg/l) with a total weight of 71 kg.

If we consider a hybrid formula with an 80 kg electric motor but a small 6 kWh battery, combined with a 30 kW combustion engine and a 5 l tank, the combination enabling autonomy of one hour, we obtain a total weight of 88 kg. This last configuration is a little heavier, but it has the advantage of being able to use 100% electric during the take-off and landing phases. It thus makes it possible to have a lower noise impact for these flight phases and to obtain a better overall efficiency, since the combustion engine can then work at an optimal operating point.

5.2. Hybridization of combustion engines: improving energy efficiency

5.2.1. *Interest in parallel hybridization*

5.2.1.1. *The case of the automobile*

By definition, a hybrid motor vehicle uses at least two different energy sources to achieve movement. To date, a hybrid vehicle has been defined mainly as a car that uses a fossil fuel via an internal combustion engine (ICE) and another complementary energy source that may require another engine (converter). For the latter, several technological families can be distinguished (Figure 5.15): electrochemical (or battery) – as for Toyota's Prius – pneumatic hybridizations combining energy storage in the form of compression of a gas such as PSA's Hybrid Air system, and mechanical, combining mechanical storage in the form of a flywheel and a combustion engine such as Volvo's KERS.

By extension, we can also define a hybrid transportation powertrain for a 100% electric propulsion system, the storage of which corresponds to the combination of two different storage modes: electrochemical and supercapacitor, for example.

The development of a hybrid system poses the problem of sizing the storage system (battery, flywheel or pressure tank) and the mechanical system, to convert the stored energy into mechanical energy (converter or motor). It also requires the analysis of the different operating modes, leading to the definition of an energy management strategy to optimize consumption, while maintaining acceptable driving performance.

In the case of land vehicles, one of the advantages of associating another energy storage system with a combustion engine and its fuel is to be able to recover kinetic energy in braking phases. The development of this phase and the sizing of the associated components are equally important.

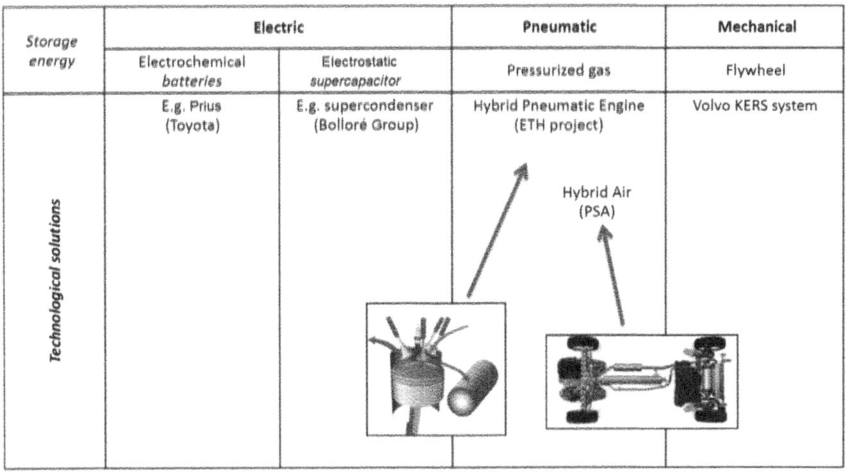

Figure 5.14. *Different hybridization technological solutions. For a color version of this figure, see www.iste.co.uk/krysinski/automotive.zip*

During the vehicle's braking or deceleration phase, the system's role is to recover a part of the kinetic energy, usually lost in heat, and store it through a reversible converter in a reusable form (battery, supercapacitor, compressed gas, flywheel). In addition to the energy gain, such a system – called the "dynamic brake" – makes it possible to limit the wear of friction brakes.

Nevertheless, it will be necessary to take into account the following elements:

– the effect of the dynamic brake is reduced when the speed becomes low;

– the amount of kinetic energy that the system can convert in a given time is limited and therefore also limits the braking capacity;

– in the case of regenerative braking, a limit is imposed by the loading state of the storage device (if full) and its maximum capacity;

– the brake distribution on the axletrees must be taken into account;

– depending on the architecture, there may be a brake failure when changing gears.

5.2.1.2. *The case of aeronautics*

For aircraft, the interest in hybridization is different from that of the automobile. The energy recovery phases, as found in land transport, do not exist. The landing phase that would allow energy to be recovered is not large enough to justify a hybrid solution on its own.

Similarly, the juxtaposition of complementary energy – electrochemical or hydrogen – with fossil, kerosene or diesel energy, is only advantageous in particular cases of aircraft operation.

Transition periods are thus phases where hybridization can be of interest without there being too much mass penalty. For example, for small helicopters with a single turbine, for safety in the event of an engine failure, the pilot must use the rotor's autorotation capability to land without damage, leaving him with little choice over his landing area. Eurocopter (Airbus Helicopters) has proposed an electric propulsion system to increase the maneuverability of this type of aircraft in the event of an autorotative landing. In the event of a turbine failure, the electric motor drives the rotor, making it easier for the pilot to control the aircraft during descent and to land it safely. This system was tested in 2011 on an AS350 Squirrel.

Hybridization can be introduced on helicopters as an equivalent to the stop-start feature of a car. On a twin-engine helicopter, fuel consumption is high due to the continuous operation of both engines. In the event of an economical cruise flight, the power of a single engine is sufficient. It may be considered to keep the second one rotating at idle. An electric motor would allow it to be restarted in a few seconds if power is required. On an aircraft such as the Airbus Helicopters Racer, where the rotor is slowed down at high speed, the power required for high-speed cruising is lower due to lower rotor drag. Single-engine operation at full power is more economical than with two engines at sub-engine speed.

The transitional take-off period is very power consuming. Electric power assist would provide additional power to a turboshaft engine. Generally, the engine is sized for high power and is therefore less efficient in cruising. On the other hand, sizing the combustion engine for cruising would reduce fuel consumption during a mission.

5.2.2. *Classification of electrical hybridization: the case of the automobile*

5.2.2.1. *Different types of hybridization*

The majority of manufacturers now offer hybrid cars in their ranges. The technical solutions, operating modes and therefore the objectives can be very different. It is then conventional to define a classification according to the degree of hybridization, with respect to an initial fuel source.

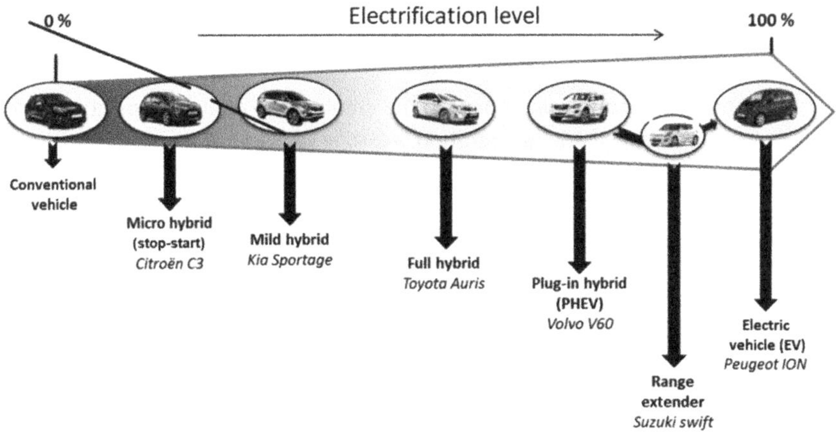

Figure 5.15. *Classification of hybridization systems*

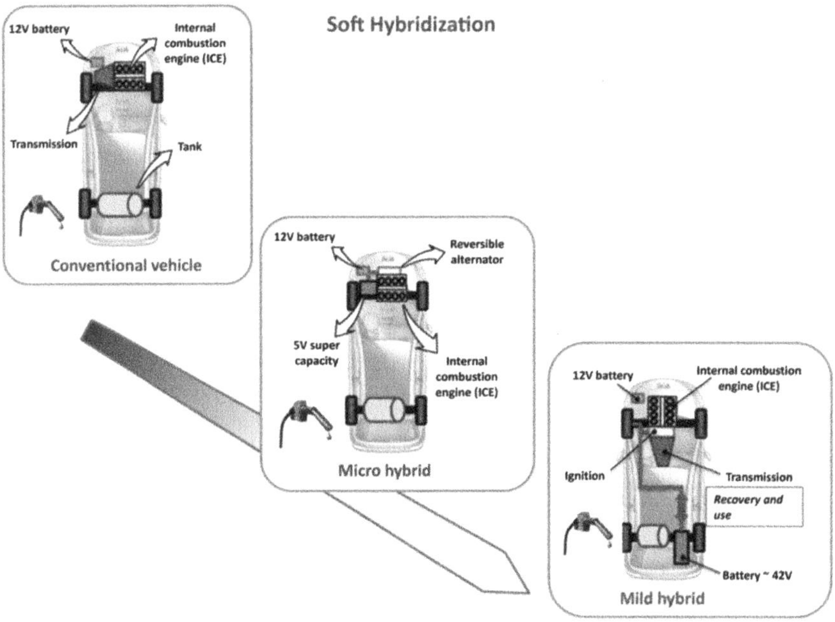

Figure 5.16. *Soft hybridization. For a color version of this figure, see www.iste.co.uk/krysinski/automotive.zip*

We can distinguish between:

– soft type hybridization:

- micro-hybridization or stop-start: the combustion engine stops automatically when the car has stopped and an "ignition" restarts the engine,

- mild-hybrid, similar to the stop-start technology: the electric device allows limited recovery during braking and assistance during take-off of the combustion engine. A very small energy storage unit (battery, super capacity) is then required;

– the more classic hybridization:

- full-hybrid: the electric machine is powerful enough to power the car, at least at a low speed. Energy recovery during deceleration and braking becomes almost mandatory. However, the quantity of batteries on board allows only a very small amount of driving: 100% zero-emission electricity (ZEV), from a few hundred meters to a few kilometers,

- the plug-in-hybrid can sometimes be seen as a variant of the full-hybrid, but with a larger quantity of battery, enough in any case to enable recharge of the sector's electricity (hence its name). In practice, the entire electric powertrain requires a larger sizing than full-hybrid, because electric driving is supposed to be able to be carried out over a few tens of kilometers with satisfactory performance, without the use of the combustion engine, except in certain exceptional cases (high power demand for example). With this architecture, it is also possible to downsize the combustion engine, or even to use thermal cycles with better efficiencies (Atkinson cycle or Miller cycle), because the electric motor and/or transmission architecture are favorable,

- if we continue to "downsize" the engine, we can then speak of a range extender, which is only a form of plug-in-hybrid with the atrophied thermal part.

Various car manufacturers wish to develop zero-emission electric or plug-in hybrid vehicles emitting less than 49 g of CO_2/km to meet standards targets by 2030. For many manufacturers, from 2019 onwards, new models will systematically have a hybrid or fully electric version.

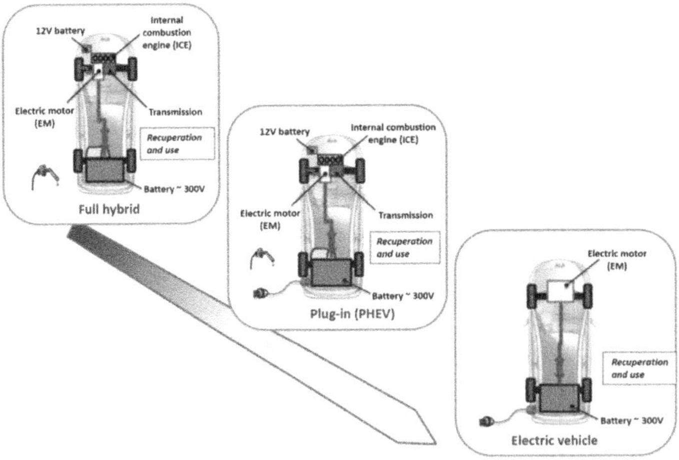

Figure 5.17. *Different types of hybridization. For a color version of this figure, see www.iste.co.uk/krysinski/automotive.zip*

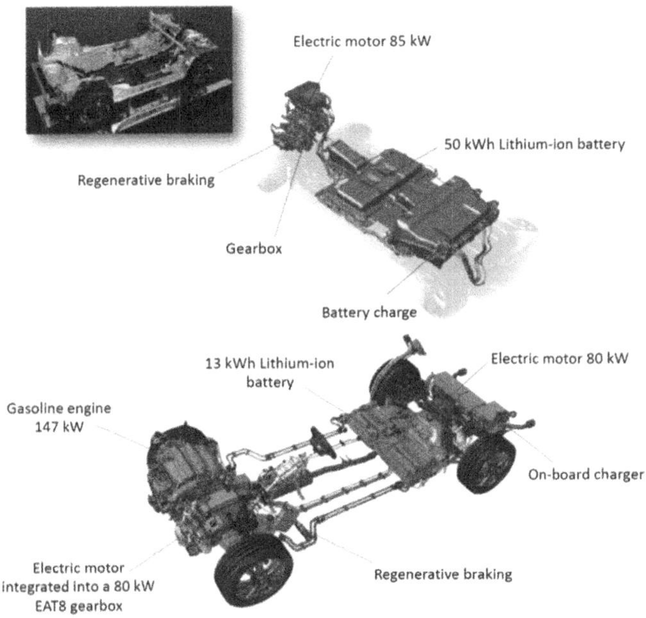

Figure 5.18. *Example of a platform for PHEV and 100% electric vehicles (PSA). For a color version of this figure, see www.iste.co.uk/krysinski/automotive.zip*

5.2.2.2. *Examples of parallel hybrid architectures*

There are several possibilities for integrating a hybrid powertrain onto a vehicle, with a combustion engine as a base. The most common are the series hybrid, the parallel hybrid and the mixed or power division hybrid.

5.2.2.2.1. Concept of kinematic coupling or torque coupling

Several mechanical devices enable mechanical coupling between different mechanical assemblies. The power balance can be expressed by equation [5.5] taking into account that there is no energy loss, that the efficiency is 1, that there is no variation in kinetic energy, therefore no variation in velocity, and that all solids are balanced:

$$P_1(1 \rightarrow S/0) + P_2(2 \rightarrow S/0) + P_3(3 \rightarrow S/0) = 0 \quad [5.5]$$

This conservation equation can be represented by a block diagram (Figure 5.19).

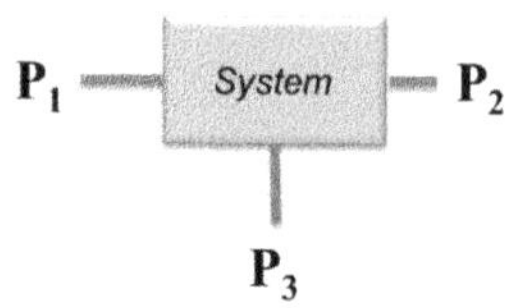

Figure 5.19. *Block diagram of a system's power conservation*

By definition, power is related to torque $C_{i \rightarrow S}$ and rotational speed $\omega_{i/0}$ by:

$$P(i \rightarrow S/0) = C_{i \rightarrow S}\ \omega_{i/0} \quad [5.6]$$

Depending on the system technology, mechanical power couplings can therefore be separated into two categories: velocity couplings or torque couplings.

Torque coupling

In this case, the technology makes it possible to obtain a linear relationship between the torques acting on the system. This relationship exists when the system technology imposes two independent relationships [5.7] linking speeds together:

$$\begin{cases} \omega_{2/0} = k_{21}\ \omega_{1/0} \\ \omega_{3/0} = k_{31}\ \omega_{1/0} \end{cases} \quad [5.7]$$

The power balance, equation [5.5] and the equation [5.7] provide the torque equation:

$$C_{1 \to S} + k_{21} C_{2 \to S} + k_{31} C_{3 \to S} = 0 \qquad [5.8]$$

This torque coupling can be achieved via several technical solutions, for example, by using a gear train or an assembly of chains or pulleys.

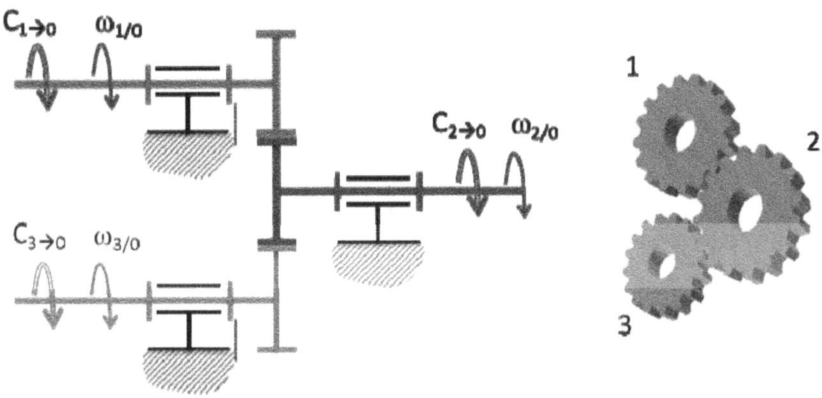

Figure 5.20. *Torque coupling by means of single trains. For a color version of this figure, see www.iste.co.uk/krysinski/automotive.zip*

Velocity coupling

In this case, the technology makes it possible to obtain two linear relationships between the torques acting on the system. This relationship exists when the system technology imposes a linear relationship between velocities:

$$\omega_{1/0} + k_{21}\,\omega_{2/0} + k_{31}\,\omega_{3/0} = 0 \qquad [5.9]$$

Then the combination of the power balance [5.5] and the velocity equation [5.9] imposes:

$$\frac{C_{1 \to S}}{1} = \frac{C_{2 \to S}}{k_{21}} = \frac{C_{3 \to S}}{k_{31}} \qquad [5.10]$$

This type of velocity coupling can be achieved by means of an epicyclic train.

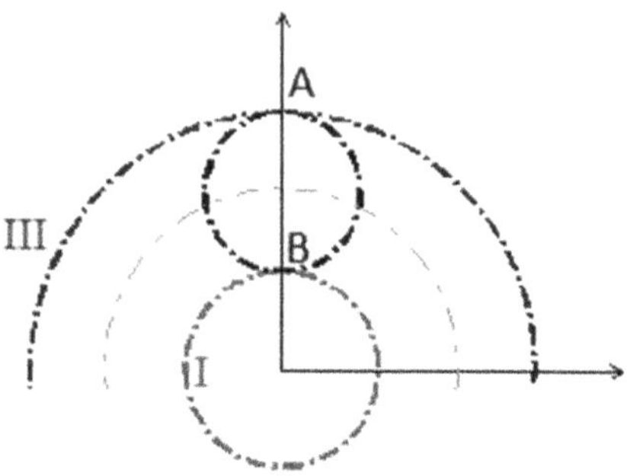

Figure 5.21. *Kinematic relationship of trains, non-slip conditions. For a color version of this figure, see www.iste.co.uk/krysinski/automotive.zip*

Non-slip relationships can be written:

$$\begin{cases} \vec{V}_{A,III/Satellite} = \vec{0} \\ \vec{V}_{B,I/Satellite} = \vec{0} \end{cases} \quad [5.11]$$

Relationships can then be expressed by:

$$\begin{aligned} \vec{V}_{A,III/Satellite} &= \vec{V}_{A,III/0} - \vec{V}_{A,Satellite/0} \\ &= r_{III}\ \omega_{III/0} + r_{II}\omega_{Satellite/IV} - r_{IV}\omega_{IV/0} = 0 \end{aligned} \quad [5.12]$$

$$\begin{aligned} \vec{V}_{B,I/Satellite} &= \vec{V}_{B,I/0} - \vec{V}_{B,Satellite/0} \\ &= r_{I}\ \omega_{I/0} - r_{II}\omega_{Satellite/IV} - r_{IV}\omega_{IV/0} = 0 \end{aligned} \quad [5.13]$$

Then, the combination of the two equations makes it possible to obtain:

$$\omega_{I/0} + \frac{r_{III}}{r_I}\ \omega_{III/0} - \left(1 + \frac{r_{III}}{r_I}\right)\omega_{IV/0} = 0 \quad [5.14]$$

The basic reason λ is defined by:

$$\lambda = -\frac{r_{III}}{r_I} \quad [5.15]$$

Thus, we can generalize the fact that – whatever the type – the two non-slip condition relationships lead to the general relation:

$$\omega_{I/0} - \lambda\, \omega_{III/0} + (\lambda - 1)\, \omega_{IV/0} = 0 \qquad [5.16]$$

With λ the basic reason for the train, which can also be presented in the form (Willis' formula):

$$\frac{\omega_{I/0} - \omega_{IV/0}}{\omega_{III/0} - \omega_{IV/0}} = \lambda \qquad [5.17]$$

In fact, we obtain the relationship concerning torques:

$$\frac{C_{I\to S}}{1} = \frac{C_{III\to S}}{-\lambda} = \frac{C_{IV\to S}}{\lambda - 1} \qquad [5.18]$$

Nomographs

The nomograph is a graphical representation of a mathematical formula that highlights the relationship between two or more factors.

In the particular case of the epicyclic train, it is possible to represent the kinematic formula [5.17] using three vertical axes.

We can write the following three equations, [5.17] and two trivial equations:

$$\begin{cases} \dfrac{\omega_{I/0} - \omega_{IV/0}}{\omega_{III/0} - \omega_{IV/0}} = \lambda \\ \dfrac{\omega_{I/0} - \omega_{IV/0}}{\omega_{I/0} - \omega_{IV/0}} = 1 \\ \dfrac{\omega_{IIV0} - \omega_{IV/0}}{\omega_{I/0} - \omega_{IV/0}} = 0 \end{cases} \qquad [5.19]$$

The three vertical axes represent ω_I, ω_{IV} and ω_{III} respectively. The three relationships [5.19] are used to place the nomograph axes.

Thanks to this graph, for a given value of λ – and if two input velocity values are set – the velocity of the third element can be measured graphically.

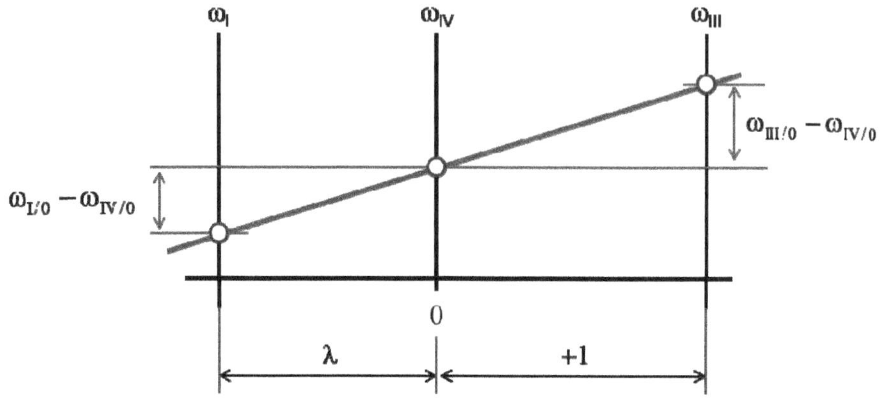

Figure 5.22. *Nomograph of an epicyclic train, λ < 0. For a color version of this figure, see www.iste.co.uk/krysinski/automotive.zip*

Figure 5.23 shows the schematic layout of an epicyclic train, to combine the combustion engine and electric motor with the wheels.

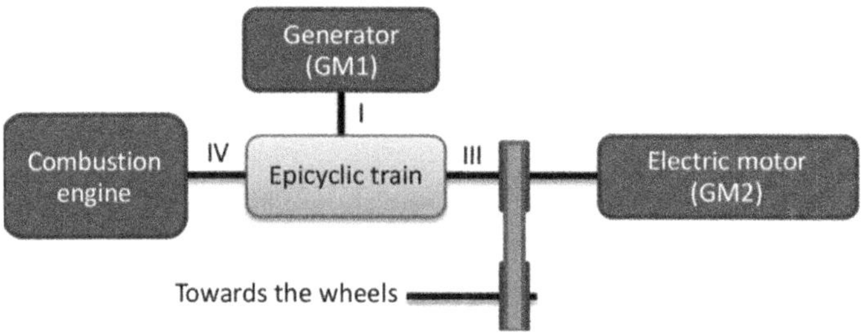

Figure 5.23. *Schematic implementation of hybridization (Toyota Prius)*

For consumption optimization purposes, the motor is maintained at a constant speed that is optimal for its own operation. The rotational speed of the wheels can then change according to the path. It can be seen that the epicyclic train then imposes a speed of rotation that can have a different sign (direct or reverse direction). The relationship of the torques remains:

$$\begin{cases} C_{III\to S} = -\lambda\, C_{I\to S} \\ C_{IV\to S} = (\lambda - 1)\, C_{I\to S} \end{cases} \quad [5.20]$$

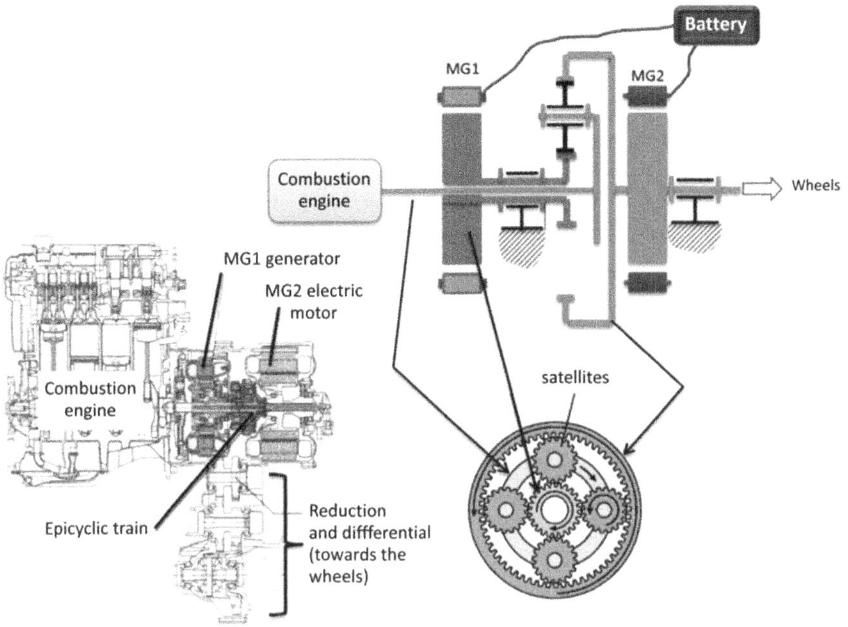

Figure 5.24. *Diagram showing the layout of the epicyclic train (Toyota Prius). For a color version of this figure, see www.iste.co.uk/krysinski/automotive.zip*

Since the torque ratio remains constant, it is observed that there may then be an inversion of the power sign. If λ is negative, the C_{IV} torque (combustion engine) is opposite to the torques of the electric motor and wheels.

As long as the wheel speed is not important, all three speeds are the same sign, then the motor power provides power to the wheels and the electric motor – the system is in bypass.

This can be illustrated by the architecture proposed by Toyota's Prius, the nomograph of which is illustrated in Figure 5.26.

Among the many operating modes of this hybrid system, a few classic situations are interesting to study, particularly with regard to power transfer.

This is the first case where the vehicle travels at a low speed (e.g. 40 km/h) for a power demand related to drag (e.g. 8 kWatt). It is assumed that the battery state of charge is low so we want to store energy (8 kWatt). The motor provides 16 kWatt of power and runs at 2,200 rpm. In this configuration, MG1 works as a generator and MG2 as a motor (Figure 5.27).

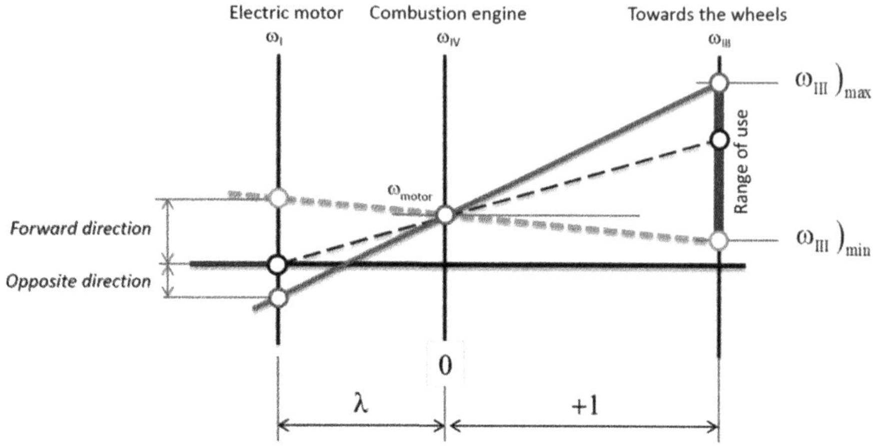

Figure 5.25. *Concept of speed inversion, nomograph. For a color version of this figure, see www.iste.co.uk/krysinski/automotive.zip*

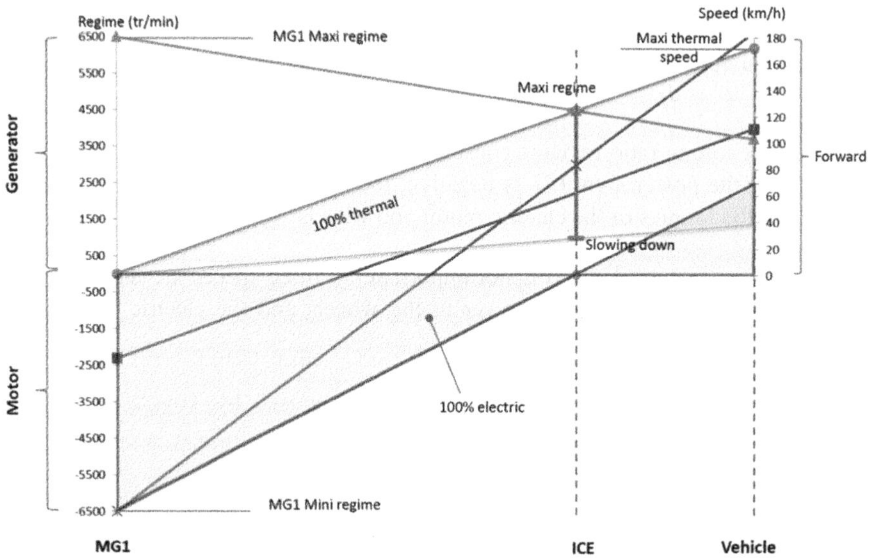

Figure 5.26. *Nomograph of a Toyota Prius. For a color version of this figure, see www.iste.co.uk/krysinski/automotive.zip*

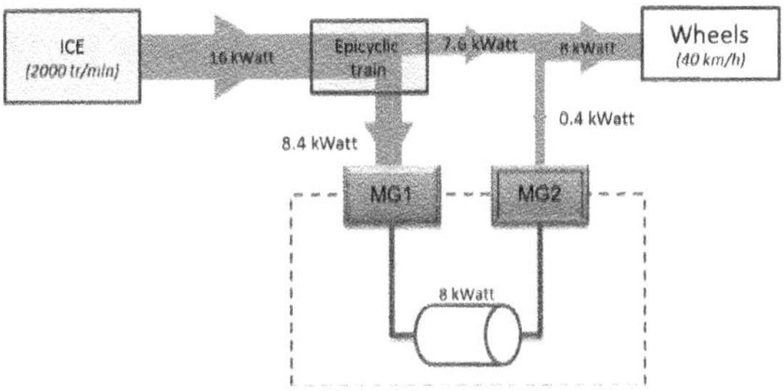

Figure 5.27. *Thermal mode with battery charge. For a color version of this figure, see www.iste.co.uk/krysinski/automotive.zip*

The second situation shown corresponds to the case where the vehicle is traveling at high speed (e.g. 80 km/h) and the driver requires significant power, for example to overtake (e.g. 48 kWatt). We choose to run the engine on an optimal efficiency point corresponding to 30 kWatt at 3,000 rpm, the battery supplying the rest of the energy. MG1 works as a generator and MG2 as a motor (Figure 5.28).

The third proposed situation should be considered on a flat road at high speed (e.g. 80 km/h). The useful power is relatively low, the motor runs at a low speed (e.g. 10 kWatt for 1,100 rpm). It can be seen that, in this part, MG2 works as a generator and MG1 as an engine, the battery is at rest (Figure 5.29). There is a recirculation of power. Taking the efficiency of electrical machines and their converters into account, losses in the bypass loop are necessarily shown. Changing the rotation of MG1 reduces the power in the bypass loop, but since this involves changing the engine rotation speed, the engine efficiency is degraded. For the assembly to be optimized, the increase in speed of MG1 must cause a reduction in losses in ICE greater than the increase in losses in the electromechanical loop.

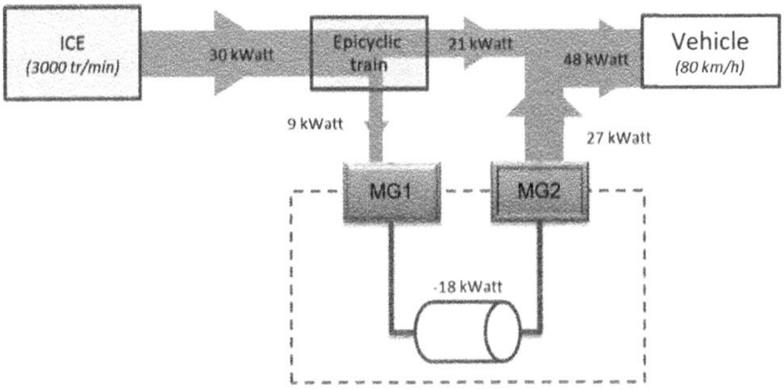

Figure 5.28. *Hybrid mode with battery discharge. For a color version of this figure, see www.iste.co.uk/krysinski/automotive.zip*

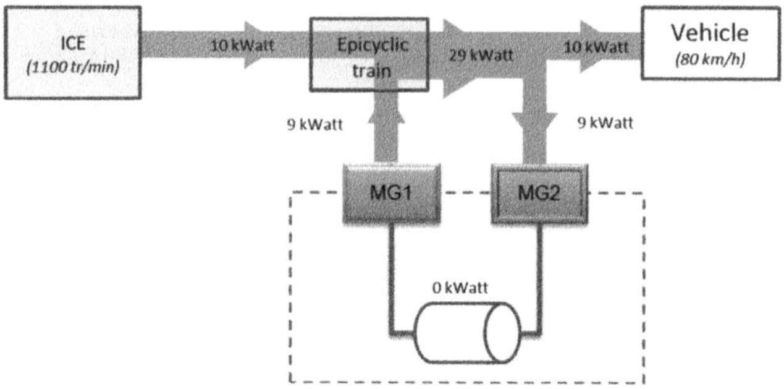

Figure 5.29. *Configuration with power recirculation. For a color version of this figure, see www.iste.co.uk/krysinski/automotive.zip*

The graph in Figure 5.30 links the impact of the engine speed and MG1 speed to the possibility of recirculation, measured by the bypass rate:

$$\left\{ \tau = -\frac{P_{III}}{P_{IV}} = -\frac{C_{III}\ \omega_{III}}{C_{IV}\ \omega_{IV}} \right. \quad \text{derivation rate} \qquad [5.21]$$

We observe that at 80 km/h, to avoid recirculation, it is necessary to accelerate the engine, which compromises its efficiency.

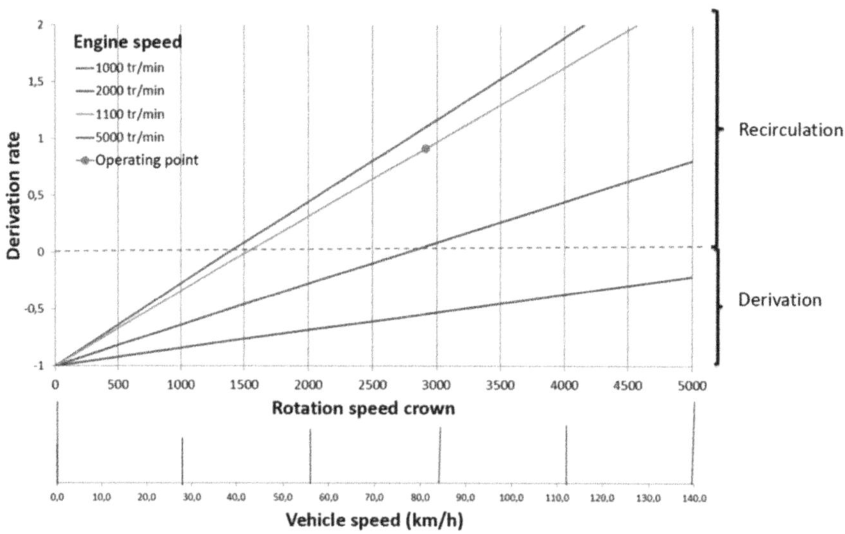

Figure 5.30. *Bypass rate according to vehicle speed for different engine speeds. For a color version of this figure, see www.iste.co.uk/krysinski/automotive.zip*

This recirculation can also be observed (Figure 5.31) by analyzing the diversion rate, according to the transmission ratio K, defined by:

$$K=\frac{\omega_I}{\omega_{IV}}=\frac{\omega_I}{\omega_{ICE}} \qquad [5.22]$$

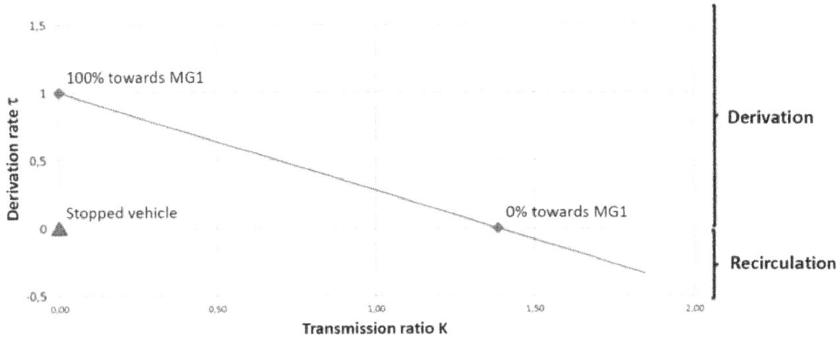

Figure 5.31. *Derivation rate according to transmission ratio*

Efficiency of power chains with epicyclic trains

The efficiency calculation only covers losses due to meshing and does not take bearing losses into account, which are generally negligible.

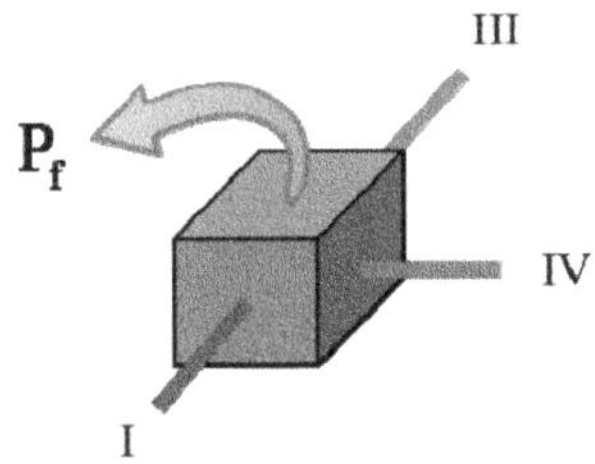

Figure 5.32. *Energy losses of the epicyclic train. For a color version of this figure, see www.iste.co.uk/krysinski/automotive.zip*

The efficiency is a function of the inputs and outputs that are selected on the epicyclic train.

The partial yield is called η, the train efficiency when the satellite carrier IV is fixed (blocked) with respect to frame 0 and when the inputs and outputs are made on planetary I and III. The system is then a fixed-axis gearbox, the efficiency of which is known. This efficiency does not depend on the number of satellites, it can be considered as the efficiency of the two gears of a one-wheel satellite with two epicyclic trains (or the efficiency of the two gears of a two-wheel satellite in complete connection).

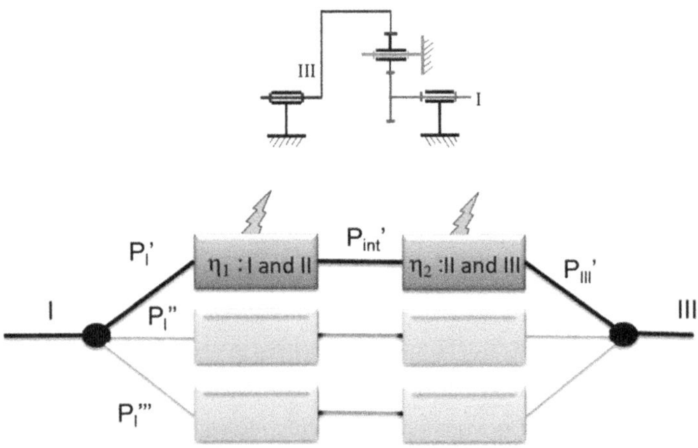

Figure 5.33. *Assessment of the performance of the blocked satellite carrier. For a color version of this figure, see www.iste.co.uk/krysinski/automotive.zip*

Under these conditions the yields can be defined by:

$$\begin{cases} \eta_1 = \dfrac{P'_{int}}{P'_I} \\ \eta_2 = \dfrac{P'_{III}}{P'_{int}} \end{cases} \Rightarrow \eta' = \eta_1\eta_2 \text{ (branch efficiency)} \quad [5.23]$$

Each branch has the same efficiency:

$$\eta' = \eta'' = \eta''' = \eta_1\ \eta_2 \quad [5.24]$$

Taking into account the different parallel branches, we obtain:

$$\eta = \frac{P'_{III} + P''_{III} + P'''_{III}}{P_I} \Rightarrow \eta = \frac{\eta' P'_I + \eta'' P''_I + \eta''' P'''_I}{P_I} \Rightarrow \eta = \eta_1\eta_2 \quad [5.25]$$

to analyze performance, regardless of the architecture and use of the epicyclic train.

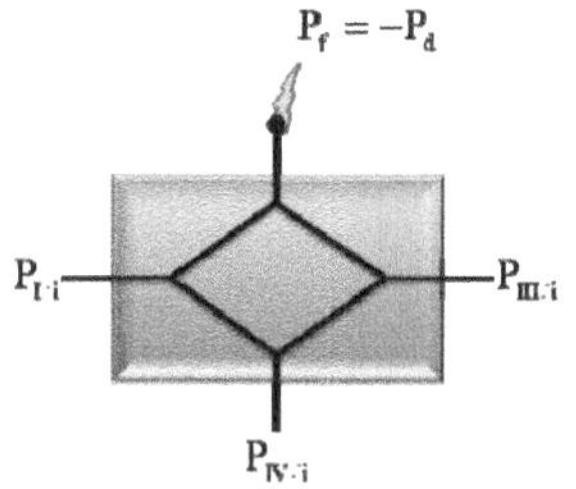

Figure 5.34. *Definition of the incoming and outgoing powers of an epicyclic train*

Case of type 1 train:

Figure 5.35. *Speed torque by means of an epicyclic train. For a color version of this figure, see www.iste.co.uk/krysinski/automotive.zip*

There are several types of trains. Type 1 – the simplest and most common – consists of single-toothed satellites, an inner and an outer planetary component (Figure 5.35). In the case of a type 1 single train, the basic reason λ is linked, assuming that the operating module is the same for all gears, by the relation:

$$\lambda = -\frac{Z_{III}}{Z_I} \qquad [5.26]$$

with ZIII and ZI the number of teeth for planetary III and planetary I, respectively.

5.2.2.2.2. Parallel architecture in 100% electric propulsion

In the proposed architecture, the powers from both motors are summarized in electrical form (in the case of an electric hybrid). The energy from the combustion engine is first converted into electricity via a generator. This electricity will either recharge the battery or power the electrical machine.

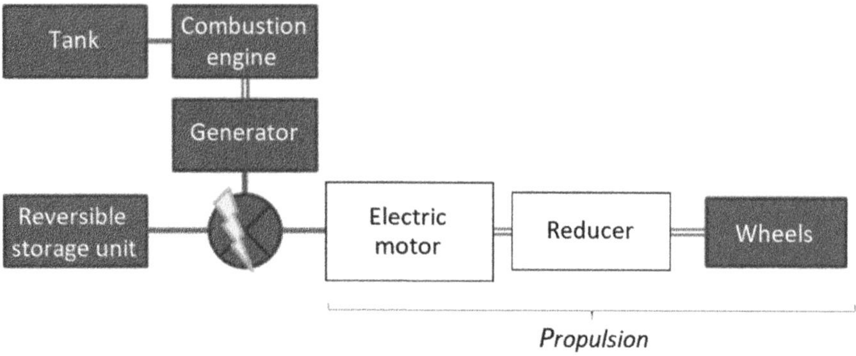

Figure 5.36. *Parallel architecture in 100% electric propulsion*

The advantages are:

– the combustion engine is completely decoupled from the wheels. It can still operate at its best efficiency point;

– the electric motor can provide a single reduction gearbox at the output. There is no need for a gearbox.

The disadvantages are:

– it requires an additional generator (weight + cost);

– the electric motor must be sized to meet all power requirements;

– there is a loss of efficiency because the mechanical energy from the combustion engine is converted into electrical energy before being converted back into mechanical energy.

A serial architecture is not *a priori* suitable for a hybrid vehicle with a conventional combustion engine, except perhaps in the case of a range extender. However, this architecture can be justified if it is another kind of combustion engine. For example, a Stirling engine (external combustion engine) has a better theoretical efficiency than a combustion engine. However, it is not very reactive and is therefore not used in a conventional vehicle. A Stirling machine would be justified in a serial architecture because it is the electric motor that generates the vehicle's reactivity. In addition, the loss of efficiency through mechanical/electrical/mechanical conversion would be compensated by a good efficiency of the Stirling machine.

5.2.2.2.3. Parallel architecture in distributed propulsion

In this parallel architecture configuration, the powers from both engines are summed in mechanical form.

Role of the gearboxes

The gearbox is essential in a parallel architecture. Indeed – as can be seen in Figure 5.37 – there is only one mechanical coupling (in torque or speed). Considering that both the torque required from the wheel and the rotational speed of the wheel are imposed, there are four variables:

$$\omega_{MTH},\ C_{MTH},\ \omega_{Mot},\ C_{Mot} \quad [5.27]$$

However, mechanical coupling is governed by three equations. In total, four variables and three equations correspond to one degree of freedom. The gearbox gives an additional degree of freedom by allowing the speed of rotation from the wheels to be varied.

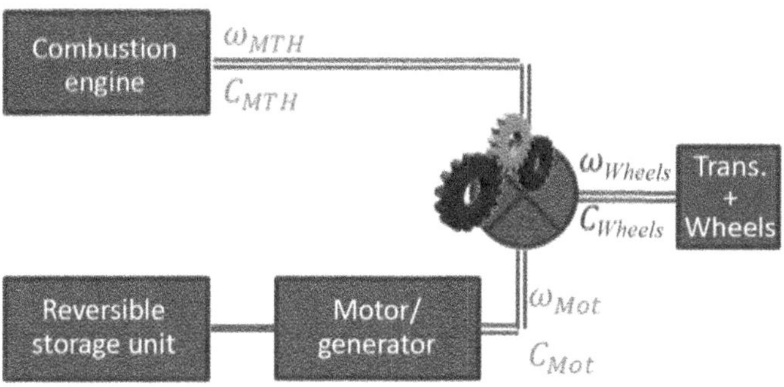

Figure 5.37. *Parallel architecture. For a color version of this figure, see www.iste.co.uk/krysinski/automotive.zip*

In a parallel architecture, the two powers are summed in mechanical form. This mechanical sum can be done at several levels (disconnections are not represented).

Implementation of the mechanical sum

At road level

For example, the thermal chain supplies the front wheels and the electric chain supplies the rear wheels. The powers are summed at road level when both sources operate simultaneously.

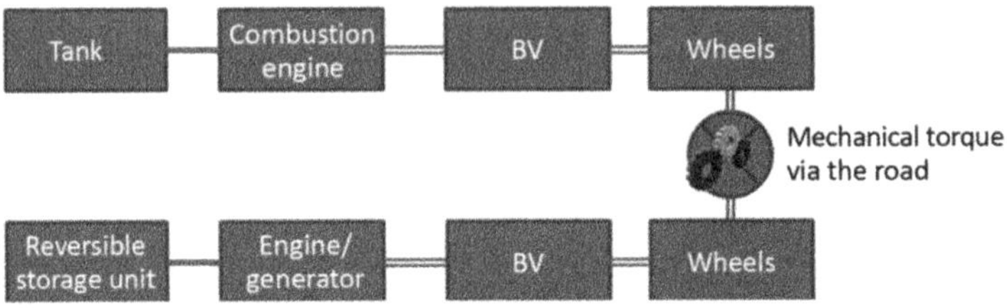

Figure 5.38. *Kinematic train with mechanical coupling with the wheels*

After gearboxes

For example, the thermal chain and the electric chain each have their own gearbox. The powers that come out of it are mechanically summed and feed the front wheels.

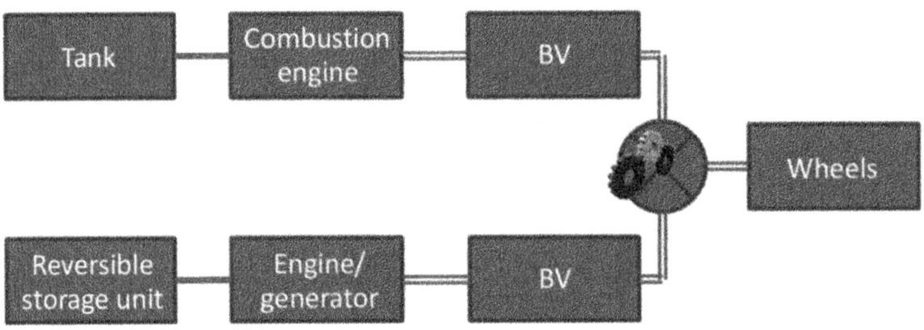

Figure 5.39. *Kinematic trains with torque at the output of the gearboxes*

Before gearboxes

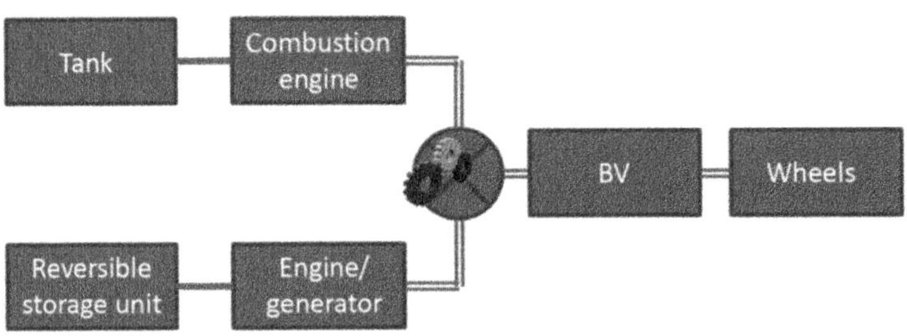

Figure 5.40. *Kinematic trains with coupling at the output of the motors*

In general, in a parallel architecture the coupling is in torque.

The advantages are:

– you can choose either the torque of the combustion engine or the rotational speed;

– there is no additional generator;

– the dimensions of the electric motor are smaller;

– there is a direct mechanical link between the combustion engine and the wheel;

– there is a speed torque, useful if this is done in combination with a Stirling engine or gas turbine that are speed sensitive.

On the other hand, the disadvantages are:

– the combustion engine cannot be operated at any point in its mapping, unless the gearbox is a CVT;

– for parallel architecture, the ignition must generally be taken care of, unlike the power division architecture. However, some parallel architectures are ignition free, especially some mild-hybrid cars (such as PSA's e-DCT transmission).

Architecture in power division: installation in vehicles

In a power division architecture, the mechanical power from the combustion engine is halved in the epicyclic train. Some of the power is transformed into electrical energy via the generator. The other part remains mechanical energy and will supply the wheels directly. If necessary, the electric motor provides additional torque.

In normal use, the power split architecture works as shown in Figure 5.42 (schematically representing the Prius). The electric chain always provides a certain power to the wheel because the MG1 generator regulates the combustion engine's speed. The power split architecture allows an infinitely variable transmission, but the counterpart is that the power at the wheels is always the sum of mechanical and electrical power.

However, it happens that MG1 operates in engine mode to regulate the engine speed. In this case, power recirculation may occur and generate additional losses.

In a power division architecture, there are two mechanical couplings: one in torque, the other in speed. The speed coupling is mainly used as a power divider (not as a summing device). Considering that the torque required from the wheel and the rotational speed of the wheel are both imposed, there are eight variables:

$$\omega_{MTH},\ C_{MTH},\ \omega_{Mot},\ C_{Mot},\ \omega_{Gen},\ C_{Gen},\ \omega_{int} \qquad [5.28]$$

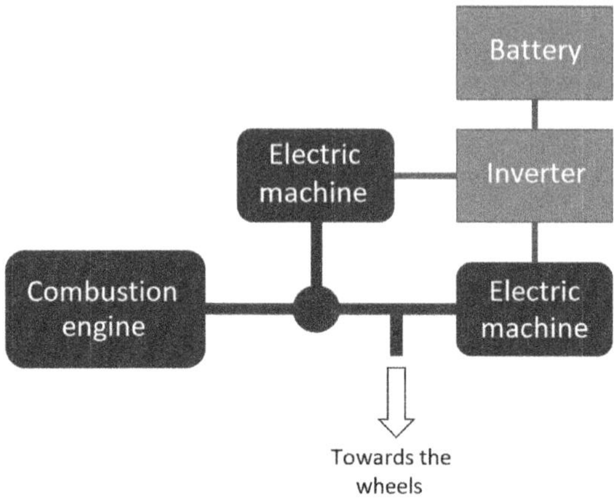

Figure 5.41. *Principle of the power chain with power split*

However, a mechanical coupling is governed by three equations (see above). Eight variables and six equations correspond to two degrees of freedom. The torque and speed of the combustion engine can therefore be chosen as desired. There is no need for a gearbox to decouple the wheels and engine.

The advantages are:

– the combustion engine can operate at any point in the engine map;

– the dimensioning of the electric motor is smaller compared to the serial architecture;

– the downsizing of the combustion engine;

– the direct mechanical link between the combustion engine and the wheel;

– there is no need for a gearbox.

The disadvantages are:

– the need to have an additional generator compared to a parallel architecture, which implies additional costs and integration difficulties (weight and cost);

– unless the epicyclic train is blocked on the highway – in general – the power from the combustion engine is divided in half. Some of it is converted into electricity. Thus, the power that reaches the wheels is the sum of a power output from the mechanical chain and a power output from the electrical chain (with the associated efficiency;

– power recirculation may occur and increase losses.

5.2.3. *Implementation of hybridization in the case of the automobile*

5.2.3.1. *General principles*

The two main criteria for sizing the propulsion system are the power to be delivered by the various converters (i.e. electric motor or generator) and the sizing of the storage system that defines the vehicle's range.

We saw in Chapter 2 that a vehicle requires power peaks during acceleration phases, particularly in urban or peri-urban areas, where there is a sequence of these power demands. On the other hand, helicopters need significant power in hover flight but on a continuous basis during this operating phase.

To size the power of the car's engine and the capacity of the storage system, approval cycles are used. In force since September 1 2017, the WLTP (Worldwide Harmonized Light Vehicles Test Procedure) approval cycle (Figure 5.42) is more restrictive than the previous cycle (NEDC) but closer to actual driving conditions.

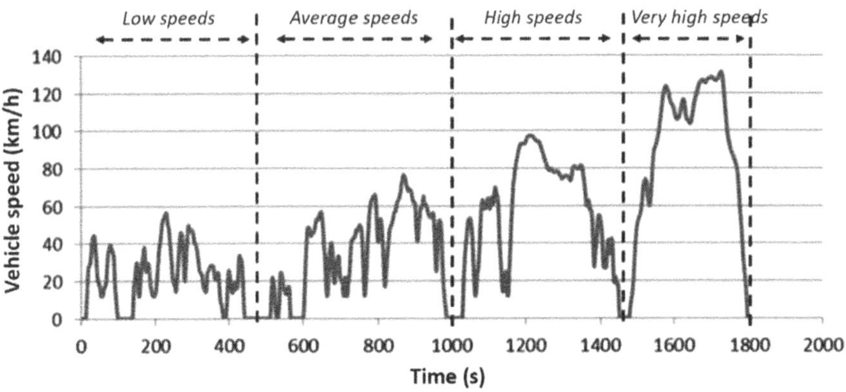

Figure 5.42. *WLTP cycle, class 3*

For example, we are interested in urban and peri-urban driving. This allows the power consumption during the low and medium speed part of the WLTP cycle to be evaluated. Figure 5.43 shows the power expended and the recoverable power for this speed profile.

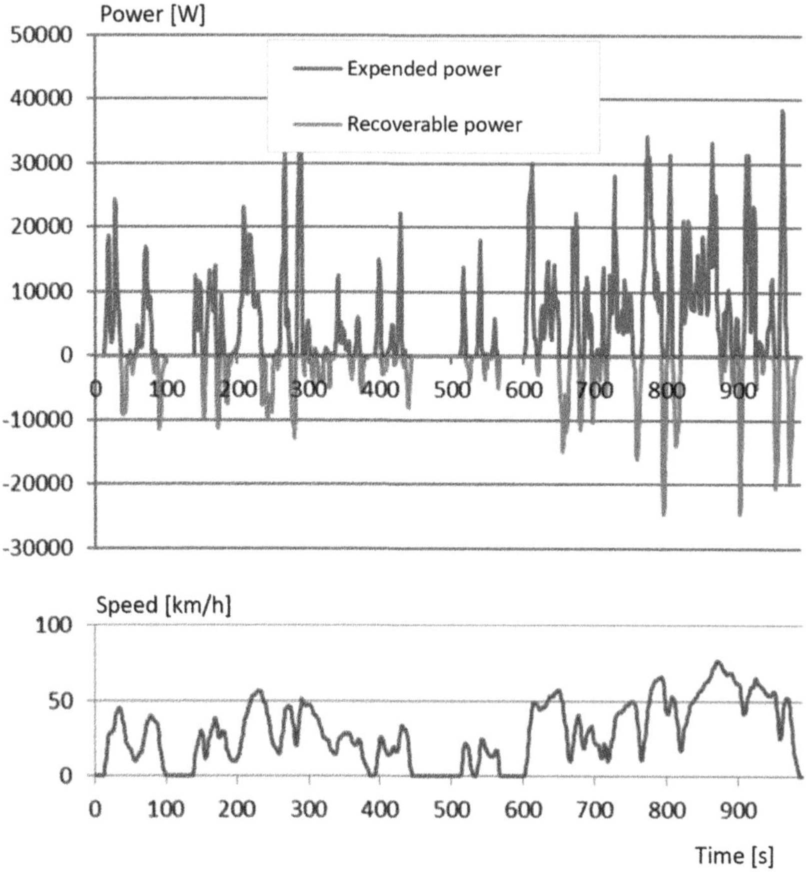

Figure 5.43. *Distribution of power over the low and medium speed parts of a WLTP cycle. For a color version of this figure, see www.iste.co.uk/krysinski/automotive.zip*

It can be seen that the requested power values range from 0 to 40 kW but a large part of the power demand is for powers below 15 kW. On the other hand, the high-power phases are very short and separated by phases of lower power. The analysis of the recoverable power shows the same type of finding.

From this, the conclusion can be drawn that it is necessary to have energy conversion systems (storage → vehicle, or vehicle → storage) with short response times.

On the other hand, in order to avoid oversizing the system, the "power requirement" and "energy requirement" functions must be separated. If we take "to be able to complete the previous 50 km journey in urban areas" as specifications, we have to size a power chain capable of supplying up to 35 kW of power and 10 kWh of energy. However, these 35 kW of power are only very rarely used, which means that the power chain is oversized in the vast majority of cases.

It may therefore make sense to work with two different chains, based on a separation of power and energy: one provides a lot of energy through storage and little power, the other provides little energy but a lot of power through a powerful motor.

5.2.3.2. *Which powertrain to choose?*

5.2.3.2.1. A single power chain, two storage units

In order to significantly reduce battery cost, this architecture proposes to separate the power source and energy source roles. The idea is to size the large energy storage unit to a minimum in power and to use another much smaller storage unit adapted to the power, to compensate for peaks in power demand. The sizing is very sensitive to the specifications: for this type of architecture to be useful, power demands must be short, as they can be on an urban adapted route.

The diagram in Figure 5.44 shows a two-storage unit architecture.

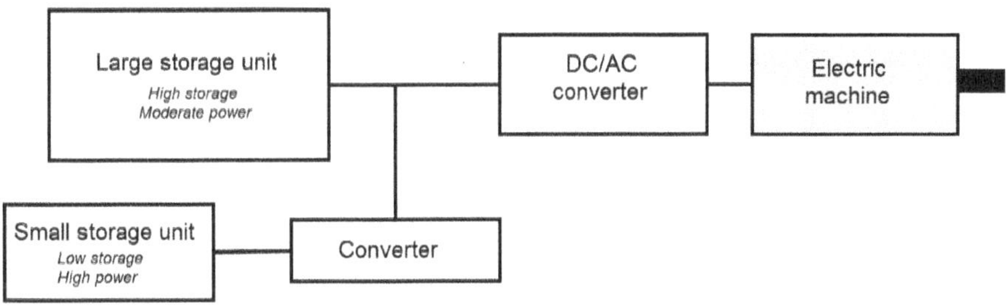

Figure 5.44. *Two storage units and a power conversion chain*

Such an architecture therefore makes it possible to significantly reduce the price of the large storage unit. However, a small storage unit with high electrical power is very expensive (usually supercapacitors are used) and the power electronics will be highly complex to manage two different storage units. In addition, high electrical powers are once again being managed, resulting in many losses.

The combination of electrochemical accumulator (battery) and supercapacitors is a solution to meet the charging profile characterized by relatively low average power and transient high-power requirements. This combination is already implemented for micro-hybrids with a 12 V battery for vehicles equipped with stop-start, as proposed by PSA's e-HDi engines. This combination is now being considered for plug-in-hybrid hybridization, where performance in terms of specific power and efficiency would be much better than that obtained by batteries alone, due to the energy efficiency of supercapacitors.

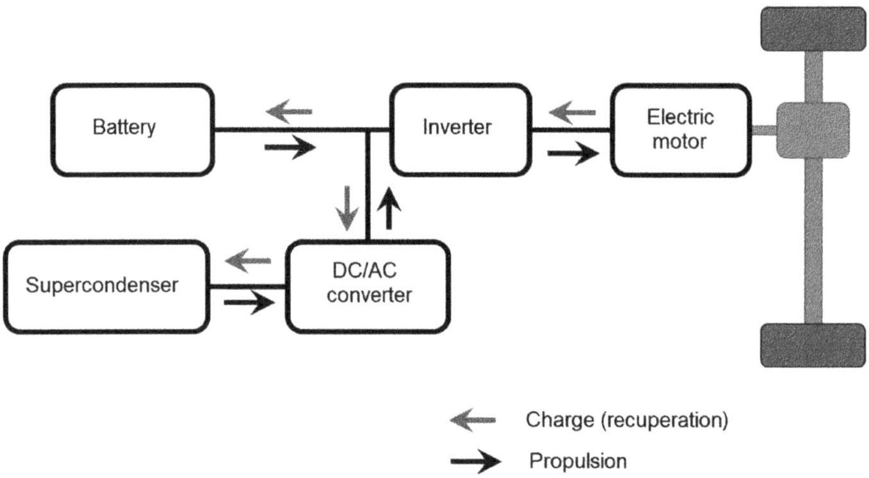

Figure 5.45. *Battery super-capacitor hybrid system for electric vehicles*

Various studies show the effectiveness of the system through demonstrators, often small vehicles in urban route configurations [BRI 16, BUR 15, KOU 18]. Manufacturers in 2018 talked about this possibility but the technologies were not yet mature enough and too expensive to be offered as standard.

A number of studies are also being carried out by combining the fuel cell (hydrogen gas storage) with supercapacitors [THO 05].

5.2.3.2.2. Two power chains, two storage units

To simplify power electronics and to give yourself more freedom in managing the two energy sources, you can also separate the two power chains: a power chain sized for low power but capable of supplying a lot of energy and a low energy power chain capable of meeting high power demands, as shown in Figure 5.46.

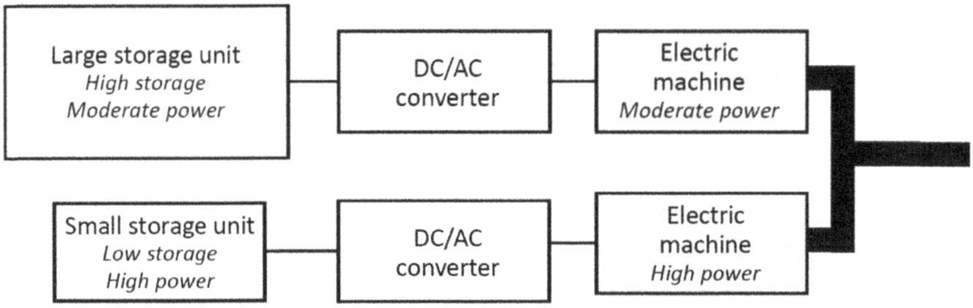

Figure 5.46. *Powertrain 3: two storage units and two electric powertrains*

This architecture has the advantage of power management and also makes it possible to size components more specifically. The low-power electric motor must be able to operate for long periods of time at low power, while the other electric motor must operate for very short periods of time at much higher power levels. The specific sizing allows us to gain a few more percent of overall efficiency. However, there remains the problem of a poor relationship between high power and electronics.

We come to the last possible plug-in power chain: a low power, high energy and all-electric part, as in the previous case, and a high power and low energy, but non-electric part. It can be mechanical (flywheel) or compressed air (Hybrid Air system). The latter two technologies have the advantage of being very efficient and almost unlimited for high powers and can also be cheaper than their electrical counterparts. The diagram in Figure 5.47 summarizes this last proposal.

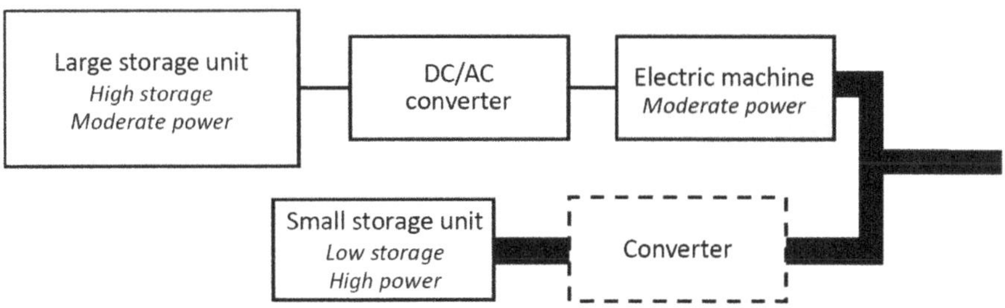

Figure 5.47. *Energy is stored and converted electrically at low power and high power is processed non-electrically*

This last proposal seems very suitable from the point of view of costs and efficiencies for a Plug-In Hybrid powertrain.

Example of the Hybrid Air system

PSA's Hybrid Air system, proposed in 2016, is based on a combination of a gasoline engine with a compressed air energy storage unit, a pump and hydraulic motor unit and an (automatic) gearbox.

An electronic management system adapts the operating mode to the user's driving style and optimizes energy efficiency. The hydraulic components (pump) recover and store energy during vehicle decelerations or braking, or in certain phases by the combustion engine, which recharges the compressed air storage. The system (via the hydraulic motor) allows the energy to be released during acceleration phases.

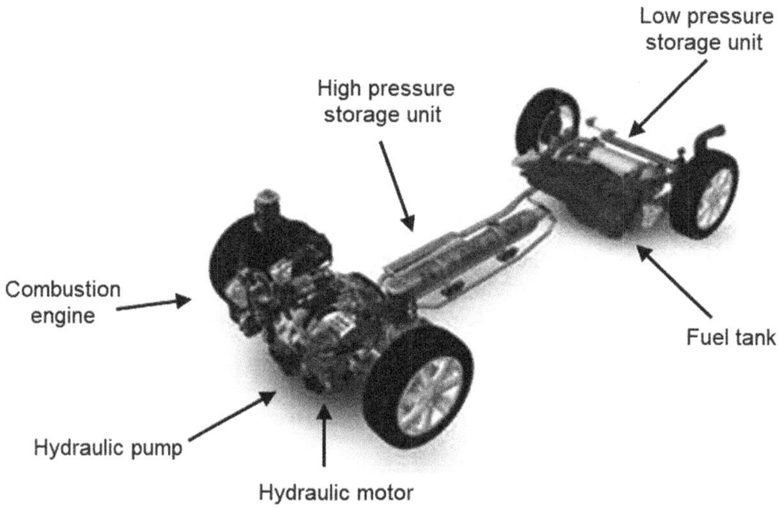

Figure 5.48. *PSA Hybrid Air system (source: PSA). For a color version of this figure, see www.iste.co.uk/krysinski/automotive.zip*

There are three distinct operating modes of the system:

– Mode 1: use of the combustion engine (fuel) at 100%. In this mode, the combustion engine provides the vehicle with traction. This mode is used when the power required by the wheel is low, typically at cruising speed when the speed is stabilized, for example in non-urban travel on highways. In this mode, fuel consumption is reduced because the combustion engine is used at an optimal operating point. This point is generally obtained for high torque and low speed which can be achieved with an appropriate transmission ratio. In this phase, the motor can also recharge the secondary storage;

– Mode 2: 100 % use of the hydraulic motor (compressed air). This mode is used when the power required by the wheel is low, that is at low speeds, typically during urban journeys with successive small acceleration and braking phases. Such a system allows very fast energy recovery and restitution, the charging and complete recharging time of the compressed air storage is only a few seconds. This mode reduces fuel consumption because the combustion engine is then at a standstill, offering this car a zero-polluting emission mode, perfectly adapted to the city. If the energy in the compressed air storage unit is completely exhausted during this phase and before the driver slows down, the combustion engine is switched on again and supplies energy, via the pump, to the compressed air tanks;

– Mode 3: combination of the two sources. This mode combines the action of the combustion engine and the compressed air system for propulsion. It is used when the power required by the wheel is high, such as in the case of high acceleration or uphill driving phases. In this mode, the combustion engine supplies energy directly to the wheels but also through the hydraulic system. This combined mode reduces fuel consumption by choosing the engine speed and torque, to ensure optimal operation. One of the particularities of this mode is also to offer a boost possibility (temporary significant increase in power). In this case, if the accumulator is full, it can deliver additional power to the hydraulic motor.

Overall, in terms of efficiency, the Hybrid Air system would have an efficiency of 95 % and a CO_2 emission of 69 grams per kilometer on the C3 Hybrid Air vehicle, which would make it very competitive with other conventional hybrid systems.

Example of the flywheel inertia system (KERS)

There are several possibilities to position a flywheel on a vehicle: either between the engine and the gearbox, or directly on the driven axletree, or on the non-driven axletree.

In 2014, Volvo tested a flywheel energy recovery system (KERS) in combination with a combustion engine. The prototype developed used a Kevlar flywheel that recovered energy during the braking phases. The system ran at a maximum speed of 60,000 rpm and offered an additional 80 hp for about ten seconds. The objective was either to increase acceleration capacities (boost) or, to reduce consumption by operating the combustion engine at a point of higher efficiency during the acceleration phase. Volvo announced that it would be able to conserve the stored energy for twenty minutes, thanks to magnetic bearings and the closing off of the steering wheel. This type of energy recovery system is effective in an urban environment, when the driver brakes a lot. Nevertheless, it can be envisaged that the engine can recharge this flywheel even when the engine acts on the wheels. Depending on the energy capacity of the steering wheel and its recharge and discharge speed, a regular relay strategy between the engine and the energy wheel would improve the efficiency of such a system.

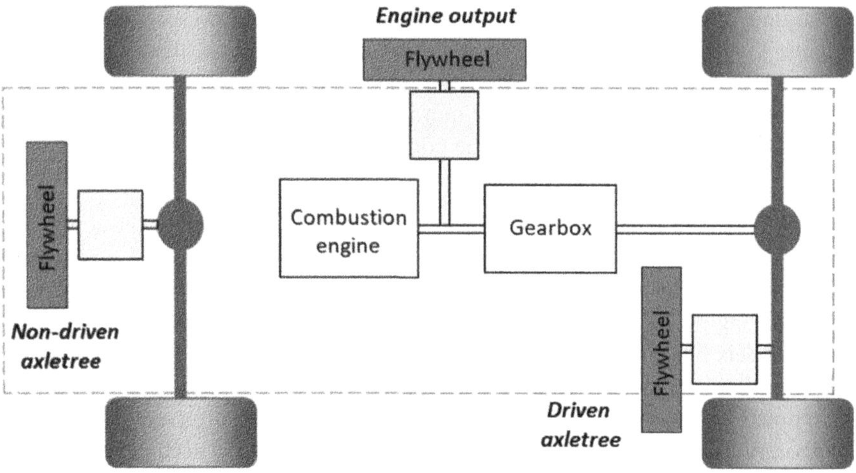

Figure 5.49. *Examples of potential installations of a flywheel on a vehicle*

5.2.3.2.3. Overview

What has just been highlighted is firstly the fact that the sizing of a hybrid system is very sensitive to the specifications that are required. However, when we add the cost issues to that, we see that there are trade-offs to be made. The best compromise seems to be a plug-in-hybrid vehicle (PHEV) the electrical part of which is designed to operate at 100% electric (ZEV) in the city with not too high speeds and limited accelerations. In this case, an architecture with separation of power and energy storage seems to be the most economically advantageous since it makes it possible to save on the price of the battery considerably, while providing services worthy of an electric vehicle.

Table 5.2 summarizes the different power chains available and the advantages and disadvantages of these chains.

Solution	Advantages	Disadvantages
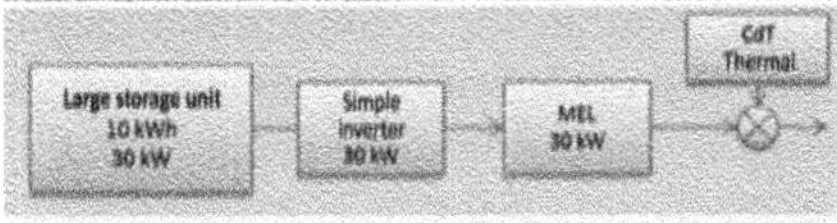	• Simplicity • Number of components	• Cost • Oversizing in 80% of cases
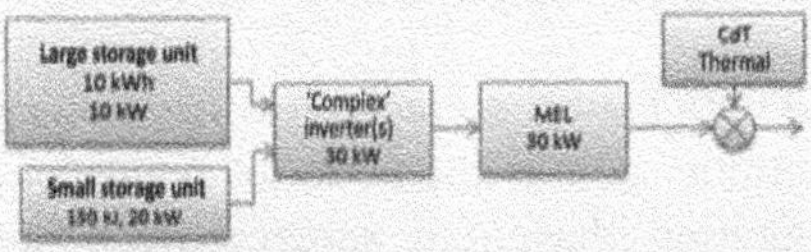	• Cost and volume of the large storage unit • Simplicity • Number of components	• 'Complex' inverter • Difficult management • High electrical power
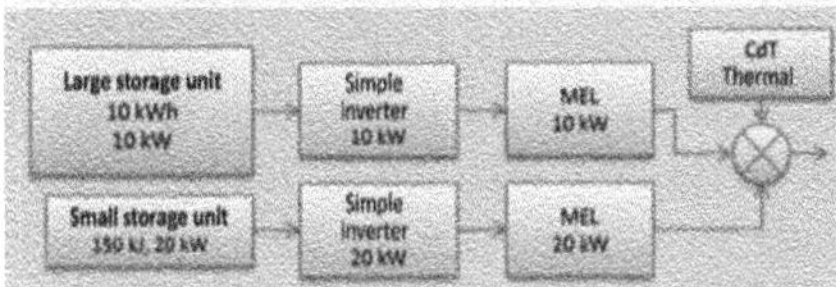	• Cost + volume of the large storage unit • Simpler management	•Complexity •High electrical power •Overall cost
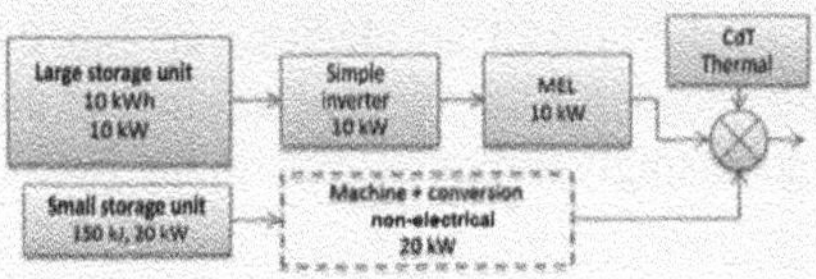	• The same as above but also: • Non-electrical high power (type AQ 35 or flywheel) • Cost	• Complexity • "Mechanical" storage problem •3 different energies

Table 5.2. *Summary of the different transmission chains*

5.3. Conclusion

Hybridization of transport systems combines at least two energy sources.

A first serial architecture consists of the second source, extending the autonomy of the installed propulsion system. This is the range extender category where – most often – the fuel storage and the combustion engine are used to power the electrochemical storage to ensure a longer autonomy than the battery alone allows. This solution allows the battery to be undersized and the combustion engine to perform at an operating point independent of the propulsion system's need, thus allowing the engine to operate at a point of highest efficiency. It also ensures the vehicle is not dependent on access to a still uncommon electric recharge, a classic limitation on the range of 100% electric cars: fuel can be filled up through the traditional distribution network.

The second type of architecture puts energy sources in parallel. A classic architecture consists of using a fossil fuel combustion engine combined with complementary energy storage (reversible for land vehicles where the recovery function is essential). This can be electrical, mechanical or pneumatic. The advantage is that the combustion engine can be operated at a point of optimal efficiency. Depending on the different operating phases, the second source provides power instead of, or in addition to, the propulsion system. The most common application for cars is the electric hybridization solution for a combustion engine (e.g. Toyota Prius). This solution allows a better cost optimization by integrating the hybridization function. Electric motors are often integrated into the engine block in combination with the combustion engine, as in the Toyota Prius.

For electric propulsion systems associated with battery storage, hybridization can be considered as a means of separating functions such as the need for power over short periods of time, such as the fast acceleration phase (e.g. supercapacitor) and autonomy needs requiring a large storage unit (electrochemical battery). Beyond the consumption and mass savings aspects, this system makes it possible to reduce the vehicle's ownership cost, the final cost being greatly affected by the cost of the battery. The increasing use of batteries leads to a reduction in their cost from \$180/kWh in 2018 to less than \$100 in 2025 (Bloomberg New Energy Finance forecast). The battery represents about 40% of the cost of an electric car, but this share should gradually decrease in the coming decades. These prospects suggest that hybrid electrification solutions will play an increasingly important role in the field of aeronautical and automotive transport.

References

[AOU 15] AOUIZERATE G., "Spécificités des moteurs aéronautiques de nouvelle génération", *Utilisation rationnelle de l'énergie et environnement, 16ème cycle de conférences Cnam/SIA*, March 17, 2015.

[BAD 13] BADIN F., *Les véhicules hybrides : des composants au système*, Technip, 2013.

[BAI 71] BAILLY M., *Thermodynamique technique*, Bordas, 1971.

[BAR 06] BAR F., "Moins pour plus…", *Le moniteur automobile*, pp. 96–97, 2006.

[BER 17] BERJOZA D., JURGENA I., "Effects of change in the weight of electric vehicles on their performance characteristics", *Agronomy Research*, vol. 15, Special Issue I, pp. 952–963, 2017.

[BEY 13] BEY R., OHREM C., BIERMANN J.W., *Downsizing Concept with Two-cylinder CNG Engine*, MTZ Worldwide, pp. 12–19, 2013.

[BIR 13] BIRCH S., "Mercedes anticipates the future with new S-Class", *SAE International Seminar*, October 2013. Available at: http://articles.sae.org/12330 (accessed January 27, 2014).

[BOI 98] BOIFFIER J.L., *The Dynamics of Flight: The Equations*, Wiley, 1998.

[BOM 06] BOMBEAU B., *Hélicoptères : la génèse, de Léonard de Vinci à Louis Breguet*, Privat, 2006.

[BOU 91] BOULET J., *Histoire de l'hélicoptère racontée par ses pionniers, 1907–1956*, France-Empire, 1991.

[BRU 81] BRUN R., *Science et technique du moteur diesel industriel et de transport*, vol. 1, Technip, 1981.

[CHA 06] CHAMBRE R., *Histoire de l'aviation*, Flammarion, 2006.

[CHA 13] CHATROUX D., "Performances of batteries technologies in vehicle applications", *Power Conversion and Intelligent Motion Europe Conference*, Nuremberg, 2013.

[DEA 16] DEAKIN A., "Kinetic energy recovery system in a bus application", *ATZ Worldwide*, vol. 118, no. 9, pp. 26–31, 2016.

[DHA 13] DHAND A., PULLEN K.R., "Review of flywheel based internal combustion engine hybrid vehicles", *International Journal of Automotive Technology*, vol. 14, no. 5, pp. 797–804, 2013.

[DIS 05] DISCOURS H., "La motorisation d'un véhicule hybride (seconde partie)", *Revue technologie*, no. 140, pp. 66–76, 2005.

[EBB 13] EBBESEN S., ELBERT P., GUZZELLA L., "Engine downsizing and electric hybridization under consideration of cost and drivability", *Oil & Gas Science and Technology*, vol. 68, no. 1, pp. 109–116, 2013.

[EHS 09] EHSANI M., GAO Y., EMADI A., *Modern Electric, Hybrid Electric, and Fuel Cell Vehicles: Fundamentals, Theory, and Design*, CRC Press, 2009.

[FAC 05] FACCHINI B., INNOCENTI L., *Refroidissement des turbines à gaz – techniques et efficacité – refroidissement interne pour aubages de turbine*, Techniques de l'ingénieur, 2005.

[FIL 10] FILIPPONE A., "Data and performances of selected aircraft and rotorcraft", *Progress in Aerospace Sciences*, vol. 36, no. 8, pp. 629–654, 2010.

[FIL 12] FILIPPONE A., *Advanced Aircraft Flight Performance*, Cambridge University Press, 2012.

[GAB 15] GABRIEL-BUENAVENTURA A., AZZOPARDI B., "Energy recovery systems for retrofitting in internal combustion engine vehicles: A review of techniques", *Renewable and Sustainable Energy Reviews*, vol. 41, pp. 955–964, 2015.

[GIA 05] GIANNELLI R., NAM E.K., HELMER K. *et al.*, "Heavy-duty diesel vehicle fuel consumption modeling based on road load and power train parameters", *SAE World Congress*, 2005. Available at: http://escholarship.org/uc/item/1mc5c0z5 (accessed January 27, 2014).

[GOM 03] GOMEZ M.M., A continuously variable power-split transmission in a hybrid-electric sport utility vehicle, PhD thesis, College of Engineering and Mineral Resources at Virginia University, 2003.

[GRA 13] GRAF F., KNORR T., JEHLE M. *et al.*, *Powertrain Management in Hybrid Vehicles with 48 V Electrical System*, MTZ Worldwide, pp. 20–24, 2013.

[GUA 04] GUALOUS H., GALLAY R., BERTHON A., "Utilisation des super-condensateurs pour le stockage de l'énergie embarquée : applications transport", *Revue de l'électricité et de l'électronique*, no. 8, pp. 83–90, 2004.

[GUZ 10] GUZZELLA L., ONDER C., *Introduction to Modeling and Control of Internal Combustion Engine Systems*, Springer Verlag, 2010.

[GUZ 13a] GUZZELLA L., SCIARRETTA A., *Vehicle Propulsion Systems: Introduction to Modeling and Optimization*, Springer Verlag, 2013.

[GUZ 13b] GUZZELA L., "Automobiles of the future and the role of automatic control in those systems", *Annual Reviews in Control*, vol. 33, no. 1, pp. 0–10, 2013. Available at: http:// 10.1016/j.arcontrol.2009.01.001 (accessed January 27, 2014).

[JEL 13] JELDEN H., LÜCK P., KRUSE G. *et al.*, *The Electric Powertrain Matrix from Volkwagen*, MTZ Worldwide, pp. 5–9, 2013.

[KIM 13] KIM N., ROUSSEAU A., DUOBA M., *GM Volt Vehicle Model Development and Validation Under Different Thermal Conditions*, VB, pp. 35–40, 2013.

[KRY 12] KRYSINSKI T., NAUZIN J.P., MALBURET F., "Lightweight design: Mass in transit", *Vehicule Dynamics International*, December 2012. Available at: www. vehiculedynamicsinternational.com (accessed January 1, 2014).

[LEB 92] LE BORZEC R., *Réducteurs de vitesse à engrenages*, Techniques de l'ingénieur, 1992. Available at: http://www.techniques-ingenieur.fr/ (accessed January 27, 2014).

[LEC 04] LECLERCQ L., Apport du stockage inertiel associé à des éoliennes dans un réseau électrique en vue d'assurer des services systèmes, Thesis, Université de Lille, 2004.

[LEP 91] LEPOURRY P., CIRYCI R., *Propulseurs aéronautiques*, Cépaduès, 1991.

[LI 13] LI X.W., SHI J.W., LU T.L. *et al.*, "A new traction control system for the vehicle with automatic transmission", *International Journal of Vehicle Design*, vol. 62, no. 2/3/4, pp. 322–350, 2013.

[LOU 01] LOUBET J.L., *Histoire de l'automobile française*, Le Seuil, 2001.

[MEI 13] MEILLAUD L., "Mode roue libre, la conso en pente douce", *Voiture écologique*, no. 7, pp. 104–105, 2013.

[MER 97] MERLIN P., *Les transports en région parisienne*, La Documentation française, 1997.

[MOR 13] MOREY B., "Learning how to optimize engine efficiency", *SAE International Seminar*, November 2013. Available at: http://articles.sae.org/12485 (accessed January 27, 2014).

[MOR 09] MORCHEOINE A., VIDALENC E., "Les transports électriques en France : un développement nécessaire sous contraintes", *ADEME & vous*, no. 21, p. 8, 2009.

[MOS 13] MOSER T., WAGNER F.X., VAN RIESEN D., *Electric Motor Technology in a High-performance Application*, MTZ Worldwide, pp. 44–48, 2013.

[NEG 08] NEGRE G., NEGRE C., "Engine with an active mono-energy and/or bi-energy chamber with compressed air and/or additional energy and thermo-dynamic cycle thereof", *U.S. Patent*, no. 7, p. 469 and p. 527, December 30, 2008.

[PEL 13] PELLETIER G., "Des solutions hybrides… et abordables", *L'argus*, pp. 28–30, November 2013.

[POG 12] POGNANT-GROS P., KETFI-CHERIF A., "Powertrain control and evaluation of hybrid powersplit systems", *IFAC Workshop on Engine and Powertrain Control Simulation and Modeling*, IFP Énergies nouvelles, October 23–25, 2012.

[RIC 07] RICHARD D., DUBEL Y., "Valeo StARS technology: A competitive solution for hybridization", *Power Conversion Conference*, Nagoya, pp. 1601–1605, 2007. Available at: http://10.1109/PCCON.2007.373178 (accessed January 27, 2014).

[RIZ 99] RIZZONI G., GUZZELLA L., BAUMANN B.M., "Unified modeling of hybrid electric vehicle drivetrains", *IEEE/ASME Transactions on MECHATRONICS*, vol. 4, no. 3, pp. 246–257, 1999.

[ROL 86] ROLLS-ROYCE PLC., *The Jet Engine*, The Technical Publications Department, 1986.

[ROU 07] ROUGIER P., *Mécanique du vol de l'hélicoptère*, Hermes-Lavoisier, 2007.

[RUA 18] RUAN J., WALKER P.D., WU J. *et al.*, "Development of continuously variable transmission and multi-speed dual-clutch transmission for pure electric vehicle", *Advances in Mechanical Engineering*, 2018. Available at: https://doi.org/10.1177/1687814018758223, (accessed March 30, 2019).

[RYD 10] RYDBERG K.E., Hydro-mechanical transmissions – Fluid power system and transmission, Course: TMMS10, Linköping University, Sweden, 2010.

[SAB 13] SABRIÉ J., CAUBET R., YARCE A. *et al.*, "Hybrid air overview – main differences with hybrid electric vehiclet", *VB*, pp. 27–33, 2013.

[SCH 13] SCHOMMERS J., SCHEIB H., HARTWEG M. *et al.*, *Minimising Friction in Combustion Engines*, MTZ Worldwide, pp. 28–35, 2013.

[TIE 13] TIE S.F., TAN C.W., "A review of energy sources and energy management system in electric vehicles", *Renewable and Sustainable Energy Reviews*, vol. 20, pp. 82–102, 2013.

[VAN 08] VAN DE VEN J.D., OLSON M.W., PERRY Y., "Development of a hydro-mechanical hydraulic hybrid drive train with independent wheel torque control for an urban passenger vehicle", *National Conference on Fluid Power*, Las Vegas Convention Center, 2008.

Index

A, B, C

D, E, F

G, H, I, J

K, L, M

N, O, P, R

S, T, U, V

Other titles from

ISTE

in

Systems and Industrial Engineering – Robotics

2019

ANDRÉ Jean-Claude
Industry 4.0: Paradoxes and Conflicts

BENSALAH Mounir, ELOUADI Abdelmajid, MHARZI Hassan
Railway Information Modeling RIM: The Track to Rail Modernization

BLUA Philippe, YALAOU Farouk, AMODEO Lionel, DE BLOCK Michaël, LAPLANCHE David
Hospital Logistics and e-Management: Digital Transition and Revolution

BRIFFAUT Jean-Pierre
From Complexity in the Natural Sciences to Complexity in Operations Management Systems
(Systems of Systems Complexity Set – Volume 1)

BUDINGER Marc, HAZYUK Ion, COÏC Clément
Multi-Physics Modeling of Technological Systems

FLAUS Jean-Marie
Cybersecurity of Industrial Systems

JAULIN Luc
Mobile Robotics – Second Edition Revised and Updated

KUMAR Kaushik, DAVIM Paulo J.
Optimization for Engineering Problems

TRIGEASSOU Jean-Claude, MAAMRI Nezha
Analysis, Modeling and Stability of Fractional Order Differential Systems 1: The Infinite State Approach
Analysis, Modeling and Stability of Fractional Order Differential Systems 2: The Infinite State Approach

VANDERHAEGEN Frédéric, MAAOUI Choubeila, SALLAK Mohamed, BERDJAG Denis
Automation Challenges of Socio-technical Systems

2018

BERRAH Lamia, CLIVILLÉ Vincent, FOULLOY Laurent
Industrial Objectives and Industrial Performance: Concepts and Fuzzy Handling

GONZALEZ-FELIU Jesus
Sustainable Urban Logistics: Planning and Evaluation

GROUS Ammar
Applied Mechanical Design

LEROY Alain
Production Availability and Reliability: Use in the Oil and Gas Industry

MARÉ Jean-Charles
Aerospace Actuators 3: European Commercial Aircraft and Tiltrotor Aircraft

MAXA Jean-Aimé, BEN MAHMOUD Mohamed Slim, LARRIEU Nicolas
Model-driven Development for Embedded Software: Application to Communications for Drone Swarm

MBIHI Jean
Analog Automation and Digital Feedback Control Techniques
Advanced Techniques and Technology of Computer-Aided Feedback Control

MORANA Joëlle
Logistics

SIMON Christophe, WEBER Philippe, SALLAK Mohamed
Data Uncertainty and Important Measures
(Systems Dependability Assessment Set – Volume 3)

TANIGUCHI Eiichi, THOMPSON Russell G.
City Logistics 1: New Opportunities and Challenges
City Logistics 2: Modeling and Planning Initiatives
City Logistics 3: Towards Sustainable and Liveable Cities

ZELM Martin, JAEKEL Frank-Walter, DOUMEINGTS Guy, WOLLSCHLAEGER Martin
Enterprise Interoperability: Smart Services and Business Impact of Enterprise Interoperability

2017

ANDRÉ Jean-Claude
From Additive Manufacturing to 3D/4D Printing 1: From Concepts to Achievements
From Additive Manufacturing to 3D/4D Printing 2: Current Techniques, Improvements and their Limitations
From Additive Manufacturing to 3D/4D Printing 3: Breakthrough Innovations: Programmable Material, 4D Printing and Bio-printing

ARCHIMÈDE Bernard, VALLESPIR Bruno
Enterprise Interoperability: INTEROP-PGSO Vision

CAMMAN Christelle, FIORE Claude, LIVOLSI Laurent, QUERRO Pascal
Supply Chain Management and Business Performance: The VASC Model

FEYEL Philippe
Robust Control, Optimization with Metaheuristics

MARÉ Jean-Charles
Aerospace Actuators 2: Signal-by-Wire and Power-by-Wire

POPESCU Dumitru, AMIRA Gharbi, STEFANOIU Dan, BORNE Pierre
Process Control Design for Industrial Applications

RÉVEILLAC Jean-Michel
Modeling and Simulation of Logistics Flows 1: Theory and Fundamentals
Modeling and Simulation of Logistics Flows 2: Dashboards, Traffic Planning and Management
Modeling and Simulation of Logistics Flows 3: Discrete and Continuous Flows in 2D/3D

2016

ANDRÉ Michel, SAMARAS Zissis
Energy and Environment
(Research for Innovative Transports Set - Volume 1)

AUBRY Jean-François, BRINZEI Nicolae, MAZOUNI Mohammed-Habib
Systems Dependability Assessment: Benefits of Petri Net Models (Systems Dependability Assessment Set - Volume 1)

BLANQUART Corinne, CLAUSEN Uwe, JACOB Bernard
Towards Innovative Freight and Logistics (Research for Innovative Transports Set - Volume 2)

COHEN Simon, YANNIS George
Traffic Management (Research for Innovative Transports Set - Volume 3)

MARÉ Jean-Charles
Aerospace Actuators 1: Needs, Reliability and Hydraulic Power Solutions

REZG Nidhal, HAJEJ Zied, BOSCHIAN-CAMPANER Valerio
Production and Maintenance Optimization Problems: Logistic Constraints and Leasing Warranty Services

TORRENTI Jean-Michel, LA TORRE Francesca
Materials and Infrastructures 1 (Research for Innovative Transports Set - Volume 5A)
Materials and Infrastructures 2 (Research for Innovative Transports Set - Volume 5B)

WEBER Philippe, SIMON Christophe
Benefits of Bayesian Network Models
(Systems Dependability Assessment Set – Volume 2)

YANNIS George, COHEN Simon
Traffic Safety (Research for Innovative Transports Set - Volume 4)

2015

AUBRY Jean-François, BRINZEI Nicolae
Systems Dependability Assessment: Modeling with Graphs and Finite State Automata

BOULANGER Jean-Louis
CENELEC 50128 and IEC 62279 Standards

BRIFFAUT Jean-Pierre
E-Enabled Operations Management

MISSIKOFF Michele, CANDUCCI Massimo, MAIDEN Neil
Enterprise Innovation

2014

CHETTO Maryline
Real-time Systems Scheduling
Volume 1 – Fundamentals
Volume 2 – Focuses

DAVIM J. Paulo
Machinability of Advanced Materials

ESTAMPE Dominique
Supply Chain Performance and Evaluation Models

FAVRE Bernard
Introduction to Sustainable Transports

GAUTHIER Michaël, ANDREFF Nicolas, DOMBRE Etienne
Intracorporeal Robotics: From Milliscale to Nanoscale

MICOUIN Patrice
Model Based Systems Engineering: Fundamentals and Methods

MILLOT Patrick
Designing Human–Machine Cooperation Systems

NI Zhenjiang, PACORET Céline, BENOSMAN Ryad, RÉGNIER Stéphane
Haptic Feedback Teleoperation of Optical Tweezers

OUSTALOUP Alain
Diversity and Non-integer Differentiation for System Dynamics

REZG Nidhal, DELLAGI Sofien, KHATAD Abdelhakim
Joint Optimization of Maintenance and Production Policies

STEFANOIU Dan, BORNE Pierre, POPESCU Dumitru, FILIP Florin Gh., EL KAMEL Abdelkader
Optimization in Engineering Sciences: Metaheuristics, Stochastic Methods and Decision Support

2013

ALAZARD Daniel
Reverse Engineering in Control Design

ARIOUI Hichem, NEHAOUA Lamri
Driving Simulation

CHADLI Mohammed, COPPIER Hervé
Command-control for Real-time Systems

DAAFOUZ Jamal, TARBOURIECH Sophie, SIGALOTTI Mario
Hybrid Systems with Constraints

FEYEL Philippe
Loop-shaping Robust Control

FLAUS Jean-Marie
Risk Analysis: Socio-technical and Industrial Systems

FRIBOURG Laurent, SOULAT Romain
Control of Switching Systems by Invariance Analysis: Application to Power Electronics

GROSSARD Mathieu, REGNIER Stéphane, CHAILLET Nicolas
Flexible Robotics: Applications to Multiscale Manipulations

GRUNN Emmanuel, PHAM Anh Tuan
Modeling of Complex Systems: Application to Aeronautical Dynamics

HABIB Maki K., DAVIM J. Paulo
Interdisciplinary Mechatronics: Engineering Science and Research Development

HAMMADI Slim, KSOURI Mekki
Multimodal Transport Systems

JARBOUI Bassem, SIARRY Patrick, TEGHEM Jacques
Metaheuristics for Production Scheduling

KIRILLOV Oleg N., PELINOVSKY Dmitry E.
Nonlinear Physical Systems

LE Vu Tuan Hieu, STOICA Cristina, ALAMO Teodoro,
CAMACHO Eduardo F., DUMUR Didier
Zonotopes: From Guaranteed State-estimation to Control

MACHADO Carolina, DAVIM J. Paulo
Management and Engineering Innovation

MORANA Joëlle
Sustainable Supply Chain Management

SANDOU Guillaume
Metaheuristic Optimization for the Design of Automatic Control Laws

STOICAN Florin, OLARU Sorin
Set-theoretic Fault Detection in Multisensor Systems

2012

AÏT-KADI Daoud, CHOUINARD Marc, MARCOTTE Suzanne, RIOPEL Diane
Sustainable Reverse Logistics Network: Engineering and Management

BORNE Pierre, POPESCU Dumitru, FILIP Florin G., STEFANOIU Dan
Optimization in Engineering Sciences: Exact Methods

CHADLI Mohammed, BORNE Pierre
Multiple Models Approach in Automation: Takagi-Sugeno Fuzzy Systems

DAVIM J. Paulo
Lasers in Manufacturing

DECLERCK Philippe
Discrete Event Systems in Dioid Algebra and Conventional Algebra

DOUMIATI Moustapha, CHARARA Ali, VICTORINO Alessandro, LECHNER Daniel
Vehicle Dynamics Estimation using Kalman Filtering: Experimental Validation

GUERRERO José A, LOZANO Rogelio
Flight Formation Control

HAMMADI Slim, KSOURI Mekki
Advanced Mobility and Transport Engineering

MAILLARD Pierre
Competitive Quality Strategies

MATTA Nada, VANDENBOOMGAERDE Yves, ARLAT Jean
Supervision and Safety of Complex Systems

POLER Raul *et al.*
Intelligent Non-hierarchical Manufacturing Networks

TROCCAZ Jocelyne
Medical Robotics

YALAOUI Alice, CHEHADE Hicham, YALAOUI Farouk, AMODEO Lionel
Optimization of Logistics

ZELM Martin *et al.*
Enterprise Interoperability –I-EASA12 Proceedings

2011

CANTOT Pascal, LUZEAUX Dominique
Simulation and Modeling of Systems of Systems

DAVIM J. Paulo
Mechatronics

DAVIM J. Paulo
Wood Machining

GROUS Ammar
Applied Metrology for Manufacturing Engineering

KOLSKI Christophe
Human–Computer Interactions in Transport

LUZEAUX Dominique, RUAULT Jean-René, WIPPLER Jean-Luc
Complex Systems and Systems of Systems Engineering

ZELM Martin, *et al.*
Enterprise Interoperability: IWEI2011 Proceedings

2010

BOTTA-GENOULAZ Valérie, CAMPAGNE Jean-Pierre, LLERENA Daniel, PELLEGRIN Claude
Supply Chain Performance / Collaboration, Alignement and Coordination

BOURLÈS Henri, GODFREY K.C. Kwan
Linear Systems

BOURRIÈRES Jean-Paul
Proceedings of CEISIE'09

CHAILLET Nicolas, REGNIER Stéphane
Microrobotics for Micromanipulation

DAVIM J. Paulo
Sustainable Manufacturing

GIORDANO Max, MATHIEU Luc, VILLENEUVE François
Product Life-Cycle Management / Geometric Variations

LOZANO Rogelio
Unmanned Aerial Vehicles / Embedded Control

LUZEAUX Dominique, RUAULT Jean-René
Systems of Systems

VILLENEUVE François, MATHIEU Luc
Geometric Tolerancing of Products

2009

DIAZ Michel
Petri Nets / Fundamental Models, Verification and Applications

OZEL Tugrul, DAVIM J. Paulo
Intelligent Machining

PITRAT Jacques
Artificial Beings

2008

ARTIGUES Christian, DEMASSEY Sophie, NERON Emmanuel
Resources–Constrained Project Scheduling

BILLAUT Jean-Charles, MOUKRIM Aziz, SANLAVILLE Eric
Flexibility and Robustness in Scheduling

DOCHAIN Denis
Bioprocess Control

LOPEZ Pierre, ROUBELLAT François
Production Scheduling

THIERRY Caroline, THOMAS André, BEL Gérard
Supply Chain Simulation and Management

2007

DE LARMINAT Philippe
Analysis and Control of Linear Systems

DOMBRE Etienne, KHALIL Wisama
Robot Manipulators

LAMNABHI Françoise *et al.*
Taming Heterogeneity and Complexity of Embedded Control

LIMNIOS Nikolaos
Fault Trees

2006

FRENCH COLLEGE OF METROLOGY
Metrology in Industry

NAJIM Kaddour
Control of Continuous Linear Systems